THE BHS COMPLETE MANUAL OF
Stable Management

OFFICIAL BHS MANUALS
published by Kenilworth Press

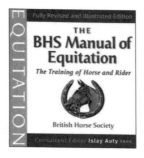

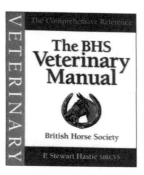

The BHS Complete Manual of
Stable Management

British Horse Society

Consultant Editor ISLAY AUTY FBHS

KENILWORTH PRESS

This book is an updated, omnibus edition, bringing together the seven titles
which previously formed *The British Horse Society Manual of Stable Management*:
Book 1: The Horse (1988 and 1991)
Book 2: Care of the Horse (1988 and 1991)
Book 3: The Horse at Grass (1988 and 1992)
Book 4: Saddlery (1989 and 1991)
Book 5: Specialist Care of the Competition Horse (1989 and 1991)
Book 6: The Stable Yard (1989 and 1992)
Book 7: Watering and Feeding (1992)

This revised, omnibus edition published in 1998 by
Kenilworth Press Ltd
Addington
Buckingham
MK18 2JR

Reprinted 1998, 1999, 2000 (twice), 2001, 2002

British Library Cataloguing in Publication Data
A catalogue record for this book is available from the British Library.

ISBN 0-872119-03-4

Typeset by Kenilworth Press

Printed in Great Britain by Bell & Bain Ltd, Glasgow

CONTENTS

Section 5 SPECIALIST CARE OF THE COMPETITION HORSE

Section 6 THE HORSE AT GRASS

Section 7 THE STABLE YARD

Appendices

PREFACE

The aim of this volume is to provide a reliable source of information and advice on all practical aspects of horse and stable management. Throughout the book emphasis is placed on the adoption of correct and safe procedures for the welfare of all who come into contact with horses, as well for the animals themselves. The book has been compiled by a panel of experts, each drawing on considerable experience and contributing specialised knowledge on his or her chosen subject.

ACKNOWLEDGEMENTS

The British Horse Society would like to acknowledge the contribution made by the late Pat Smallwood FBHS, who undertook much of the initial work on this project.

The British Horse Society is also grateful to the following individuals for their invaluable assistance:

Joan Allen; Islay Auty FBHS; Goran Breisner FBHS; Dr James L. Duncan; Eric Ellis; Sidney Free; Judy Harvey FBHS; Stewart Hastie MRCVS; Jeremy Houghton-Brown BPhilEd; Jane Kidd; Elizabeth Launder MSc; James Lord; Deborah Lucas MSc; Gillian McCarthy BSc; Jane McHugh; Dr David Marlin; Tessa Martin-Bird FBHS; Clive Scott; Richard Shepherd; Michael Simons MRCVS; Barbara Slane Fleming FBHS; Claire Tomlinson; Gillian Watson FBHS; Helen Webber FBHS; and members of the 1997 BHS Training Committee.

Illustrations
The line drawings are all by Dianne Breeze, except for the foot detail diagrams on pages 13 and 87 (left), which are by Christine Bousfield.

NOTE: Throughout the book the term 'horses' is used and it will often include ponies.

THE HORSE

CHAPTER 1

CONFORMATION

The conformation of a horse or pony affects his soundness, ability to perform and comfort as a ride. If the horse is in good condition, the following features should be sought:

General Impression

The general impression should be that he is built in proportion, with all sectors matching and his outlook alert, bold and confident.

The 'top line' – the neck, withers, back, loins and dock – should form a succession of well-developed outlines, each of which blends smoothly into the other. He should have a good sloping shoulder, a relatively short back and a long croup, i.e. length from hip to point of buttock, giving an appearance of 'standing over a lot of ground', whilst being well balanced over his legs, which are four square underneath him.

If the horse is in poor condition, only an experienced person with a 'good eye' can assess his potential. It takes skill to recognise whether he has the right make and shape; and to see that when given suitable care, food and work, he will make up into a worthwhile or even top-class animal.

Similarly, too much fat can disguise a horse's outline, and can make it difficult to visualise what he will be like when the excess weight is lost and he is fit.

A horse with good skeletal structure, but poor muscle development, can be transformed by being given suitable work and the opportunity to use and develop his muscles in the correct way.

Feet

The much-quoted saying 'no foot, no horse' is very true. The make and shape of feet are vital to the soundness and functioning of the horse.

Front feet and hind feet should be matching pairs. Any difference obvious in outline, angle of foot to the ground, or size of frog, should be viewed with suspicion. The only exception is the horse whose foot has been worn down through losing a shoe.

- The front feet should slope at an angle of 45–50° from the ground.

- The hoof wall should continue at the same angle as the pastern.

- The hind feet should have a slightly steeper slope, and should be longer and narrower than the front feet.

- All feet should point straight forwards. Any deviation from this is usually caused by poor conformation of the leg, and will result in faulty action.

The heels should be wide, with a well-developed frog to help absorb concussion. The sole should be slightly concave. Contracted heels and poorly developed frogs restrict the blood supply to the foot, and increase the vulnerability to navicular disease

Large flat feet can cause problems, particularly on stony ground. They are liable to develop corns and are easily bruised.

Small feet are a disadvantage, as they give a smaller

The points of the horse

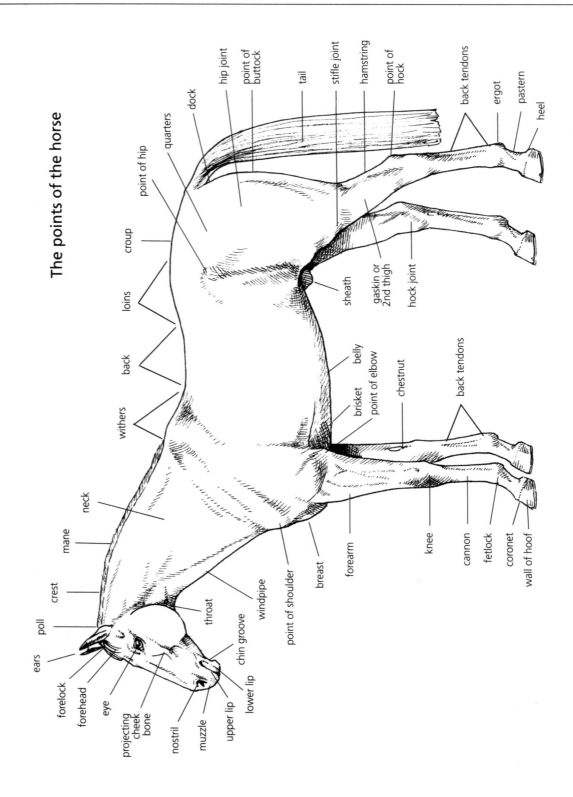

ears
poll
crest
mane
neck
forelock
forehead
eye
projecting cheek bone
nostril
muzzle
upper lip
lower lip
chin groove
throat
windpipe
point of shoulder
breast
forearm
knee
cannon
fetlock
coronet
wall of hoof
back tendons
chestnut
point of elbow
brisket
belly
withers
back
loins
croup
point of hip
quarters
dock
hip joint
point of buttock
tail
stifle joint
hamstring
point of hock
back tendons
ergot
pastern
heel
sheath
gaskin or 2nd thigh
hock joint

The skeleton

Bones of the foot

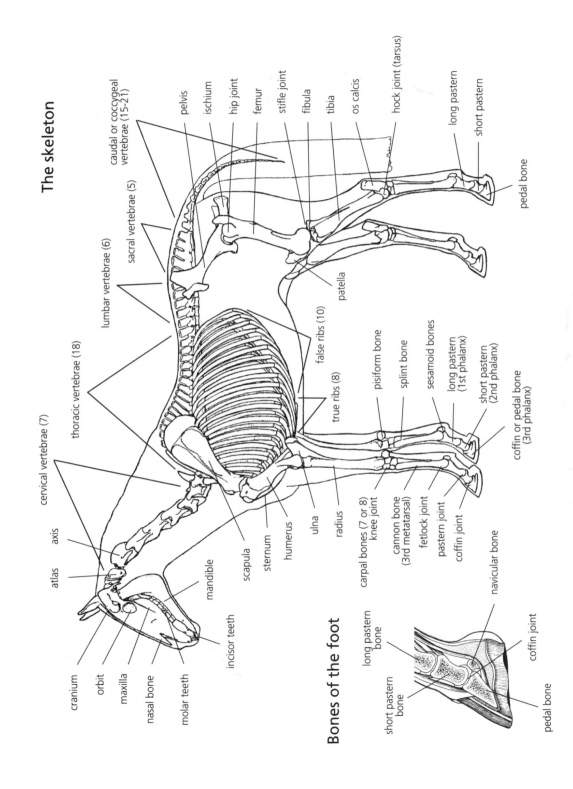

caudal or coccygeal vertebrae (15–21)

sacral vertebrae (5)

lumbar vertebrae (6)

thoracic vertebrae (18)

cervical vertebrae (7)

axis

atlas

cranium

orbit

maxilla

nasal bone

molar teeth

incisor teeth

mandible

scapula

sternum

humerus

ulna

radius

pelvis

ischium

hip joint

femur

stifle joint

fibula

tibia

os calcis

hock joint (tarsus)

long pastern

short pastern

pedal bone

patella

false ribs (10)

true ribs (8)

pisiform bone

splint bone

sesamoid bones

long pastern (1st phalanx)

short pastern (2nd phalanx)

coffin or pedal bone (3rd phalanx)

carpal bones (7 or 8)

knee joint

cannon bone (3rd metatarsal)

fetlock joint

pastern joint

coffin joint

navicular bone

long pastern bone

short pastern bone

coffin joint

pedal bone

The main superficial external muscles

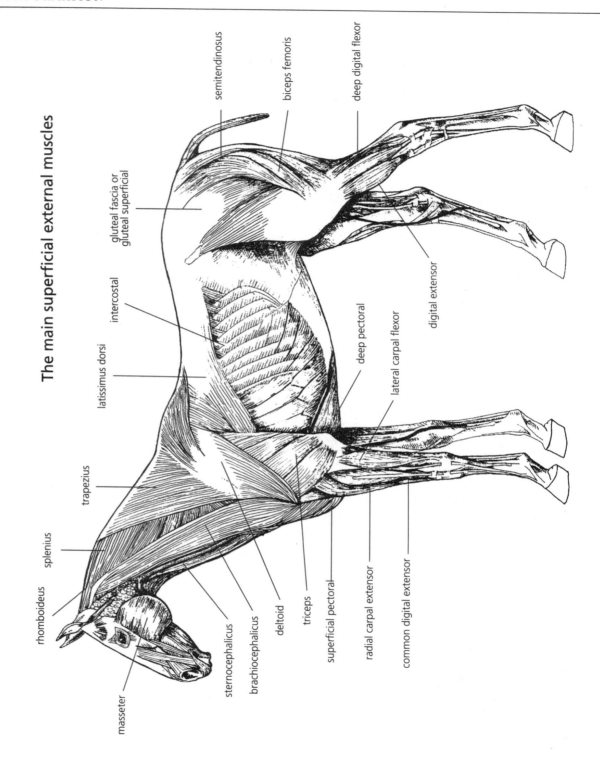

rhomboideus

splenius

masseter

sternocephalicus

brachiocephalicus

deltoid

triceps

superficial pectoral

radial carpal extensor

common digital extensor

trapezius

latissimus dorsi

intercostal

gluteal fascia or gluteal superficial

semitendinosus

biceps femoris

deep digital flexor

digital extensor

lateral carpal flexor

deep pectoral

weight-bearing surface, and there is a greater tendency towards unsoundness in the foot.

The wall of the foot should be smooth and free from cracks. Rings and grooves are a sign of change of management or diet; pronounced rings can signify laminitis (see Shoeing, page 86 onwards).

The texture of the horn is of importance. Some breeds of horses have naturally hard feet, especially native ponies, Arabs and Hackneys. Poor horn texture can be improved by careful feeding, particularly the use of the additive biotin, and by the use of horn stimulants.

A skilled farrier can greatly improve a horse's feet through regular attention. Equally, lack of care or an inexperienced farrier can spoil good feet.

Head

The head should be lean, well set on to the neck, and in proportion to the size of the horse. There should be sufficient width between the branches of the lower jaw to allow ample room for the top of the windpipe. The jaw bone itself should not be too large. There must be room between the lower edge of the jaw bone and the jugular furrow, to allow the horse to flex.

This set of the head on the neck is very important, as it affects respiration, and ease of flexion, which in its turn affects the control and balance of the horse.

Eyes

The eyes should be of a good size and expression, and widely set to give broad vision. Their expression is a key to temperament and reliability, and should be bold and generous, not mean and sharp, if the horse is to work well.

Muzzle

The muzzle should be fine with a well-defined nostril. The jaws should be of equal length. If overshot (parrot-mouthed) or undershot, the horse may have problems when biting grass, although his ability to chew will not be affected. This is an hereditary fault which could be passed on through breeding. Horses with this conformational fault should not be used for breeding.

Ears

The ears should be relaxed, mobile and of a good size. When pricked, they should be carried forward. Lop ears droop forwards and down, or to the side and down, but they have no ill effects other than to give an impression of dejection.

Neck

The neck should be muscular and of a length and substance proportional to the body. The top line should be convex, with a definite arch between the withers and poll. A heavy crest, expected in a stallion, is to be regretted in a mare or gelding. The muscle under the neck should appear to slip into the shoulder without any definite division. The neck should not be set into the shoulder so low that there is a big dip between it and the withers.

Neck muscles, and thus outline, can be improved by correct training.

Shoulder

The shoulder should be deep with a definite slope forward from the withers to its point. This shape gives a better ride, and longer, more flowing strides. The top of the shoulder blades should be close together; if wide apart they make a lumpy or 'loaded' shoulder, which gives an uncomfortable ride and poor movement. Upright shoulders allow the saddle to slip forward, make the rider feel that he has little in front of him, and restrict the freedom of movement of the horse. This type of shoulder is more suitable for the harness horse, as it provides greater pulling power.

Withers

The withers should be clearly defined and of a sufficient

height to provide room for the attachment of the covering muscles of the shoulders. If too high, they will make it difficult to fit a saddle; if too low the saddle tends to work forward – which may result in girth galls.

In a mature horse, the withers and croup should be of a similar height. If the croup is higher, known as 'croup high', the horse appears to be standing downhill and tends to be on his forehand when ridden. There are, however, some good jumpers and racehorses with this conformation.

Chest

The chest should be of medium width, giving plenty of heart room. If too narrow, the horse moves very close in front with the likelihood of brushing. If too wide, he tends to roll in canter and have limited ability to gallop.

Forelegs

The forelegs' positioning will depend on the length of the humerus bone, lying between the elbow and the point of shoulder. If this is too long, the legs are too far under the body, and the elbow is likely to be tied in with a consequent lack of freedom of movement. The elbow should stand well away from the ribs.

The forelegs should be straight from the top of the leg to the foot when looked at from the front, and straight from the top of the leg to the front of the fetlock when looked at from the side. Any deviation from a straight line when looked at from the front is a serious fault, which affects the straightness of the action and gives extra strain on the tendons, ligaments and joints.

The forearm should be well developed, and longer than the leg below the knee, which should be relatively short so as to minimise strain on the ligaments and tendons.

Knees

The knees should be broad, flat and deep to give room for the tendons that are attached over the top of the knee, and also to give room for the attachment of tendons and ligaments at the back of the knee.

A horse **'back at the knee'** (concave when viewed from the side) or 'calf kneed' will be suitable for slower work, but is more likely to strain tendons when galloping and jumping.

A horse a little **'over at the knee'** (convex when viewed from the side) is less likely to sprain tendons. This conformation, in many people's opinion, is no drawback.

Swollen knees caused by injury are always suspect, and the cause should be investigated.

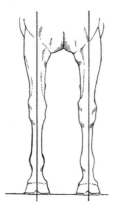

Good column of support through centre of limb.

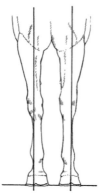

'Out of one hole.'

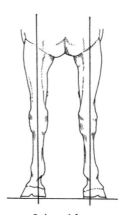

Splayed feet.

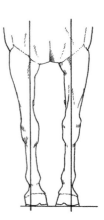

Pigeon-toed.

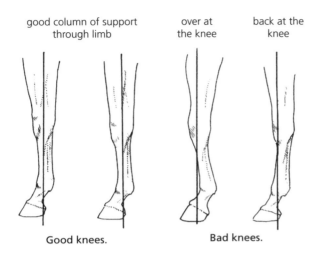

good column of support through limb

over at the knee

back at the knee

Good knees. Bad knees.

Cannon Bones

The cannon bones should be short and straight. If narrower below the knee than at the fetlock, the horse is said to be 'tied in below the knee'. There is less room for tendons and ligaments, and therefore a greater risk of strain.

Bone is measured immediately below the knee and is the circumference of the leg at that point. By placing the thumb and first finger round the leg an approximate measurement can be taken: 20cm (8ins) is sufficient bone for a riding horse or lightweight hunter, although this will depend on breeding; 21.5cm (8½ins) for a mediumweight hunter and 23–23.8cm (9–9½ins) for a heavyweight hunter.

A well-known saying is that 'blood carries weight'. Thoroughbred and Arab horses have much denser bone structure than a horse of common breeding, i.e. of cart-horse blood. Horses with this denser bone are capable of carrying more weight, relative to their size, than the commoner breeds.

Fetlock

The fetlock joint should give an appearance of flatness rather than roundness. Lumps on the front or back of the joint are a sign of work and age; on the inside they indicate that the horse moves close and brushes.

Pasterns

Pasterns should be of medium length. Long, sloping pasterns make for a springy ride, but are liable to strain. Short, upright pasterns are strong, but give a bumpy ride, and tend to cause lameness through the extra jar and concussion.

Body

The body should be deep through the heart, with well-sprung ribs and plenty of room for the lungs. This gives a natural girth line.

A horse who is shallow through the heart is described as 'showing a lot of daylight' or 'on the leg'. The measurement from the lowest part of the girth to the withers should approximately equal that from the girth to the ground. Many young horses have a 'leggy' appearance, but lose this look as they mature.

Ribs

There are eight true ribs attached to both the vertebrae and the sternum, and ten false ribs attached to the vertebrae at the top and at the bottom, interconnected by cartilage. The first rib is only slightly curved, the curvature of each succeeding rib gradually increasing to give a well-rounded appearance and internal space for many of the main organs of the body.

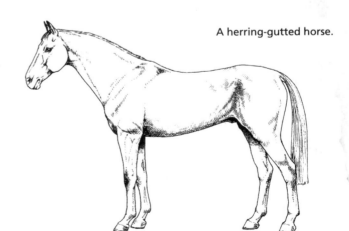

A herring-gutted horse.

An animal which is flat-sided, which runs up light, and which is 'herring gutted', is usually a 'poor doer', lacking in stamina. There should be only a short gap between the last rib and the point or hip, i.e. 5–8cm (2–3ins).

Back

The back, having to carry the weight, should be of medium length and almost level.

- LONG BACKS give a comfortable ride but are liable to strain. It is usual for mares to be longer in the back than stallions or geldings.

- SHORT BACKS are strong, but give a less comfortable ride.

- HOLLOW BACKS are a sign of weakness in young horses, but are also a sign of age in an older horse. Weak- or dip-backed horses are often suffering from progressive arthritis.

- ROACH BACKS – i.e. arched upwards – are strong but give an uncomfortable ride. This outline can develop with age, and may then signify arthritis.

The correct muscular development of the back and loins area is essential if the horse is to perform well under the weight of the rider.

Loins

The loins are immediately behind the saddle and on either side of the spinal processes. They should be broad and well developed. Their strength and correct muscular development play a major part in the horse's balance and ability to perform. A slack loin goes with a weak back and should be avoided.

Quarters

The quarters provide power and should be muscular, the hips broad, rounded and a pair.

    '**Hip down**' (a fracture of the point of the hip caused by a fall or hitting a door post) is shown by the flatter

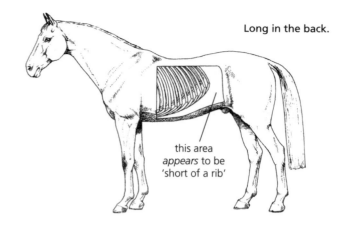

Long in the back.

this area *appears* to be 'short of a rib'

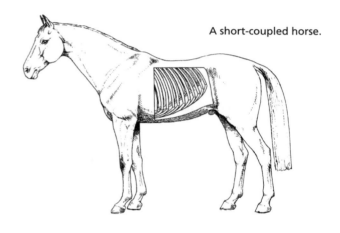

A short-coupled horse.

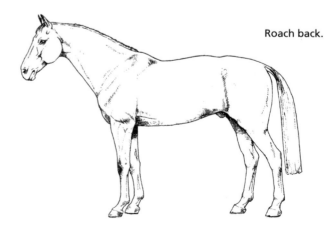

Roach back.

outline of the injured hip. It can be unsightly but often goes unobserved, and once the initial bruising has subsided, it rarely causes lameness. It can sometimes cause problems in pregnancy.

A **'goose-rumped'** horse has quarters which slope sharply from croup to dock, and this often accompanies jumping ability. A prominence on the croup is called a 'jumping bump'.

The dock should not be set on too low. There should be plenty of length between the hip and the point of buttock and the point of the hip and point of hock.

Viewed from behind, the impression should be one of strength, with rounded quarters and upper or first thighs sloping down to a well-developed second thigh or gaskin. 'Split up behind' means a poorly developed upper thigh.

Hind Legs

When the horse is standing naturally and the hind leg is viewed both from the side and from immediately behind, there should be a straight line from the point of buttock through the point of hock down to the fetlock and so to the ground. Any deviation from this shows a weakness, and places greater strain both on the stifle joint, or the patella above, and the hock joint below it. These two joints of the stifle and hock work in harmony, and provide the main propelling force for galloping and jumping. When in movement, the stifle joint should not be thrown outwards, but should stay in line with the body. The stifle joint is the equivalent of the human knee.

Hocks

The hocks should be large, and the outline should be clean and flat with a prominent point at the back. There should be plenty of bone below the hock. There should be length from hip to hock. The hock should be 'well let down', i.e. the section of the limb from hock to fetlock should appear short.

There are various weak shapes of hock:

- **BENT OR SICKLE HOCKS** If viewed from the side, there

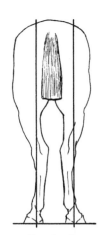

'Split up behind'. Cow hocks.

Wide behind. Good hocks.

is a more acute angle to the hocks, which are too much under the horse, usually with a poor second thigh and cut-in above the hock.

- **COW HOCKS** If viewed from the rear, the points of the hocks come together and the toes turn out. Such a horse is liable to brush.

- **BOWED HOCKS** If viewed from behind, the points of the hocks are wide apart, the toes turn in and the foot is likely to screw as it comes to the ground.

- **STRAIGHT HOCKS** These are good for galloping, but have less leverage for jumping. There is more

Hocks out behind.	Straight hocks.	Good hocks well supported under the body.	Sickle-shaped hocks.	Hocks too high.

concussion unless compensated for by a long pastern. They are liable to strain, and young horses with straight hocks may suffer from a slipped stifle joint. This is usually rectified as the horse matures and the muscles and ligaments strengthen.

- **HOCKS OUT BEHIND** Viewed from the side, such hocks are way out behind the horse. There is a big angle above the hock over the hamstring. Such horses often jump well, but rarely gallop.

DEVELOPING AN EYE FOR A HORSE

It takes considerable experience and constant practice to 'develop an eye' for a horse and to be able to assess a horse's conformation correctly. It can only be achieved by looking at many different types and sizes of horses and learning mentally to compare and evaluate what has been seen.

Assessing Conformation

First look at the horse at rest in the stable. Then ask the attendant to bring it out and stand it up for inspection.

General Impression

- Does everything match?

- Assess the size of head in relation to body, and parts of the body in relation to each other.
- Consider the size of body in relation to the amount of 'bone' and to the length or shortness of legs.
- Assess structure and size of forehand in relation to hindquarters.
- Look at the size and shape of the feet.

Head, Neck, Forehand, Legs and Feet

- Check teeth for age, wolf teeth, and bit injuries.
- Check shape of the upper and lower jaw, and width between the branches of the lower jaw.
- Check the set of the head on the neck.

Stand in front of the horse and check:

- Width of the chest.
- Straightness of the legs.
- Size and shape of the knees.

Stand on the left or near-side and:

- Check the shape and feel of the withers.
- View the angle and length of the shoulders.
- Check by eye and right hand the position and set of the elbow joint.

Face towards the tail and:

- Run the left hand down the back of the forearm,

knees, tendons and fetlock joint of the left leg.

- Move to the right or off-side and compare the look and feel of the right leg.

Face the horse and:

- View the size and shape of the knee, the cannon bone, fetlock joint and pastern.
- Crouch down and using both hands check the front and back of the knee and leg to the coronet.

Face towards the tail and:

- Run the left hand down the shoulder and leg and pick up the foot. If necessary, pick out the foot. Check the size and shape of the foot, sole and frog.
- If shod, check the shoes for fit, type and wear.

Repeat the above procedure with the right leg.

The Middle

Stand back and view:

- The general outline.
- Top line.
- Length of back.
- Depth of girth.
- Shape of ribs.
- Strength of loins.
- Run your left hand over the withers and along the back. Then feel the ribs and loin area.
- Feel for muscle development and condition.

The Hindquarters, Legs and Feet

Stand back and view the outline, particularly:

- Length from hip joint to hock.
- Stifle to hock.
- Hock to fetlock.
- The size and shape of the hock, and the angle at which it is set on. The development of the muscles of the second thigh.

Run your hand along the body and over the quarters.

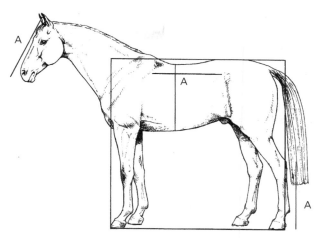

As a rule of thumb, in a well-proportioned horse the length of the horse's head (A) is approximately equal to: the distance between the back of the shoulder blade and the point of hip; the depth of body at the girth; and the distance between the point of hock and the ground. The horse's body from shoulder to point of hip should roughly fit into an imaginary square whose sides are A x 2.5.

Take hold of the tail, and standing behind the horse check:

- The levelness of the points of the hip, and the width between the points of the hip.
- The muscular development of the buttocks and upper thighs.
- The straightness of the hind leg from point of buttock through the hock joint to the ground.
- The angle of the foot.

Standing to the side:

- Run your hand down the quarters and with both hands examine the hock joint for signs of strain or injury.

Feel down the hind leg in the same way as for the front leg, particularly the inside of the fetlock joint.

Check the hind feet in the same way as the front feet.

Repeat the last three procedures with the other leg.

Standing in front of the horse:

- Look through the front legs and, diagonally, at the inner outline of the hock. Compare the outline of each hock joint.

Assessing Condition

Consider:

- The **general impression,** from expression of eyes and appearance of skin, coat and colour of gums. A running nose or eyes should be noted and mentioned.

- The feel of the **coat and skin**: staring coat and tight skin may imply worms or other illness and lack of condition.

- The feel of the horse's **body**.

- **Neck**. The crest should feel tough and strong. Poor crest means no reserves of fat.

- **Back and ribs**. These should be well covered. In old age or pregnancy the weight of belly may give an impression of poorness above it. A second feel along the backbone should verify condition.

- **Quarters**. Their shape varies according to breed and type. A fit horse has firm, well-developed muscles. A fat horse will show a round outline. A poor horse shows poverty lines down the back of the quarters, and the muscles will feel underdeveloped and soft.

- The **feet**. These reflect general condition and management. The outward evidence of changes – either for the better or worse – may take nine to eighteen months to show.

CHAPTER

2 ACTION

The ability of a horse to move well greatly depends on his conformation, breeding and type. A good mover uses himself equally well in front and behind.

To move well in front, a horse must actively use his shoulders as well as his knees and fetlock joints. He should move straight, with a length of stride suitable to his make and shape. He should flow, giving an impression of moving with ease, which will be more likely if the elbow is well positioned, allowing plenty of room between elbow and ribs. There should be slight knee elevation, but if this is exaggerated the horse may be better suited to driving rather than riding. A horse with an upright shoulder rarely moves well: he has a short stride and is a moderate ride. A show horse should always move straight. Horses destined for most competitive disciplines will usually be less susceptible to strains and injury if their action is 'straight'.

The movement of the hind leg is equally important. Good movement comes from well-developed loins, quarters and thighs, and from active hocks. The appearance should be of strength and power. The action should again be straight, with the hocks brought well forward underneath the body.

In trot, if as the horse moves away from you he is sound and has full use of his joints, it should be possible to see briefly the whole sole of each foot when the foot is at its highest point.

Some allowance should be made for young horses, particularly when trotted in-hand, and for four-year-olds when ridden. Lack of balance and muscular development may affect their action. As the horse matures and comes off his forehand, small defects in his action often disappear. A skilled farrier can do much to improve a horse's movement.

DEFECTIVE ACTION

Dishing
The front foot is thrown outwards, particularly in trot, and although this action appears to come from below the knee it is related to the flexion of the elbow. When still, the horse will often stand with toes turned in – 'pigeon-toed'. The movement can occur in one or both front legs. As long as it is only moderate dishing it is of little detriment to the horse, except for showing. An exaggerated movement puts great strain on the fetlock joints, and this may eventually cause unsoundness.

Plaiting or Lacing
At walk and trot the horse places one foot in front of the other. This can apply to one or both feet. Narrow-chested horses are prone to this. Excessive plaiting is likely to make the horse stumble and even fall. A slight deviation can be acceptable. The knee should be carefully checked for any scarring or bruising.

Toes Turned Out
Horses who stand with either front or hind feet pointing outwards usually brush their legs, knocking the

inside of their fetlock joint with the opposite foot. Evidence of this is an enlarged joint from constant bruising and/or rubbing off of the hair over the joint. It is a serious fault, as it necessitates always using protective boots when the horse is ridden, which can cause other problems.

Going Wide Behind

This often relates to stiffness in the back, and the hind legs are frequently bowed outwards. It is unsightly.

Forging

Forging is not considered to be a defect, as usually it can be corrected. It occurs in trot when the toe of the hind shoe strikes the underneath surface of the front shoe on the same side. The front foot lingers and is caught by the back shoe.

It is sometimes found in free-moving young horses, not yet sufficiently balanced and strong under the weight of the rider, and thus not able to bring the weight off the forehand. The problem ceases when the horse is stronger and better balanced. Another cause is trotting too fast. This is usually the rider's fault – if the trot is slowed down, the forging ceases.

Poor conformation can sometimes lead to forging. The remedy is schooling to improve and build up muscles. If the feet are too long at the toe this can also cause forging, but this can be corrected by a farrier.

Over-reaching

Over-reaching occurs when the hind limb over-extends and the toe of the hind shoe interferes or strikes into the forelimb between the knee and the bulbs of the heel. Usually the lower part of the limb is most affected, particularly the bulbs of the heels. A high over-reach can cause severe damage to the tendons in the back of

Pigeon-toed – weight not evenly distributed down the centre of the limb.

Plaiting.

Dishing.

the leg. Over-reaching usually occurs when galloping or jumping, or through loss of balance.

ASSESSING ACTION AND SOUNDNESS

- Ask the attendant to walk the horse away, turn and walk back. Check on evenness and length of stride, and head movement.

- Ask the attendant to trot the horse away, turn and trot back, keeping to as straight a line as possible. Check on the straightness of movement, the activity and levelness of the trot, and the evenness of the movement of the horse's head.

- Ask for the horse to be trotted up again, and view from the side. Observe the general attitude of the horse.

Forging.

Over-reaching.

CHAPTER 3

THE PSYCHOLOGY OF THE HORSE

When dealing with horses and ponies it is important to bear in mind their natural lifestyle and their instinctive defences. For many centuries they were nomadic, grass-eating, herd animals, and speed was their greatest protection against enemies. Even a newly born foal was soon on its feet, ready to keep up with his mother and, if the necessity arose, to join the rest of the herd in flight. Bucking was an instinctive, defence reaction to dislodge a predator leaping on the back. Thus, if a horse is frightened, his first thought is to gallop off, bucking as he goes.

If a frightened horse is cornered, his instinct is to kick his way out. He may also stamp and strike out with his front feet. Allied with this protective behaviour is very acute hearing, good eyesight both forwards and to the side – and a nose sensitive to any strange or foreign smell. Horses therefore react very quickly when anything disturbs or alarms them. Their first thought is flight; if this is made impossible they resist and fight. It is essential that on these occasions any person dealing with the horse is of a calm temperament and unafraid.

Horses have little reasoning power but excellent memories. Their training from the earliest age must establish confidence, trust and good habits, which in time become instinctive. Their memory can, however, also work against them. If startled or frightened by circumstances or people, a horse never forgets. Given the same situation he will always be likely to think back and react in a similar way.

Most bad behaviour by horses, both in and out of the stable, is caused by incorrect handling, particularly when the horse is young. The horse is a creature of habit, and appreciates a regular routine. He can be easily upset by sudden and unexpected change.

PROBLEMS AND POSSIBLE CAUSES

Pulling back when tied up
Cause: Tying up before the horse is calm. Tying up too short, so that the horse feels restricted.

Kicking or biting when being groomed
Cause: Nervous groom, sensitive skin, rough handling.

Bad to shoe
Cause: Insufficient handling of feet when young. Hasty or rough farrier.

Bad to box
Cause: Hasty or rough handling, insufficient time taken to box. Frightening the horse by driving too fast, particularly on corners.

Kicking when travelling
Cause: Insufficient room to balance, resulting in restriction and fear.

Pulling back when in trailer
Cause: Tying up before back strap and back are put up.

Resisting examination and treatment
Cause: Fear, pain, nervous handler.

THE HERD INSTINCT

This is exhibited in the following ways:

- Young horses often show unwillingness to leave others or reluctance to work alone.

- Horses left on their own when others are taken away will often gallop about, and may even jump over a fence or gate.

- A group of horses or ponies in a field will usually have a herd leader and a very distinctive pecking order. This is very obvious when hay is fed loose on the ground in winter.

As the herd instinct is very strong it can be used to help control and manage horses in such instances as:

- When horses are stabled they are always happier and more settled if other horses are within sight and hearing.

- When travelling, a young or nervous horse often settles if allowed a quiet older pony as a travelling companion.

- Horses at grass are happier when in company. Ponies who are difficult to catch will often give in if encouraged to follow others to the gate.

- A horse who gets loose when away from his stable or field feels insecure and worried and tends to return to familiar surroundings. He usually goes back to either the stable or outside the field gate.

THE TEETH AND AGEING

The horse has three types of teeth. These are:

(1) molars or grinding teeth;
(2) the incisors or biting teeth; and
(3) the tushes.

The **molars** are situated to the rear of the mouth, on either side of the face and in the upper and lower jaws. The **incisors** are in the upper and lower jaws in the front of the mouth. The **tushes** are found in the space between the other two groups of teeth. To open the mouth safely the fingers may be inserted into this space, which is where the bit lies.

The horse grows two sets of teeth during his lifetime: temporary milk or deciduous teeth, and permanent teeth. The temporary teeth are small and white with a distinct neck. Permanent teeth are larger, stronger, pale fawn or yellow in colour, with no distinct neck.

TYPES AND NUMBER OF TEETH

Molars or Grinding Teeth

These are in the long head bones and jaw bones. Their function is to grind food. There are twelve temporary molars (also called premolars): three in each side of the upper and lower jaws. There are twelve permanent molars (also called true molars): three in each side of the upper and lower jaws. The twelve temporary molars are called premolars because each one is replaced by a permanent tooth. In addition, there are twelve true

molars, which are permanent from the outset, not replaced like the premolars. By the time the horse is three and a half to four years old, it has a second set of twelve permanent premolars plus all twelve true molars, making a total of twenty-four permanent cheek teeth.

The upper jaw is wider than the lower jaw and this, in conjunction with the criss-cross movement of the jaws as the horse eats, causes the teeth to wear unevenly and require regular rasping to maintain a level surface. It is the outside of the upper teeth and the inside of the lower teeth which will need attention. If the sharp edges are left, mastication is affected, and the horse may also cut the inside of his mouth. The teeth should be checked every six months.

Incisors or Biting Teeth

These form a rather flat arch in the front of the top jaw or maxilla, and in the front of the lower jaw, the mandible. Their function is to cut through growing herbage and to collect food.

There are twelve temporary incisors and twelve permanent incisors, six in each jaw. The centre teeth are called 'centrals', on either side of them are the 'laterals' and the remaining ones are called 'corner' teeth.

Tushes

These are found in the mouths of adult male horses. They appear between the incisor and molar teeth, one

2 years

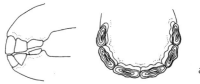

all baby teeth

2½ years

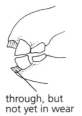

through, but not yet in wear

central adult teeth

3 years

baby teeth

adult teeth

3½ years

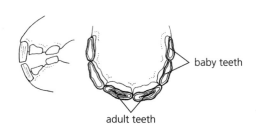

adult lateral teeth come through towards four years old

4 years

baby teeth

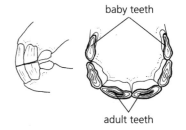

adult teeth

4½ years

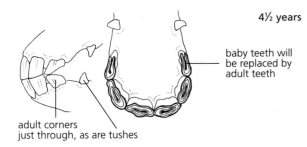

baby teeth will be replaced by adult teeth

adult corners just through, as are tushes

5 years

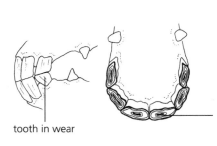

corners not in wear

oval teeth; well-defined infundibulum

young teeth; corners not in wear

6 years

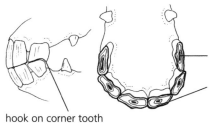

full mouth in wear; infundibulum still visible in all incisors

tooth in wear

7 years

infundibulum well defined in corners, fading in centrals

hook on corner tooth

8 years

remains of infundibulum

fang hole

9 years

slightly longer teeth

10 years

infundibulum almost gone   fang hole

10–12 years

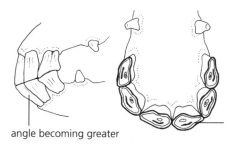

angle becoming greater   less oval; infundibulum gone

15 years

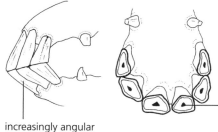

increasingly angular   more triangular in shape; pulp cavity or fang hole shows alone

Milk Teeth

Birth or shortly after	Two central incisors appear in each jaw
4–6 weeks	Lateral incisors appear
9–10 months	Corner incisors appear
12 months	All incisor teeth are in full wear
2 years	Incisor teeth show signs of wear

Permanent Teeth

2½ years	The central milk teeth are gradually replaced by the permanent incisors which at 3 years are in full wear
3½ years	The lateral permanent incisors appear and at 4 years are in full wear
4½ years	The corner incisors appear and in the male, the tush
5 years	The corner incisors are well up but smaller than their neighbours. The tables are not quite level or in full wear. The tush is well grown
6 years	The corner incisors are in full wear but the infundibulum or 'mark' is less distinct
7 years	The mark on the central incisors may have gone, and that on the lateral incisors may be less obvious. A distinct hook shows on the edge of the upper corner incisor: the 'seven-year hook'
8 years	The mark may have disappeared from the lateral incisor and is less distinct on the corner incisors. The seven-year hook has been worn level. The fang hole or 'dental star' is seen in the central incisors as a line in front of the mark
9 years	The mark has all but gone from all the incisors. The dental star appears on the lateral incisors
10–12 years	The dental star gradually becomes a spot rather than a line and is present in all incisors

in each jaw, four in all. They appear at the age of three-and-a-half to four years, and are fully developed twelve months later. They were originally 'weapons', but now serve no purpose in the mouth of the domesticated horse.

If awkwardly placed, they can sometimes cause discomfort when a bit is in the mouth. As the horse grows older they often become encrusted with tartar, which should be removed to prevent the gum from becoming inflamed.

Tushes can be present in the mouth of mares, but they are small and often barely show through the gum.

Wolf Teeth

These small extra teeth which may erupt between the ages of two and five years old are usually found more commonly in the upper jaw, close up against the first molar tooth. They may be of varying size and root depth, but are always smaller than the true molar. They are usually discarded with the milk teeth. There are occasions when they remain in the jaw and cause the horse considerable discomfort when bitted. Their presence often goes unnoticed, and they can be the cause of mouth problems.

It is almost always advisable to have wolf teeth removed by your veterinary surgeon. This should be a simple task, but occasionally may require surgery if deep roots are encountered.

Before starting to train young horses or retrain older horses their mouths should be inspected in case of teeth problems. When inspecting the mouth, always untie the horse. Most horses will show some resentment when having their mouths looked at, so a tactful approach is needed and this should be carried out by an experienced person.

THE PARTS OF THE TEETH

The surface of the tooth which bites and/or grinds the food is called the **table,** also known as the tabular surface, masticating surface or occlusal surface; it is this which gets worn down.

The **crown** of the tooth is that part which is above the gum. In milk teeth, the point where gum and tooth meet is the **neck.**

The **root** part of the tooth lies within the jaw. It is hollow and its cavity contains the blood vessels and nerves which nourish the tooth.

As the incisor tooth grows older, this cavity, the **fang hole,** fills up with a substance of a lighter colour. As the table of the tooth wears down, this will eventually appear as a small white area on the centre of the tooth in front of the mark.

The **mark** or **infundibulum** is the blackened depression seen on the table of the permanent incisors. It is lined by a ring of enamel. In the young tooth the mark is broad and deep, but as the table of the tooth wears down, the mark becomes shallower and eventually disappears. The blackening of the mark is due to discoloration from food and is not present in the new tooth.

TEETHING

When new teeth are coming through, the horse's gums become inflamed and painful, and if the animal is on feed other than grass they may make eating uncomfortable. Such animals should be put on a soft diet and meadow hay. Ridden horses can become more sensitive in their mouths, and if the problem is not recognised this 'mouthiness' can turn into a permanent habit. Horses should not have a bit in their mouths until the teeth are through and the gum healed, so the mouths of all young horses should be regularly checked, particularly when they are likely to be cutting teeth. Should there be any unusual tooth formation or wolf teeth present, veterinary advice should be sought.

AGEING THE HORSE

The official age of Thoroughbred and warmblood horses is taken from 1 January.

The age of a horse can be assessed by inspection of his incisor teeth. This assessment can be reasonably accurate up to the age of eight years. After this, changes are visible, both in the shape of the teeth and the tables, but the time of their occurrence varies among different horses, largely according to their management and diet. Experienced observation of many different horses enables an approximate opinion to be given. When offering this opinion, consideration should be given to all aspects of the mouth and teeth.

Inexperienced people may mistake a two-year-old mouth for that of a five- or six-year-old. Both mouths will have a full set of incisor teeth, but the teeth of a two-year-old are smaller and whiter. The molar teeth of a two-year-old consist of three temporary molars and two permanent molars in each jaw. The five-year-old has six molars in each jaw. A five-year-old male horse will have a fully grown tush.

Looked at from the side, the teeth meet approximately at a right angle. From now on changes appear on the tables, first of the central incisors, followed by the lateral and then the corner incisors. Changes in the mouths of undershot or overshot horses do not necessarily follow the normal pattern.

Galvayne's Groove

This is a depression on the outer side of the upper corner incisor. It is seen mainly in cold-blood heavy horses. It is often absent in warmblood light horses or ponies. It appears just below the gum at nine to ten years, has grown half way down by fifteen years, and to the bottom by twenty years. It is half grown out at the top by twenty-five years and disappears at thirty years.

Eleven-Year Hook

Although this normally appears at eleven years, it can appear any time from nine years on. By reference to the tables and the length and shape of the teeth, it should be possible to distinguish it from the seven-year hook. It usually persists throughout the rest of the horse's life.

Shape of the Table and Angle of the Teeth

From the age of seven years onward, the tables of the teeth will change from oval to round and then to triangular. The back of the tooth forms the apex of the triangle. As the horse ages, his gums recede, his teeth appear to be longer and they project more forward and at a more acute angle.

Bishoping

This is the practice of attempting to make an old mouth look young. The teeth are filed short, and a false mark is gouged out of the centre of the table. The practice can be recognised, as the teeth do not meet naturally, and there is no enamel lining to the false mark.

Crib Biting

When examining a horse for age it is easy to recognise signs of crib biting. The central, and possibly lateral, upper teeth are worn on their outer edges.

5 BREEDS

Breed societies keep a stud register. For a horse or pony to be recognised as belonging to a certain breed and to be recorded in the stud book of that breed society they must be qualified. Most breed societies are affiliated to the British Horse Society. An up-to-date list can be found in the current **Horseman's Year Book** published by the British Horse Society.

The following are some of the well-known breeds:

Thoroughbred

The word Thoroughbred is used to describe horses registered in the **General Stud Book**, which was first published in 1791 and is generally referred to as the GSB. All these horses can trace their ancestry in the male line to three Arab stallions imported into Britain in the late seventeenth and early eighteenth centuries: the Byerley Turk, the Darley Arabian and the Godolphin Arabian. On the female side, all horses in the GSB trace back to some thirty mares, mostly of Arab blood, originally bred or imported by King Charles II in the early seventeenth century and known as the 'Royal Mares'.

The accepted abbreviation for Thoroughbred is TB.

The General Stud Book is kept by Messrs Weatherby of Wellingborough, Northamptonshire, and TB types are registered by reference to them.

Foals should be registered before they are four months old. The information required is sex, age, colour and markings, plus the name of the previous owner and/or breeder and the sire and the dam.

Since 1974 Weatherby's have kept a **Non-Thoroughbred Register,** previously published by Miss F. M. Prior as the HB (half-bred) Stud Book. Up to 1987 this was for the produce of TB stallions out of mares who had been identified by Weatherby's. Entry has now been widened to include the produce of certain non-TB stallions registered with Weatherby's.

British Sport Horse

The National Light Horse Breeding Society or Hunters' Improvement Society (HIS) now has a new name – The Sport Horse Breeders of Great Britain (SHB). Its aim is to identify and promote breeding of the range and type of horse for the future, largely using Thoroughbred blood. It will also assist in the registration, promotion and marketing of horses bred in the UK and establish British-bred horses internationally.

British Warmblood

This is a new breed group started in the 1970s. The foundation stock are the Continental warmbloods (Hanoverian, Danish, Dutch, etc.), many of which have been mated with the Thoroughbreds in Britain. Mares and stallions are eligible for the stud book on the basis of pedigree, plus conformation, action and veterinary tests. The aim is to produce a British warmblood with correct conformation, athletic movement and a tractable temperament, which is particularly suitable for dressage, jumping and general riding.

Arab

The Arab is a foundation breed, crossing with which has improved other breeds of horses. Many countries now have their own Arab breed societies and stud books. In the United Kingdom the Arab Horse Society produces three stud books:

- for pure-bred Arabs.
- for Anglo-Arabs – a cross between Arabs and Thoroughbreds.
- for part-bred Arabs – horses having a minimum of 25% Arab blood crossed with any breed other than Thoroughbred.

Cleveland Bay

This is a breed of great antiquity originating in Yorkshire. It has been used mostly as a carriage horse, and with the revival of driving is still popular. It is also in demand as a show jumper, eventer and hunter. In recent years Ferdi Eilberg has produced the Cleveland Bay Arun Tor as a renowned successful international dressage horse.

Irish Draught

This breed has been used in Ireland as a general farm horse. A stud book was started in 1917, and in the 1970s steps were taken to maintain the breed which, because of the popularity of its cross-bred, was diminishing in numbers. When mated with Thorough-bred or Arab blood the mares produce quality weight-carrying stock, which has achieved great success in competitive sports and as hunters.

Hackney

The Hackney horse is descended from the Norfolk Roadster, which until the mid-nineteenth century was used as a general utility animal on farms in the eastern counties. The first stud book was published in 1883. The modern Hackney, or Hackney Pony, ranges in height from 13–15.3hh (145–155cm). It has a fast and spectacular trotting gait, with good use of the shoulders and great flexion and thrust from the hock. The Hackney is now mostly seen in the show ring and in driving competitions. It can be a bouncy riding horse.

Cross-breds

The pure-breds do not fulfil all the demands of riders or drivers, and in Britain there has been considerable cross-breeding to produce cobs, hacks, hunters and show jumpers. As mentioned above, the mountain and moorland breeds, particularly the Welsh Cob and Connemara, are popular for this purpose.

Heavy Horse Breeds

These are also used for cross-breeding. Many good heavyweight riding horses are produced by putting a Shire or Clydesdale mare to a TB horse. If Percheron or Suffolk Punch mares are used, the offspring grow less feather on the legs. The Percheron-cross tends to have the better action.

MOUNTAIN AND MOORLAND BREEDS

Nine of these breeds are indigenous to the British Isles. Many of the smaller breeds have some Arab blood, whereas large ponies such as the Dales were crossed with Clydesdales to give the offspring greater size and strength. The severity of the native environment has made the breeds very hardy. For many centuries they have had to survive in winter with the minimum of food and shelter, which has ensured that only the toughest of them have been left to carry on the breed.

The popularity of trekking holidays has increased the demand for the larger ponies and cobs. They make admirable mounts for novice adults, and are capable of carrying heavy riders.

Welsh

There are five types of ponies and cobs registered with the Welsh Pony and Cob Society:

SECTION A is the Welsh Mountain Pony, the foundation stock of all the other sections. The ponies make excellent mounts for children, and have been very successful in the show ring. They have also been used to improve many other breeds. They are well-made, pretty ponies with good movement. They should not exceed 12hh (122cm). Grey is the dominant colour, but all colours except piebald and skewbald are accepted.

SECTION B is the Welsh Pony up to 13.2hh (134cm). This pony has more bone and substance than the Welsh Mountain Pony. Colour as for Section A.

SECTION C is the Welsh Pony cob type over 12hh (122cm) not exceeding 13.2hh (134cm). This pony is an excellent ride and drive type. Colour as Section A.

SECTION D is the Welsh Cob. There is no height limit, but 15hh (152cm) is favoured. This is a general utility animal, which makes an excellent riding or driving horse and is a good jumper. It has been crossed very successfully with other breeds to give greater substance and stamina. When crossed with suitable Thoroughbred stock it has produced successful dressage horses, show jumpers and driving horses. Colour as for Section A, but excessive white markings are not popular.

SECTION E consists of geldings only. Heights are as for the Sections above, except Section D cobs, where height is unlimited.

Exmoor

These ponies have bred on Exmoor for centuries. They are distinguished by their heavy mane, 'mealy' nose and 'toad' (i.e. hooded) eyes. They are about 12.2hh (124cm), and are bay, brown or dun in colour, with 'mealy' shading under the belly and inside the forearms and legs. There should be no white markings or white hairs.

Exmoors make good riding ponies for children, but need careful training. They are also capable of carrying a small adult.

They are now bred off the moor, but the moor pony is still the foundation stock. Exmoors can always be identified by brand marks, which are given when the pony is examined as a foal by Exmoor Pony Society inspectors. If of suitable standard, the pony is allocated a number which is branded on the near flank; the Exmoor star and the herd number are then branded on the near shoulder. Most unbranded ponies bred on the moor are not of true Exmoor stock.

Connemara

This pony breed originated in western Eire, and there are now numerous private studs in England. It makes an excellent mount for a teenager or a light adult, and is known for its good temperament and jumping ability. A Connemara-cross-Thoroughbred makes a good all-round competition horse. The pure-bred stands 13–14.2hh (132–144cm). The predominant colour is grey. There are also blacks, bays, browns and various shades of dun, some verging on palomino.

New Forest

These ponies have bred in the New Forest for many centuries. There are now many private studs, where selective breeding has improved the stock. They make a good type of riding pony, ranging in height from about 12–14.2hh (122–144cm). They can be any colour except piebald, skewbald or blue-eyed cream. They have an equable temperament and are easy to train. Crossed with TB or Arab blood they make excellent small riding horses of great versatility.

Dartmoor

These ponies are indigenous to Dartmoor. They breed wild on the moor but are also bred privately off the moor. They make good children's ponies and are also used successfully for private driving turnouts. In height they should not exceed 12.2hh (127cm). They are usually brown, bay or black. White markings are discouraged. Piebald or skewbald colouring is barred. They make good foundation stock and, when crossed with other breeds such as TB or Arab, have produced many successful show animals.

Dales

These ponies are native to north-eastern England. They are a well-built cob-type of animal, with active paces suited to driving. In height they should not exceed 14.2hh (144cm). In colour they should be black, brown, bay or grey, with limited white markings. The local term 'Heckberry Brown' means a deep chestnut with black dapples on the coat. They grow a considerable amount of fine 'feather' on the legs and heels. They are quiet, docile ponies, much favoured for trekking.

Fell

This breed is smaller than its neighbour the Dales, and is not over 14hh (141cm). It is bred on the western side of the Pennines. It is strongly built, active, and brown or black in colour. It makes an excellent ride-and-drive pony of cob type. White markings are discouraged. Broken colours or chestnut are barred.

Highland

The Highland pony is a breed of great antiquity. There were three types ranging in height from 12.2–14.2hh (124–144cm), but cross-breeding has merged the differences. Well built and sure-footed, they were originally used on small highland farms and for carrying deer. They make good docile riding ponies, and the larger ones are capable of carrying adults. In colour, grey and various shades of dun predominate. There should be no white markings.

Shetland

These ponies originated in the Shetland Isles, where many are still bred. They are a very ancient breed,

exceptionally strong for their size, and should not exceed 42ins (106.6cm). In former years they were employed as pit ponies. They are now used very successfully for driving. Their small height also makes them popular as riding ponies for children – although they are wide and can be strong, requiring firm handling. There are no limitations as to colour.

DESCRIPTIVE TERMS

HALF-BRED Denotes a horse of whom one parent is a Thoroughbred.

THREEQUARTER-BRED Denotes that one parent is a Thoroughbred and the other a half-bred. These animals can now be registered.

TYPE These include hunters, hacks, polo ponies, cobs and vanners. They are usually cross-breds, and are distinguished from breeds as they are not registered in a stud book.

COLD-BLOOD Applies to heavy work-horses such as Shires, Clydesdales and Percherons.

HOT-BLOOD OR FULL-BLOOD Applies to Eastern breeds, i.e. Arab, Barbs, etc, and to Thoroughbreds.

WARMBLOOD Applies to horses with a mixture of blood in their foundation stock. These breeds are used for riding and driving.

PONY The difference between a horse and a pony lies more in build and movement than in exact height, but a show pony should not exceed 14.2hh (144cm).

HORSE A horse is generally described as 15hh (152cm) and over, but any animal over 14.2hh (144cm) may be deemed a horse. Arabs, whatever their size, are always referred to as a horse.

HACK A lightweight horse of TB type. Show hacks range from 14.2–15.3hh (144–155cm).

COB A sturdy weight-carrying type, not exceeding 15.3hh (155cm).

HUNTER A type and usually a half- or threequarter-bred animal, ranging from 15.3hh (155cm) upwards.

POLO PONY A thoroughbred type under 16hh (162cm) with weight-carrying ability and suitable for adult riders.

MULE A cross between a donkey stallion and a pony mare (female horse).

HINNY Cross between horse stallion and female donkey. The name 'Jennet' is also used to denote this cross.

6 BREEDING

Breeding a foal is an expensive and labour-intensive business and should not be undertaken without considerable thought and planning. Unless either the stallion or the mare have a proven record, it is unlikely that the breeder will make a profit on the eventual progeny, whether it is a small pony or a racehorse.

Facilities

It is essential to provide safe facilities for rearing a foal. Good clean grazing of not less than three acres is required. 'Clean' means that it has not been grazed by horses with a worm burden; that is not 'horse sick' (see Grassland Management, page 297 onwards). Ideally, it should have been cross-grazed by cattle or sheep during the previous twelve months. Fertilisers, if used, should be organic. The grazing should have very safe fencing, as foals have little sense of self-preservation, and shelter should be available from the prevailing winds.

A loosebox or barn large enough to hold the mare and foal is another requirement: 4.25 x 4.25m (14 x 14ft) is the minimum size for Thoroughbreds, but 4.6 x 4.6m (15 x 15ft) is more suitable. A smaller size is adequate for ponies.

Selection of Mare and Stallion

It is important to breed from tough, strong mares of good conformation, who are free from known hereditary disease, and who, if possible, have stood up to work.

Consideration must be given to the type of young horse required and, if it is to be sold, as to whether there is a market for it. It is pointless to breed a foal whom no one wants. Proven ability and soundness in both mare and stallion should be considered. The eligibility of the foal to obtain registration papers is becoming more important, so both parents should be registered or eligible to be registered with a society.

A stallion who complements the mare should be chosen. The breed societies have lists of the various stallions standing at stud. The National Light Horse Breeding Society runs a Premium Scheme which chooses selected Thoroughbred horses to stand at stud at reduced rates in most districts. The British Warm-Blood Society has a list of their graded stallions. The stallion should be seen and the quality of the stable management at the stud taken into consideration.

In many cases it is sensible to delay sending the mare for service until late April or May. The foal will then be bred in the warmer weather, and the fresh grass will encourage a good supply of milk from the mare. Foals destined to be shown in-hand are better born earlier, as they are more mature for the first shows.

MANAGEMENT OF THE MARE

Before Being Sent to the Stallion

- The mare should be in good condition but not fat or over-fit. She should be improving in condition rather than being let down.

- She should have been wormed regularly, preferably at four- to six-week intervals.

- She should have been immunised against influenza within the last two months. Her tetanus vaccinations should be up to date.

- She should have had her hind shoes removed (some studs prefer the front shoes to be removed as well).

- Note should have been taken of the dates she came in season over the previous six weeks.

- Most studs insist that mares are swabbed by a veterinary surgeon, and passed clean before they are served. This can be carried out at home, but some studs insist that it is done at the stud.

The stallion owner/stud groom should be informed of all the above facts. In addition, they should be told if the mare is a maiden or barren and – if she has previously had a foal – whether there were difficulties getting her in foal. If she is due to foal at the stud, details of the last date of service and anything known about her previous breeding history should be supplied.

She should be sent to the stud with the following:

- A well-fitting comfortable headcollar, preferably with her name or the owner's name on it.

- Her passport, if she has one.

- Her flu and tetanus certificate if she has no passport.

- A veterinary certificate of cleanliness if she has been swabbed at home.

- A foal slip or headcollar if she is to foal at the stud.

At Stud

She may be out all the time or in at night. She is 'tried' to determine whether she is coming in season. When she is ready, she is served by the stallion. She is 'tried' again two to three weeks later to see whether she has 'held' to the first service. Depending upon the arrangements with her owner, she is returned home.

After Returning Home

- If she has not been tested 'in foal' at the stud, this has to be done (see Pregnancy Testing, page 45).

- She should be kept in good condition, but not allowed to become too fat.

- She should be wormed regularly with a wormer which is safe for pregnant mares.

- She may be ridden gently for a few months if desired. Mares who look well and have shelter from the bad weather may remain out at grass until a short time before foaling. This type of programme is certainly best for mountain and moorland ponies.

- Stabled mares should be turned out as much as possible, but only with other mares or alone, not with colts or geldings. Most mares need to be brought in at night from November onwards, but should still be turned out for exercise during the day. If the fields are too wet, the mare should be allowed to go loose in an indoor or outdoor school.

- After September the grass becomes less nutritious, and extra feed should be given as required.

- Whether in at night or out all the time, as soon as the grass loses its feed value she should be fed good hay and 1.8–3.6kg (4–8lbs) of hard food. Ponies require less hard food. During the last months before foaling, the hard food needs to be increased (see below). A feed compound specifically formulated for broodmares is desirable.

The Last Four Months Before Foaling

- The mare carries the foal for approximately eleven months, and during the last four months the foetus grows very fast, its size and weight increasing three-fold. A vitamin and mineral supplement is needed, and energy requirements should be gradually increased by about 12% over this time, while the roughage intake is somewhat reduced.

- A month before she is due to foal she should have a tetanus booster. She should also have a flu inoculation, if this has not been done in the last six months. Antibodies are passed to the foal in the colostrum (first milk).

- The mare must continue to be regularly wormed. Worms are lethal to young foals, and can be passed in the mare's milk, or picked up by the foal eating the mare's droppings. It is important that the wormer is safe for pregnant mares, and that the instructions are carefully followed.

FOALING (PARTURITION)

The mare carries the foal for between 320 and 350 days. The average is about 335 days.

If the person in charge of the mare is experienced, and if assistance is available, the mare can be foaled at home. If not, and especially if the mare is valuable and has been to an expensive stallion, it is safer to send her to a recommended stud to foal. If she is to be put in foal again she can go to the stud where the stallion stands.

Foaling Away From Home

It is better to send the mare at least three weeks before the expected foaling date, so that the upsets of the journey do not bring on the foaling, and she can settle down and acquire some immunity to the germs at the stud.

Foaling at Home – Facilities

- A large, safe, disinfected loosebox with as few fittings as possible. A good size is 4.6m x 4.6m (15ft x 15ft).

- A deep, clean straw bed, well banked up on the walls.

- Good lighting.

- Some means of viewing the mare without disturbing her.

- A safe available power point, which can be used if necessary for heating.

If she is foaling at home and is the only in-foal mare, it is better for her to sleep in the foaling box a week or two before the expected foaling date.

Signs of Foaling

BAGGING UP During the last few days the udder grows larger, stiffer and rounder. During this time the udder may be gently handled, so that the mare does not resent being suckled. This is particularly beneficial with maiden mares.

WAXING Usually between six and forty-eight hours before foaling starts, the teats exude a wax-like substance. This is not a reliable sign, as it may be closer to the foaling, or the mare may not wax at all.

SLACKENING The muscles on either side of the root of the tail slacken; this occurs usually twelve to eighteen hours before foaling.

When Foaling is Imminent

At this stage, it is wise to put a light, well-fitting headcollar and a clean tail bandage on the mare. The likely signs, which may be spread over a period of about four hours, are:

- Walking the box.

- Pawing the bedding.

- Looking round at flanks.

- Kicking at the belly.

- Swishing the tail.

- Lying down and getting up again.

- Starting to sweat.

At this time the mare should be watched constantly, but without disturbing her. At night, the lights should be left switched on.

The veterinary surgeon's telephone number and an antibiotic powder for the foal's navel should be at hand.

The assistant's hands should be well scrubbed before handling the foal or mare.

Normal Foaling

Usually, but not always, the mare lies down.

1. The foetal fluids are discharged as the bag breaks. The foal should then be born within 20–40 minutes.

2. Within about five minutes, the front legs appear, still inside the membrane, with one foot slightly in advance of the other.

3. The head appears, lying along the front legs.

4. The shoulders appear, and then the mare usually has a short rest.

5. The membrane over the nostrils (caul) should break. It is essential that this is done before the navel cord is broken, otherwise the foal suffocates. If the foal has not broken this for himself, the watcher may do so, but not until the foal's ribs are clear of the mare.

6. The hind legs slip out, or are withdrawn by the foal.

7. The foal may struggle to get up, or the mare may get up and the cord will be broken naturally. Antibiotic powder can then be sprinkled over the stump.

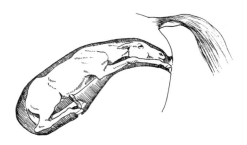

Position of foetus immediately prior to foaling.

After the first discharge (Stage 1 above), the foal has only one to three hours to live inside the mare. If there is any deviation from the sequence just described, veterinary help should be sought immediately. The vet requires hot water, soap and clean towels.

After Foaling

If all has gone well, which in the majority of cases it does, the mare and foal should be allowed to rest, but should be watched unobtrusively. Usually within an hour the foal is on his feet and trying to suckle. It is very important for him to get this early milk (colostrum), as it contains the antibodies which protect him for the first few months of his life. After that time he starts to manufacture his own. Some foals are slow to find their way to the udder and have to be helped.

Sometimes a maiden mare resents the foal and has to be restrained. The assistant can hold up a front leg; a tranquilliser may be given; or in very severe cases the mare can be twitched. As the application of the twitch can cause a rise in blood pressure, care must be taken. It is possible to cause a haemorrhage in a newly foaled mare.

Sometimes the foal does not get up. This may be because the cord has broken early, and the foal has not obtained a full blood supply. If he otherwise seems bright and healthy, two people may support him with a clean sack under his belly, so that he can get to the mare's udder. There are cases where it may be necessary to milk the mare and let the foal drink from a sterilised bottle with a calf's teat on the end. If the foal is not bright and alert, or begins to behave strangely, veterinary help should be sought immediately.

The **afterbirth** (foetal membranes) gradually separates from the uterus and is expelled by the mare, often within an hour of foaling. If it is retained for more than three hours, veterinary help should be sought. On no account should any attempt be made to pull it away. If possible, it should be inspected to make sure that nothing has been retained inside the mare. The placenta has a main 'body' which filled the womb and enclosed the foetus, with two 'horns' at the top.

As soon as the mare is on her feet, she should be offered a chilled drink and (if she will take them) electrolytes may be added. Hay may be given, and after an hour or two a small feed. Hay should always be fed loose, so that there is no danger of the foal becoming entangled in the net.

Foaling Outside

ADVANTAGES
- It is nearer to nature.
- There is less risk of infection.
- More natural for native breeds.

DISADVANTAGES
- In an emergency it is difficult to provide human assistance, especially at night.
- The mare may choose a dangerous place to foal, e.g. too near a fence.
- In bad weather the foal may suffer.
- Once the foal is on his feet the mare may refuse to be caught.

If this practice is to be followed, a safe, level, well-fenced field with no ditches, ponds or other hazards should be chosen. There should be no horses or cattle in the field and it should be adjacent to the stables. There should be shelter, e.g. a thick hedge, or shed.

Usually the system works well with ponies, but for valuable horses it is not advisable.

The mare and foal should be caught and handled each day. In bad weather, driving rain and wind, they should be brought in.

MANAGEMENT AFTER FOALING

The First 24 Hours

Some people like their foals to have an antibiotic injection within the first 24 hours. In any case, it may be a wise precaution for the vet to check the mare, the foal, and the afterbirth. He can also check whether the **meconium** (foetal dung) has been expelled correctly .

The mare and foal should be looked at frequently. The foal must be seen to be suckling successfully. He should pass the meconium, which is blackish and may be hard, within thirty-six hours. If this is not passed, veterinary assistance may be sought, or an experienced person can give an enema.

The mare may remain on the same diet as before foaling – i.e. high protein mash-type feeds – but smaller quantities of energy food should be given for the first few days. It is unwise suddenly to introduce bran mashes, as the mare's digestion will not be used to bran. If she has been having flaked barley, mashes may be made from this.

After the First 24 Hours

If the weather is good, the mare can be led out with the foal (see Handling the Foal, page 43) and given an hour's grazing. With an unknown mare it is wise to graze her in-hand the first time she goes out. She should have a headcollar on when she is turned loose with the foal. When the mare is released the foal should be in front of her, where she can see him, so that she will not kick him by mistake as she gallops off. The foal should be led out and not allowed to go loose. He should be released before the mare.

March foals should come in at night with their dams until the weather improves. If the weather is suitable, May foals (after they are one week old) can be left out at night, but they should be caught and handled daily.

When flies become troublesome the mare and foal may be brought in during the daytime. They should also be brought in during bad weather. If a foal is very cold and shivery he should be rubbed down, and if possible put under a heat lamp. It is also wise to bring foals in before a thunderstorm, as horses shelter under trees and are therefore vulnerable to lightning. During storms young horses can panic and may gallop through fences.

Scouring

The foal is likely to scour when the mare first comes into season on the foaling heat. This is quite normal.

However, should he start scouring at any other time, an antibiotic injection may be necessary to deal with any infection. Severe scouring can be serious owing to the danger of dehydration.

The foal's hindquarters should be kept clean by washing and drying them carefully. Baby oil applied on either side of the tail can help to prevent the coat being lost.

Worming

Most modern wormers are safe for foals. The instructions should be carefully followed. Worming should be carried out as early as the product allows, usually at four weeks old. Keeping foals as worm-free as possible is a vital safety measure. Much damage, if not death, can result from a worm burden and – even if not evident at the time – can result in serious trouble later in life. Many cases of colic and subsequent death are attributable to worm infestation when young.

The mare should be wormed at the same time as the foal. Thereafter it is sensible to worm them both every four weeks.

FEEDING THE MARE

Feeding the mare depends on the quantity and quality of the grass. In May and June, when on good grazing, no supplementary feeding may be necessary. At other times, or if on poor, restricted grazing, the mare will need additional food to maintain her own condition and the milk supply for her foal.

Quantities must vary according to circumstances, and a close watch should be kept on the condition of both mare and foal. The foal may or may not look thin and ribby when he is born, but should quickly start to look round and well.

The mare's condition has no bearing on her foal's well-being. Very fat, good doers often make the worst mothers, both in producing a foal in good condition and in keeping it looking well.

When the grass is highest in protein and at its most digestible – i.e. in May and June – the diet must not be too high in protein. Before the grass starts to grow, and, again, when it starts to lose its value (after it has seeded) the protein/energy food should be increased. If the mare is in foal again, this should be taken into consideration.

Specimen Daily Feed

DURING WINTER AND THE MONTH BEFORE FOALING:

> Best meadow hay ad lib.
> 1.8–3.6kg (4–8lbs) oats, flaked barley or stud cubes or mixture.
> 0.45–1.8kg (1–4lbs) dry sugar beet soaked for 24 hours.
> 57g (2oz) milk pellets or 113g (1/4lb) soya-bean meal.
> 28g (1oz) salt.
> Calcium – ground limestone or equivalent: 28g (1oz) (This is particularly important on land with a low lime count.)
> Additives – selenium and vitamin E if the land is deficient in selenium.

Two weeks before foaling, you can, if you wish, begin to substitute flaked barley for part of the oat ration. The oats at this stage should be rolled and both they and the sugar beet may be made into mashes if required, both before and after foaling. Bran mashes should not be suddenly introduced.

MARCH AND APRIL: As above.

MAY AND JUNE: Depending on the grazing, no supplementary feed at all or up to half of the above without milk or soya.

JULY ONWARDS: Working gradually towards the March ration, but without milk or soya.

OCTOBER ONWARDS: As the grass begins to lose its value, dried milk or soya meal may be re-introduced to ensure that the essential protein content of the diet is maintained.

FEEDING THE FOAL

Encourage him to eat some of the mare's rations by putting the food in two separate feed bowls.

Foals requiring extra food because they are not doing well should be fed by hand on a nutritious and easily digested diet.

Too high a protein diet causes the bones to grow too fast, and knuckling over may result, but foals require a higher protein diet than mature animals to enable them to grow. If too little calcium is fed, the end of the bones will not harden and epiphysitis may result. A modern specialist rearing diet would be the best recommendation.

Feeding the Weaned Foal

The object is not to allow the foal to go back in condition. Before being weaned he should be eating grass/hay and hard food. The constituents of the diet should not be changed but, if he has still been getting some milk from the mare, his food quantity should be stepped up. Depending on his management during the winter he may require more energy food as the weather deteriorates.

HANDLING THE FOAL

Foals should at all times be treated quietly but with firmness. The foal should be handled within twelve hours of his birth. Some foals are very friendly, others very shy.

To catch the foal, two people are required: one to hold and reassure the mare, the other to catch the foal. The foal may try to escape by going in front of or underneath the mare. He should be approached from behind, never from the front. He should be caught with one hand round the front of the chest and the other round the quarters. Once caught, he should be stroked and soothed.

Within 24 hours, a foal slip should be fitted. An assistant to hold the foal makes this easier.

The Foal Slip or Headcollar

Opinion is divided as to whether the foal slip should be left on all the time. Many people consider this to be a dangerous practice, as it is easy for the foal to get caught up in it. Whether or not it is worn all the time it should be adjustable over the head and around the nose, and should be made of leather, not nylon, unless fitted with 'panic breaks'.

To put on the headcollar, fasten the headpiece round the neck first, push it forwards to behind the ears, and then fasten the noseband. To begin with it will need adjustment for length and tightness every three or four days. It must be carefully fitted, so that it is not too tight, but must never be loose enough for the foal to trap a hind leg when scratching his face.

Each day the foal should be handled all over, and a hand run down his legs. After a few days his feet may be picked up.

Teaching to Lead

Leading lessons should start in the box. With a very young foal, put a stable rubber round his neck and move him along with a hand around his quarters. As soon as he understands this, a 1.8m (6ft) light rope may be threaded through the foal slip to guide and steady him, but a hand should be kept round his quarters to push him along and to prevent him running back, as he may fall over. The foal should be 'played' like a fish. The rope must be threaded through rather than clipped on, so that should he inadvertently escape, the rope will not be left dangling, which will frighten him.

Foals should not be allowed to go to and from the field loose; they may easily gallop into something and damage themselves. If the leading lessons are started early enough, they learn very quickly. If the training is left too long they become much stronger and, being unused to discipline, will be difficult to control.

When leading them out to the field, the mare should be in front with the foal close to her side. The person leading the mare should watch the foal, and stop immediately if he is left behind.

Feet

When the farrier comes to dress the mare's feet he should handle the foal and pick up his feet, even if they do not need attention. Foals handled only by women can become quite nervous with men. It is important to study the foal's action and to keep his feet correctly trimmed. This will prevent unlevelness of the foot caused by faulty action.

WEANING

When deciding on the best time to wean, several considerations must be taken into account.

If the mare is not in foal again there is no hurry, and the foal may be left with the mare until she weans him naturally. If the mare is in foal again, the best time for weaning is usually when the foal is five to six months old.

For the last month, the foal should have had a companion, ideally another foal, or, failing that, a small, good-natured pony or worm-free donkey. The foal (and his companion) should be put in a box which he knows, and both the bottom and the top door should be shut.

The mare should be removed to a safe place out of earshot of the foal. A box a mile away may be necessary. They should be kept apart until the mare's milk dries up and the foal begins to settle down. They should not be put near one another for at least four weeks.

After Weaning

The mare must be given a low protein diet and plenty of exercise. If she has a lot of milk, her bag should be examined carefully each day to make sure that she is not developing mastitis. It is not advisable to draw off the milk, as this encourages further production.

Pony foals may winter out as long as there is shelter, but young horses usually lose too much condition. The latter are best turned out in the daytime, and kept in at night. This means that they lie warm and dry without expending food to keep themselves warm, and they are able to take the daily exercise which is necessary to ensure that they grow up strong and active.

STALLIONS

Stallions who are used solely as stud horses have a rather different lifestyle from those who are ridden in equestrian activities, though some do both.

The annual programme for Thoroughbreds, or similar stallions, used solely for stud work begins three to four months before the start of the stud season. At this time, daily exercise must begin, so that the horses are fit before starting their duties. At Thoroughbred studs, mares are covered in February, but at other studs usually not before March.

After the season is over, which may be June, or even later, the stallions should be let down and roughed off. Stallions who are ridden as well as being used at stud, should have a particularly strict regime so that they understand what they are required to do in the differing circumstances. It is important to use a different bridle and to choose a special place for serving the mares.

When dealing with stallions, discipline is of paramount importance. Their natural pride and arrogance must not be spoiled, although they must understand that their handler is the boss.

When being attended to, the horse should always be tied up. A chain, rather than a rope is essential, as stallions by their nature pick up everything with their teeth. People dealing with them should not wear scent or aftershave, as the smell is likely to arouse them.

When holding or leading a stallion it is advisable to carry a stick or whip. If he is being led in a headcollar, a clip chain under the jaw or over the nose, attached to a long lead rope, is a sensible precaution. If he is being led in a bridle, a coupling to the bit and a lead rein 2.4–3.6m (8–12ft long) should be used. The attributes for dealing with all stallions are patience, firmness and common sense. There should be no fear and no loss of temper.

PREGNANCY TESTING

MANUAL This is carried out by a veterinary surgeon, from 35 days after service up to the time of foaling. Some experienced vets can detect pregnancy earlier than this.

BLOOD-TESTING This is carried out between 45 and 100 days after service. A blood sample is collected from the jugular vein and sent to a laboratory for analysis. The test is a biological one, for gonatrophic hormones.

URINE TESTING This is carried out at any time after 130 days. A urine sample is collected by the owner and sent to a laboratory to be tested for oestrogenic components.

ULTRA-SOUND SCANNING This method of testing is now available from most veterinary surgeons who specialise in treating equine breeding stock. It can be used from 17 to 20 days after the last service.

IDENTIFICATION

Brand Marks

Brand marks are used to denote breed or ownership. A hot branding iron carrying the appropriate mark is applied to the horse's skin. Horses imported from Europe often carry a breed mark, which is usually found on the left hindquarter but can be on the shoulder and/or neck.

Many native ponies breeding wild on the hills and in the forests are branded, to denote ownership. A general round-up takes place in the autumn, and ponies in their first year and then still running with their mothers can be easily identified and branded. The brand is usually on the quarters, under the saddle, or on the shoulder.

Freeze-Marking

To deter thieves, many owners now have their horses freeze-marked. The process has been patented by Farm Key of Banbury, who keep a national register of horses, their owners, and their allocated numbers. The process is quite painless and leaves a permanent white mark. It is applied either under the saddle or under the mane.

Describing a Horse

Consider the following:

- **Colour.**

- **Sex.**

- **Markings.**

- **Height:**
  (a) With a measuring stick (see below).

  (b) By eye or by touch. Use your own eyes, mouth or chin as a marker, e.g. for a 5ft 3in. person the chin is about 14.2hh (147cm). Then stand up to the horse's wither to gauge the approximate height.

- **Age** – See Chapter 4, page 27 onwards. Examine teeth: length, tables, shape and condition.

- **Type** – Stand back and check build and bone shape.

Sex and Age

The word 'horse' can be used to describe the whole equine species without reference to sex. The following terms are more precise descriptions:

- STALLION/ENTIRE Uncastrated male horse of any age.

- GELDING Castrated male of any age.

- COLT Uncastrated male under four years of age.

- MARE Female horse of any age.

- FILLY Female under four years of age.

- FOAL Described either as a 'colt' or 'filly' foal and under one year old.

- RIG Male horse that has one testicle retained in the

abdomen. The condition is often difficult to diagnose and may well escape veterinary inspection. It can often be assumed from misbehaviour in company, or reactions when turned out with other mares and geldings. A rig is capable of putting a mare in foal. The condition can be rectified, but would probably involve a major operation. Rigs are not suitable mounts for children. If known at the time of a sale the condition must be declared. If established at a later date, the sale can be declared null and void.

Height

Traditionally horses are measured in hands (4ins/ 10.16cm). All height measurement for competitions and documentation is expressed in centimeters. The measurement is taken from the highest point of the withers in a perpendicular line to the ground. For accurate measurement, the following items and conditions are essential:

- A measuring **stick** with a spirit level on the cross bar.

- The horse should be standing on a smooth and level surface.

- The horse must stand squarely on all four feet. His head should be lowered so that the eyebrows are in line with the withers.

- The horse must be calm and must remain still. If he is tense he should be allowed 15–20 minutes to get used to having the stick placed on the withers.

For the purposes of the Joint Measurement Scheme, the horse must be measured without shoes. For general purposes, the shoes – providing they are of normal weight – may remain on and 12mm (½in.) is taken off the measured height.

Life measurement certificates are now obtainable. The horse must be six years of age or over, measured without shoes, and the measurement must be taken by a member of a special panel appointed by the Joint Measurement Scheme.

Correct height measurement is important because:

- It forms part of a correct description of the horse for sale.

- It provides for correct division and sub-division of horses in showing and jumping classes, and in other competitive sports.

- It serves as some indication of size when buying saddlery and clothing, although type and build are of equal importance.

Colours

For the purposes of identification, it is important for a horse's colour and marking to be described correctly. His precise colour can sometimes be difficult to determine, but reference to the colour of his 'points' – muzzle, eyelids, tips of the ears, legs and the mane and tail – should clarify the problem. The legs below the knee can be black or a darker version of the body, but they also may have white markings. A horse with no white markings is said to be **whole coloured**. When describing horses, any patches of white hair caused by pressure or injury should be noted. These are most likely to occur in the saddle and bridle area or on the legs. Scars should also be noted.

- A **black** horse is black in colour with black points and a black muzzle.

- A **brown** horse is dark brown in colour with black limbs, mane and tail.

- A **dark bay** horse is mid-brown in colour with black limbs, mane and tail.

- A **bright bay** horse is mahogany in colour with black limbs, mane and tail.

- A **light bay** horse is a paler shade of brown, or pale mahogany, with black limbs, mane and tail.

- A **bay-brown** is a horse where the predominating colour is brown, with bay muzzle and black points.

- A **dark chestnut** is a rich red colour with matching

points. He sometimes has small patches of black hair on the body. He may be whole coloured as in a Suffolk Punch, but is most likely to have white markings on his legs.

- A **chestnut** is a paler version of the above and may have a flaxen mane and tail.

- A **liver chestnut** is a darker shade verging on brown with darker points.

- A **grey** horse is one with both black and white hair growing in the coat, with matching points and mane and tail. The skin is black.

- An **iron grey** has predominantly black hairs and can appear nearly black.

- A **light grey** has predominantly white hairs.

- A **flea-bitten grey** has dark hair growing in speckles over the body.

- A **dapple grey** has circles of black hair over the body.

NOTE: All grey horses become lighter with age, but are never described as white. On examination, the skin will be seen to be dark coloured.

- A **white** horse is one whose skin is white or pale pink. The skin is lacking in pigmentation, and the coat is white. It is a very unusual condition.

- A **blue dun** is a diluted black colour evenly distributed. The mane and tail are black. There may be a dorsal stripe. The skin is black.

- A **yellow dun** horse varies from mouse colour to dark gold with black points. He may show a 'list', a dark line along the back bone. The skin is black.

- A **palomino** varies from light cream to bright gold with similar coloured points, and lighter or silver coloured mane and tail.

- A **cream** horse has a light cream coat verging to white in the muzzle area and legs. The muzzle is white. The skin lacks pigment. The eyes may also have a pinkish or bluish appearance.

- A **roan** has a mixture of white and other colours growing in his coat. There is a tendency to get whiter with age.

- A **strawberry roan** is white and red with similar points.

- A **blue roan** is white and black with black points. The coat has a blue tinge.

- A **red roan** or a bay roan is white and bay or bay-brown with black limbs, mane and tail.

- A **chestnut roan** or **sorrel** is white and light chestnut with matching points. Mane and tail similar or chestnut in colour.

- A **piebald** is a mixture of large irregular patches of black and white. The mane and tail may also be black and white.

- A **skewbald** is a mixture of large irregular patches of white and any other colour or colours.

- A **spotted** horse often has pink or mottled skin. It may have: (a) **leopard spot** markings, when dark spots are distributed over a lighter background; (b) **blanket markings**, when there are dark spots on the rump of a lighter coloured horse; (c) **snowflake markings**, when white spots appear on a darker background.

- An **Appaloosa** has a pink skin and a silky white or grey coat with darker coloured spots on the coat. These markings can include leopard, blanket or snowflake markings.

- **Odd-coloured** horses are those whose body coat consists of large irregular patches of more than two colours, which may merge at the edges.

Markings

The Head

- A **star** is a white mark on the forehead. It can be further described as large, small, irregular etc.

Star.

Star and stripe.

Snip.

Blaze, extending to both nostrils.

White face.

- A **stripe** is a narrow white marking down the face. It may be a continuation of a star, and can be further described as narrow, irregular, etc.

- A **blaze** is a broad white marking extending from between the eyes and down the face over the nasal bones.

- A **white face** is an exaggerated blaze. It covers the whole of the forehead, front of the face to the mouth.

- A **snip** is a white mark between the nostrils. If extending to a right or left nostril it should be so described.

- A **white upper lip** and **underlip** describes skin at the edges of the lips.

- A **white muzzle** describes where white skin is found on both lips and up to nostrils.

- A **wall eye** is one which shows a lack of colouring matter. It has a greyish-white or blue appearance. The sight is not affected.

A white wall eye.

White upper and lower lips.

White muzzle.

The Body

- **List, dorsal stripe** and **ray** describe the dark lines along the back of dun horses. They are also found on donkeys.

- **Zebra marks** describe any stripes on the body. They occur more frequently on donkeys.

- **Salmon marks** are fine lines of white hair found on the loins and quarters.

- The **prophet's thumb mark** is a pronounced dimple sometimes found on the neck of Thoroughbred or Arab horses. It can also appear in the shoulder or

Sample description of a horse

NAME SAM

SIRE Scottish Venture.

DAM Breeding unknown.

Chestnut gelding. Foaled 1980. Height 16.1hh with shoes. Mane and tail on. Freeze-marked 6783 under saddle.

HEAD Whorl in centre of forehead above eye level. Small star, narrow stripe and snip to left nostril.

LEFT FRONT Small mark on front of coronet. White stripe on front of hoof. Whorl on posterior-lateral mid-forearm.

RIGHT FRONT Half cannon. White hoof.

LEFT HIND Half pastern to mid-fetlock behind. White hoof.

RIGHT HIND Threequarter cannon. White hoof.

BODY Whorl left anterior crest. Whorl right mid-crest. Whorl left anterior jugular. Whorl mid-trachea.

ACQUIRED Scattered saddle marks on both sides. Girth mark on left side. Scar anterior mid-cannon left hind.

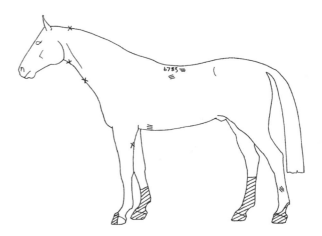

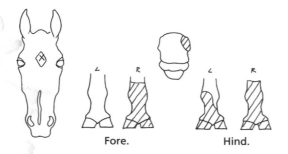

Fore.   Hind.

hindquarters. It is said to be a sign of good luck.

• **Flesh marks** are patches of skin devoid of colouring and care should be taken to use this term instead of white wherever appropriate.

The Legs

• White markings on the legs are defined by reference to the anatomy. The traditional terms 'sock' or 'stocking' are now used only for casual description.

Nowadays the terms 'right' and 'left' are used instead of 'off' and 'near'. For example, typical descriptions are: right pastern; left heel; right leg to above knee.

• The term 'ermine' is used where black spots occur on white markings.

Hooves

Any variation in colour of the hooves should be noted, in particular a white stripe or line down the hoof.

Whorls

Whorls are formed by changes in the direction of hair growth. They are an established method of identifying horses, as they vary to some extent in every animal. They occur on the head, neck, body and upper limbs. When describing horses for identification it is the ones on the head and neck which should be noted. Grey and whole-coloured horses, or those with few markings, should have at least five whorls noted, either on the head, neck or body.

Chestnuts

For horses with fewer than five identifying whorls a print can be made of the chestnuts on each front leg. These are like human fingerprints and show a different shape and/or design in every horse.

Certification of Identity

Official certificates must be completed and signed by a veterinary surgeon. Horses are identified by means of a written description of the horse and a sketch map of his body, together with separate sketches of the front view of the head, muzzle and legs. (See sample description, page 50.)

- **White markings** on the horse should be outlined in red and filled in with diagonal red lines.

- **Whorl**s are shown by an X marked in black.

- **Scars** are shown by a tick marked in black.

- **White marks on a grey** are not noted except for flesh marks.

- The written description of a horse must be in black ink, and should exactly describe the markings, whorls, and scars as shown on the sketch map. The description should start with the forehead.

- The position of head whorls should be exactly described in relation to eye level, centre of forehead, and any white markings. Crest whorls on the neck should be divided into anterior, middle and posterior. Body whorls should be designated by reference to the particular anatomical position.

8 BUYING AND SELLING A HORSE

BUYING

When looking at horses or ponies with a view to buying them, inexperienced persons should always take a more knowledgable person with them. Even experienced people find it helpful to have another opinion.

The owner of the horse should be asked to provide information as to age, any vices and/or bad habits, and suitability for intended use. It can be helpful if this information is written down in the presence of another unbiased person – who can, if necessary, corroborate what has been said.

If the horse is registered for competition he will be registered with the British Horse Database. His record can be verified with the appropriate society and the BHD.

Breed papers should be available for inspection when the horse is first seen. If he is a show animal he may have a height certificate. If not, and if you intend to show him in height-restricted classes, the vendor can be asked to warrant the height.

Before coming to a final decision it is always advisable to have the animal examined by a veterinary surgeon – preferably one of the purchaser's own choice. Veterinary surgeons will not usually give a warranty as to soundness, and cannot certify height unless able to measure the animal under Joint Measurement Scheme rules. If they consider the animal suitable they will provide a certificate giving a description (including approximate height) and stating that the said animal has no abnormality or injury likely to affect his use for the purpose for which he is being bought. This certificate should be sufficient for insurance purposes.

If possible, the intending purchaser should attend the vetting so that any queries can be answered on the spot. If this is not possible, it is certainly helpful to have a verbal report and to discuss the examination with the veterinary surgeon. Veterinary certificates can sometimes be difficult to interpret, especially for a layman. For an expensive horse required for competition, X-rays of the lower limbs and feet plus blood-tests at the time of purchase are a worthwhile investment.

Viewing the Horse

Procedure

First, insist on seeing the horse loose in his stable. Then have him brought out and stood still for inspection, before being walked and trotted up on a hard, level surface. Ask for the horse to be saddled up, and watch whilst this is being carried out. Ask for someone to ride and jump the horse before riding him yourself. Remember, some horses will jump well over coloured fences, some will be bolder over natural fences, and some will perform very much better over known fences at home than they will over an unknown course. It is up to the purchaser to test the horse or to see it perform according to its proposed use.

Information Needed

According to intended use, it is advisable to obtain the following information about the horse in question:

- Is he good in traffic? This question should be expanded to cover reactions to heavy traffic, high lorries, main roads, narrow side roads etc.

- Is he quiet to handle, groom, shoe, and clip?

- Is he good to box and travel, both in a trailer and horsebox?

- Is he free from any stable vices? These should be itemised. (See page 120.)

- Is he well behaved in company, e.g. with hounds, other horses, dogs, children?

- Is he safe in a field with other horses, and/or happy to be turned out alone?

- Is he free from any known allergy?

- In the case of a child's pony or a horse for a novice rider, is he easy to catch, both on his own or in company, good to tack up, and well behaved on roads, particularly with regard to shying?

- In the case of a hunter it is preferable to see him ridden out hunting, or to talk to independent people who can vouch for the horse's reliability as a hunter.

- Young event horses should prove their willingness to go through or jump over water. Older horses can often be vouched for by independent testimony or by competition results.

Procedure After Deciding to Purchase

Confirm your decision in writing and keep a copy of your letter. Agree to complete the sale at the stated price, subject to a satisfactory opinion from your veterinary surgeon. Should the veterinary surgeon give an unfavourable or qualified report, the buyer may still decide to buy, subject to an appropriate reduction in price.

List the information about the horse which has been supplied by the vendor.

State any agreed arrangements which will apply if the horse should prove unsuitable, e.g. return or exchange. (There may be no such arrangements.)

State that no responsibility is accepted for the horse until payment is made.

Confirm arrangements as to payment, and as to when the horse will be collected or accepted.

After payment for the horse, the buyer should take possession of the passport, breed papers and height certificate. The horse may have no official papers, but it is helpful to know records of flu and tetanus inoculations and worming programme.

The vendor may agree to this letter verbally, or he may write his acceptance, or he may ignore it. In any case, should the horse prove to be unsuitable, the letter may help if the buyer wishes to return the horse and reclaim his money.

There are advantages in buying from reputable dealers, as they are usually willing to exchange a horse or pony should he prove to be unsuitable. With a private owner this is unlikely.

In most cases the buyer should remember the rule of common law caveat emptor – let the buyer beware. In the event of problems, the Trades Description Act and/or the Sale of Goods Act may apply. Legal advice will clarify this and its relevance, as the situation can differ according to whether the vendor is a dealer or private owner.

SELLING

The vendor can protect his interests in the following ways:

- On meeting the buyer and seeing the horse ridden, he should indicate if he considers that the horse is not suitable.

- He may state that he believes the horse sound, but should not declare him sound. This is a matter for a veterinary surgeon.

- He should state the time limit that he will allow:

  for trial

  to arrange the veterinary examination

  to come to a decision.

If paid by cheque he should ensure that the cheque is cleared or guaranteed before allowing the buyer to take possession.

He should state clearly that possession passes when the cash/cheque is paid, and that the buyer is responsible for the horse from that time.

A Trial

From a seller's point of view, this is rarely a sensible procedure. Should any accident or illness befall the animal, the question of responsibility is always a problem. It is wiser to allow the purchaser to make several visits, but not to take the horse home. If necessary, the horse can be taken to an independent yard and tried in new surroundings and over unfamiliar jumps.

If other buyers are interested, the purchaser must be given a definite time limit in which to decide.

AUCTION SALES

When buying or selling a horse or pony at an auction, both the purchaser and vendor are normally bound by the 'Contract of Sale Conditions' as laid down by the auctioneers in their catalogue. All prospective purchasers are advised to study these conditions before bidding in the sale. It is advisable for the seller or his agent to ensure that any statements made about the horse – or information given to the auctioneer to facilitate the sale of the horse – are true and demonstrable. Should the animal concerned fail in respect of some part of the information given by the seller, the purchaser may be entitled to return the animal to the seller, and to have his money refunded. Auctioneers usually retain any monies paid for at least 24 hours, and they should be informed immediately of

any breach of sale conditions. Should the purchaser suspect that pain-killing drugs have been administered to the horse (the analgesic effect usually wears off in 24 hours but may take up to 14 days), he should inform the auctioneer and arrange to have the blood tested.

ALLERGIES AND SEASONAL DISORDERS

Because of prevalence of '**allergic coughing**' it is essential to obtain information about the general management, feed and bedding of any intended purchase.

Head shaking or '**summer asthma**' is another allergy which may escape veterinary detection in the winter months. The animal does not exhibit the tell-tale signs until a spell of warm, humid weather occurs. He will then wave his head up and down and strike at his face with a foreleg. There is no known cure for this ailment. At best, the horse will be uncomfortable to ride, and depressing to compete on, particularly in dressage; at worst he will be dangerous.

Sweet itch is another summer complaint which can go undetected in the winter, particularly if the mane is hogged.

Laminitis, though not an allergy, is a very common cause of unsoundness, particularly in ponies. It is a recurring disease which if diagnosed in its early stages can, with careful management, be avoided or controlled. Confirmed sufferers are easily diagnosed by changes in the shape of their feet. However, a mild attack may leave no visible evidence, although the tendency to recur will remain.

Blood-testing

As pain-killing drugs can be used to disguise unsoundness, many buyers and their veterinary surgeons now ask for a blood-test before a sale is agreed. There can be no valid objection to this from the seller, provided that all expenses are paid by the purchaser, and that the testing is completed without delay and in a proper manner.

X-rays

In recent years it has become common practice for purchasers of valuable horses to ask to have the horse's feet X-rayed. In some cases they will ask for the leg and the knee or hock joint to be included. A foot X-ray can be a wise precaution. Other, more extensive X-rays are justified if an old injury needs investigation.

Many sellers are reluctant to have their horses X-rayed. Should the sale of the horse fall through because of the results of an X-ray, the seller is put in a difficult position with regard to future purchasers. Often an X-ray will reveal old injuries, many of which may never cause unsoundness. Also, a reasonable assessment can only be made by someone who has had considerable practice and experience in the actual reading of radiographs (X-ray plates).

CARE OF THE HORSE

9

HANDLING THE HORSE

In the following pages, advice is given on how to handle, groom and move around a stabled horse. The advice is well tried and proven. When young people are being trained it is essential that they get into the habit of working in a safe way. It is in their own interest and those of the horses to do so. During training they usually look after quiet animals, who are easy to handle, but if correct procedure is practised from the beginning, it will become instinctive to do the right thing, even when they are faced with more difficult animals.

Horses are by nature nervous creatures. Most become docile and quiet if they are well handled; others always remain on edge and require quiet, confident handling. The expression in the horse's eye will indicate his state of mind, revealing whether he is calm and relaxed, or worried and frightened. A worried, frightened horse must always be treated with care. He requires firm but kind handling if he is not to become a danger both to himself and to his attendants.

Horses have to be taught good stable manners and behaviour. Well-trained grooms are able to do this. They are also capable of advising and helping other less-experienced staff. Horses are creatures of habit, and they become more relaxed if there is a routine to their life.

Horses respond to the tone of a voice, and in this way they can be rewarded for good behaviour, or corrected for bad behaviour. When entering the stable, always speak first. Go up to the horse's shoulder and pat him on the neck. Before handling him further, put on a headcollar and tie him up with a quick-release knot (see diagram). He should be tied up sufficiently short to prevent him turning and nipping or biting, but not so short that he feels constrained, is frightened and pulls back.

When dealing with the legs and feet, never kneel or sit, always crouch. If they are frightened or worried, some horses will use their legs as defence and they may:

- Strike out with the front feet - so when holding a horse always stand to the side and never in front.

- Deliver a cow kick with a hind foot.

- Kick forward and out to the side.

- Kick out to the back with one or both hind feet.

To minimise the danger of being kicked, never stand behind the horse. When attending to legs or feet, always stand close to the horse. Doing so minimises the impact of any kicks.

SECURING THE HORSE

In a Stable

Horses should always be tied up when a groom is either working in the stable or actually grooming. They should wear a headcollar; the rope or chain should be attached to the back of the headcollar and thence tied to a string loop on the tie-ring with a quick-release knot (see diagram). The purpose of the string loop is that it will

string loop

How to tie a quick-release knot.

break if the horse pulls back violently. It is important therefore to use string that will break under stress. Modern, strong binder twine usually needs splitting to make it breakable under duress.

In a Stall

Horses kept in stalls are tied up by means of a headcollar, a clip-on rope and a log. This method allows them sufficient room to lie down, but the heavy log takes up the slack of the rope, so that there is no danger of the horse getting his foot caught. The log is a piece of heavy wood (preferably circular), with a hole drilled through the middle large enough to take the rope. The headcollar rope is passed through a ring or an opening in the fixed manger, and is then fastened to the log by means of a slip knot. This is made to lie across the base of the hole, so that it cannot pull through. Straw placed in this knot will prevent it pulling too tight and becoming difficult to undo. An alternative is to put a short length of twisted string through the log and secure the rope to this. A tie-ring and string loop above the

manger give a shorter tie-up when the horse is being groomed.

PICKING UP THE FEET

To pick up the near fore, stand by the near-side shoulder, facing the tail. Pat the neck, run the hand down the shoulder and leg to the fetlock joint, squeeze above the joint, and at the same time slightly push the horse over with the shoulder so that he puts weight on the off fore. When not picking out the feet, face the horse's shoulder, support the weight of the horse's foot by placing the toe of the foot in the palm of the right hand, and hold it with the fingers. When picking out

How to pick up a near fore.

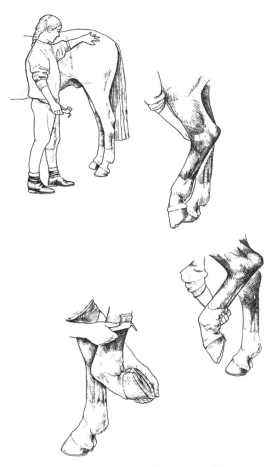

How to pick up a near hind.

the left hand round to the front of the leg, and hold the foot in the palm. Do not hold the toe high, as this will unbalance the horse. When picking out the foot, take up the foot as above, then hold it in the left hand. Allow the horse's leg to move to the rear, and pick out with the right hand.

(2) Slide the left hand around the inner side of the hock, and run it down the fetlock. Pick up the foot as before (see diagram on left).

Never leave the arm supporting the hind leg behind the horse as a sudden kick could result in its being broken.

GROOMING

Sensitive horses, particularly when fit, are often tricky to groom. They require firm but tactful handling. Corrections can be given by the voice or a slap, but only in exceptional circumstances should a horse be hit, and never by junior or inexperienced staff, who are liable to punish the horse at the wrong moment for the wrong reason. Rough handling of horses in the stable is a sign of bad temper or nerves on the part of the groom, and can only have a bad effect on the horse.

When being groomed, horses may bite through irritation, discomfort or over-freshness. Should this occur:

- A muzzle may be put over the headcollar and the horse tied up in the usual way.

- The rope can be put through the stable ring and held with one hand while the other uses the brush. In this way, the horse is kept under control.

- The horse can be tied from both sides (cross-tied).

RESTRAINT

Restraint is most likely to succeed if it is applied in a kind but firm manner by a person whom the horse knows and respects.

the feet, face the rear, hold the foot in the left hand, and use the hoof pick in the right hand. The process is reversed for the off fore.

To pick up a hind foot, either of the following methods may be used, but the first is safer if the horse is likely to kick:

(1) Approach the shoulder and pat the neck. When picking up the near hind, run the left hand along the body, quarters, and down the back and outside of the hind leg. Squeeze the fetlock, and ease the leg a little forward and up. Keep an arm between the face and the horse's leg. When not picking out the foot, use the right hand to support the foot; slip

Additional aids:

• Fidgety horses often stand if they are given a small feed or are fed by hand with sliced carrots or apples.

• If treatment allows, an attendant can hold up a front leg.

• The horse's attention can be distracted by:

Patting him on the neck.

Scratching him between the jaw bones, under his throat and round his ears.

Gripping loose skin on his neck and giving it a twist.

• A cavesson headcollar or a snaffle bridle gives more control than a stable headcollar. Care must be taken that the handler does not attempt to restrain the horse by jerking the reins of the bridle and thus damaging his mouth.

• Holding the upper lip in the fingers can be effective. It makes a mild form of twitch. Take care not to interfere with the breathing.

Using a Twitch

If the above are not successful, a twitch is the quickest and least traumatic means of restraint for most horses.

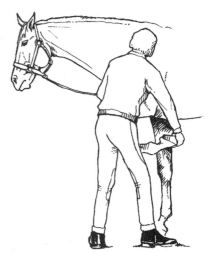

Holding up a front leg as a method of restraint.

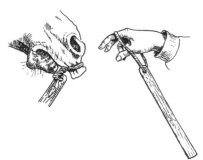

How to put on a cord twitch.

This can be made using a piece of smoothly finished wood about 46cm (18ins) long (a broken-off hay fork handle, well-planed down, is ideal). A hole is drilled in one end, through which is threaded 30cm (12ins) of soft, thick cord made into a loop. Thin string must not be used, as it is likely to cut into the horse's nose.

Various forms of patented twitches are available. One type takes the form of two metal handles hinged together at the top. The device is placed on the horse's upper lip and held firm by squeezing together the bottom of the handles. The main asset of this type of twitch is its quick release, but it is more difficult to keep in place should the horse resist.

Once the twitch has been placed on the nose and the horse has submitted to it, sensory stimulation of muzzle reflexes releases pain-killing agents called endorphins into the bloodstream. These have an analgesic effect so that the horse becomes quiet and less aware of pain and other stimuli.

A twitch should never be placed on a horse's ear. Do not hold the horse by the twitch alone and never leave it on for more than 15 to 20 minutes. It is important that the person handling the twitch keeps a careful watch to see that the horse does not become so becalmed that he appears to fall asleep. This state can precede a sudden and violent reaction, which could be dangerous. The handler of the twitch should keep the horse alert.

Putting on a Cord Twitch

An assistant should hold the horse. The handler takes the end of the twitch in one hand and places the other hand through the loop of cord. He then grips the horse's upper lip with this hand, slips the loop over the horse's lip and twists the end of the handle until the cord is tight. The twitch should then be held at the end of the handle. This is the easiest way to hold it securely. When the twitch is released, the horse's upper lip should be firmly rubbed to help restore circulation. A twitch left on too long can cause a permanent scar.

Some horses become adept at getting out of a twitch, particularly if the handler, in an effort to be kind, has failed to put it on sufficiently tightly. They can then often be difficult to re-twitch.

Most horses accept the twitch, and their treatment can be concluded without any more trouble. However, certain horses become violent and dangerous; such animals should never be twitched.

It is possible for one person to put on a twitch and to secure it by tying the handle to the headcollar. However, this is a dangerous procedure, as the twitch can become loose, the horse may then throw his head about, and the handle of the twitch may injure both horse and handler.

If other methods of restraint fail, it may be necessary to sedate or anaesthetise the animal (see also Clipping, page 80).

Attending to the Mouth or Teeth

Some veterinary surgeons use a gag, which is a device placed on the horse's lower face and adjusted to hold his jaws apart so that his teeth can be reached for rasping,

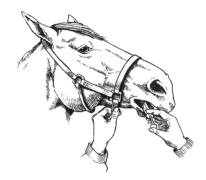

Administering a worm dose via a syringe.

scaling or the removal of wolf teeth. When attending to or examining the mouth and teeth it is often necessary to hold the tongue and bring it out to the side of the mouth. Care should be taken that the tongue is not injured by pulling or twisting, or the use of unnecessary strength. If the horse struggles, let go of the tongue immediately.

Never pass the tongue from one side of the mouth to the other. Let it go and re-take it from the other side. If the tongue is pulled out to the front, there is a danger of the incisor teeth cutting into it. If it is pulled out and backwards, the upper first molar may cut it.

Protective Restraints

Horses often attempt to chew at a wound covering, especially one applied over a knee. Preventive measures include:

CRADLE This is a framework of wood which is placed on the horse's neck, and prevents him putting his head down. It is secured with leather straps. The crest and withers should be protected with a pad.

BIB Attached to the underside of the headcollar (leather) prevents the horse chewing rugs or reaching his knees. See illustration, page 122.

A KNEE CAP will sometimes solve the problem if protection is only required for the knee.

ADMINISTERING MEDICINES AND POWDERS

By Syringe

Large, used, well-cleaned, wormer syringes are suitable. Most powders can be mixed with water, placed in the syringe and squirted down the horse's throat. This is the most efficient method of dosing the horse, other than by injection or stomach tube. The syringe is placed in the corner of the horse's mouth, pointing down over the tongue, and firmly squirted. The horse's muzzle can be raised to assist swallowing. If any liquid remains in the mouth, food should not be allowed for several minutes.

In Food

Powders can be mixed with a damp feed. Soaked sugar beet or molasses help to disguise the taste. Take care not to spoil the horse's appetite by using this method.

In Water

Certain soluble medicines can be given in the drinking water. Care should be taken that the horse is not put off drinking, as this could cause other problems. Should the horse choose not to drink, the medication may be wasted.

As an Electuary

The powder is mixed with treacle or honey and is placed to the back of the horse's tongue with a smooth wooden spatula or a wooden spoon.

Drenching

Modern veterinary practice is generally not in favour of drenching. Efficient drenching requires skill and practice. Inexperienced administration can cause choking and death. Horses with respiratory problems should never be drenched. Drenching should only be undertaken with veterinary advice.

By Stomach Tube

This method should only be used by a veterinary surgeon. A flexible rubber tube is introduced through the nasal tract and passed down the oesophagus and into the stomach. This may be used to introduce liquid paraffin, large amounts of fluid and vitamin supplements, e.g. for colic or dehydration.

By Injection

Intra-muscular injections, if authorised by a veterinary surgeon, may be given by an experienced horse owner or trained groom. To give an injection efficiently requires practice and skill. Intra-venous injections must only be given by a qualified veterinary surgeon. It must be the responsibility of the stable manager to ensure that difficult horses are handled only by experienced people. If they are mishandled by nervous or inexperienced staff, they may at best become playful, and at worst cause an accident. Consult the Health and Safety at Work regulations.

LEADING THE HORSE IN HAND

A horse is normally led from the near side, except when on a road. He should, however, be taught to lead equally well from both sides.

In a Headcollar

Put on the headcollar and rope or halter. Tie a knot in the end of the rope. Hold the end of the rope in the left hand, but never wrap the rope around the hand. Put the right hand on the rope, about 30cm (12ins) from the horse's head. With a fresh horse, put the hand over the top of the rope to help restrain him. With a lazy horse, put the hand under the rope to help push him forward. Never attempt to go ahead of the horse and try to pull him along, as his reaction will then be to pull back. The person leading should always be by the horse's shoulder, never in front; the handler should look ahead, not at the horse.

In the case of a fresh or strange horse, it is advisable to lead the horse from a cavesson headcollar with a lunge rein or long rope attached. Horses should never be led from a headcollar without a lead rope. If the horse plays up it will be impossible to hold him and the leader may injure an arm or hand.

When on foot and leading a young or fresh horse, gloves should be worn. These give a safer grip and prevent rein or rope scalds, which could result in a dropped rein and a loose horse. Such animals may well rear in play, and since the head of the leader is vulnerable, it is advisable to wear an approved hard hat, securely fastened. Strong leather boots or shoes are recommended, to give extra protection to the feet, should the horse jump or tread on them.

In a Bridle

Take the reins of the bridle over the head, and hold them in the right hand, with the end of the reins in the left hand, as for a rope, but with the reins separated by the first finger. The stick is carried in the left hand, pointing to the rear. If the horse is lazy, the left hand should be moved back behind the body and the horse tapped on the lower rib cage, just to the rear of where a rider's leg would rest.

In a Saddle and Bridle

Ensure that the stirrups are run up and secured, as they are for lungeing. The girths should be done up, but not too tightly. A running martingale should be detached from the reins and secured to the neck strap. Then proceed as above. The point of the stirrup bars may be turned up to give extra security to the leathers. These points should never be turned up when the horse is being ridden.

Leading the Horse on a Road

Ponies being brought in from the field to the stable may be led back in halters or headcollars. However, if it is known that a pony is difficult to control, or is traffic-shy, he should be led in a cavesson or bridle. If a horse has to be led along a main road, a bridle should always be used for safety and to fulfil insurance requirements in the event of an accident. The person leading the horse should be between the horse and the traffic and on the left-hand side of the road. It is advisable to wear some kind of fluorescent clothing, either a hat cover or a tabard, so as to be more visible to traffic.

Horses Being Led out for Exercise

Whether beside another horse or on foot, horses should wear a snaffle bridle or cavesson. They should be led in the direction of the flow or traffic with the led horse on the inside of the leader, whether the latter is on a horse or on foot. The near-side rein of the bridle can be put through the ring to the bit on the off side, and then both reins should be held with the left hand on top of the rein about 5cm (2ins) from the horse's head. The slack of the rein may be held in the right hand.

The reins of the ridden horse should be held in the right hand. The led horse's head should be kept level with the rider's knee. If the led horse is allowed to move too far forward, or too far behind, he becomes more difficult to control. A cavesson with a short lunge rein attached may be used. The reins of the bridle can be twisted and secured in the throatlash. This keeps them out of the way, and makes the horse more comfortable.

Leading a Horse in and out of a Loosebox or Through Doorways

Because the horse is wider across the hips than the shoulders, it is very easy for him, when led by a careless attendant, to knock and bruise his off-side hip on a doorway. After such a mishap he will be nervous and will try to hurry or rush through doorways, thus risking a repeat injury. In bad cases, a fragment of bone can be displaced. It is important when leading a horse in and out of a loosebox to ensure that:

- The horse is securely held and steadied as he reaches the doorway and goes through.

- The horse walks on a straight track through the doorway and into the box and does not approach it at an angle.

- The groom checks that the door will not swing to and hit the horse as he goes through. On windy days, doors should be hooked open.

- The horse is not allowed to go into the box on his own, with the reins thrown over his neck. This will tend to make him hurry.

Trotting out for Soundness

This is best done with the horse wearing a headcollar and no saddle. If the horse is fresh, put on a cavesson headcollar for greater control, and if this is not available, use a snaffle bridle. Check that the yard gate is shut. Trot the horse out slowly, and make allowance for the fact that a cambered yard or corner makes many older horses go unlevel.

To turn the horse, steady him to a walk and push him away from you and round to the right. This gives greater control, and the horse stays better balanced and is less likely to step on the leader's foot. If he is allowed to come round to the left, his quarters will tend to swing out, and he may well kick.

Standing up the Horse for Inspection

For inspection or show, square the horse up and show the near side. The front legs should be level, but the near hind should be a little to the rear of the off hind, not vice versa. The attendant stands at the front, but a little to the side of the horse, with the left hand holding the reins, divided by the fingers, about 30cm (12ins) from the horse's head. The slack of the reins, and the stick, if one is carried, are held in the right hand. To trot the horse up, reverse the hand hold on the reins and stick, and proceed as for leading in a saddle and bridle.

SAFETY PRECAUTIONS

- Never lead the horse out of a stable with the rug undone at the front, and wearing a roller with no breastplate. The rug and roller may slip back and frighten the horse, who may then play up and get away.

- Always tie the horse up when working in the stable or grooming.

- In the case of horses who tend to play with their stable-door bolt and open the door, preferably fit a horse-proof bolt, or at least a spring clip on the bolt.

- When tying up a horse outside the box, make sure that he cannot touch or sniff at another horse, as this will cause squealing and possibly kicking. If two or more horses are tied outside, make sure that they are not within sniffing or kicking distance of each other.

- Playful horses, who are likely to chew wood or string, or to pull at stable fixtures, should have their stable checked for likely hazards. If necessary, move the horse to a more suitable box.

- Yard gates leading to public roads should be kept closed.

- Should a horse slip up and not be able to rise easily, lean your knee on his head and neck to keep him still. Straw, manure, sacks or old rugs spread on the ground will help him to get to his feet.

- Keep a sharp knife in a predetermined, accessible place in the stable yard. Should a horse become tangled up in a rope, or if he pulls back and is held fast because the safety string fails to break, rapid action will be essential to prevent injury.

CHAPTER

10 GROOMING

To maintain his health, the stabled horse requires regular daily grooming. The fit horse on a high concentrate diet excretes a considerable amount of waste products through the skin, so thorough grooming is essential.

However, there are circumstances when grooming should be reduced. The sick horse, who has to be kept quiet, needs only minimal grooming, while one with an infectious disease, for example ringworm, should not be groomed, since there is a risk of spreading the infection over the horse's body.

When not in work, the grass-kept horse does not require grooming. The natural oils in his coat help to keep him warm.

When in work, he requires only modified grooming: remove surface mud, check his feet and generally tidy him up.

The Objectives of Grooming

The objectives of grooming are:

- To keep the skin in good condition by removing dirt, sweat and waste products.

- To stimulate circulation of blood and to tone muscle.

- To improve the horse's appearance.

- To provide an opportunity to check for signs of heat or swelling in the legs, soreness in the saddle, girth or mouth areas, heat in the foot, or skin infections.

GROOMING EQUIPMENT

- Container for holding equipment.

- Hoof pick for picking out the feet. This should have a length of coloured knotted string attached so that it can easily be seen if it is dropped in the horse's bed.

- Rubber curry comb for removing mud and sweat (see water brush). It is also used to massage the horse, and to remove loose hairs when a stabled horse is changing his coat.

- Dandy brush. This can be used on the coat of unclipped or coarse-coated horses to remove mud and sweat. It should not be used on a thin-coated, sensitive horse.

- Body brush for cleaning the coat and for brushing out the mane and tail. Avoid using it on horses that live at grass as it removes the natural grease from the coat, which is essential for warmth and waterproofing.

- Metal curry comb for cleaning the body brush. It should not be used on the horse.

- Hay wisp or leather pad for stimulating the circulation and improving muscle tone.

- Sponges for cleaning the eyes, nose, mouth and dock area.

- Water brush for laying the mane and damping down the tail. A dry water brush is also useful for removing mud or sweat from thin-skinned horses.

- Old dandy brush for washing the feet.
- Stable rubber for removing surplus dust and giving a final polish to the coat.
- Hoof oil and brush for oiling the feet.
- Plastic curry comb, for use on the long-coated or unclipped horse to remove mud and sweat. It can also be used on a thick mane.

All brushes and curry combs should fit the hand of the groom; the retaining straps should be adjusted as required.

Extra Equipment

- Spare stable rubber.
- Cactus cloth, a rough, coarse-weave cloth imported from Mexico. It is a very useful addition to grooming equipment for removing dried mud and sweat in place of a rubber curry comb.
- Large sponge for use in washing the horse down and removing stable stains.
- Sweat scraper, preferably rubber or plastic, to be used to remove surplus water after washing the horse down.
- Rubber glove with soft bristles for use when washing down.
- Towels for drying off the horse.
- Pair of blunt-ended scissors.
- Tail bandage.
- Tail comb for pulling the mane and tail and for plaiting.
- Mane comb to comb out the mane (this is rarely required).

NB: Neither comb should be used to comb out the tail, as they may break the hairs.

METHODS OF GROOMING

Quartering or Brushing Off

This is done in the morning before exercise. If the horse is not to be saddled immediately, the rugs should be left on and simply turned back to allow stable stains to be removed.

Method

- Pick out the feet.
- Brush off any stable stains. If necessary, wash clean and towel dry. Pay particular attention to the hocks, knees and flank, and under the stomach.
- Sponge the eyes, nose and dock.
- Brush out the mane and tail.

Strapping

This involves a thorough grooming of the horse, as described below, but in addition he is wisped or banged for 10 to 15 minutes to stimulate circulation of the blood and muscle tone. In many yards strapping is no longer practised. Each horse is groomed for approximately 20 minutes, or for as long as it takes to get him clean. The procedure is the same as for strapping, but the wisping or banging is omitted. It is felt that balanced feeding and correct work will build up muscle where it is needed, and that the time formerly spent wisping a horse can be put to better use.

Wisp Over

Competition or racehorses are often wisped over again before their tea-time feed, when rugs are checked and the box set fair. This unsettles some sensitive horses, so it is often better to finish the grooming in one session.

The time taken to groom a horse is not necessarily the criterion of expert work. The appearance and feel of the horse must be the final factor when assessing the quality of the work done. An experienced worker should groom his horse in half an hour. After hunting or competing in muddy conditions, it may well take longer. Extra wisping may take 20 minutes.

Grooming

This is the thorough grooming best carried out after exercise, when the horse is warm and the pores of the skin are open. On hunting or competition days, it is done in the morning instead of quartering. The time taken is 20 to 30 minutes. Wisping or banging is not carried out.

Method
The horse should be dry and cool.

- Collect the headcollar and rope (never use a frayed or worn rope as it can tighten and prove impossible to undo), and a bucket of water. A horse who chews his rope can be tied up with a rack chain attached to string.

- Put on the headcollar. Place the rope through a short length of string (called a safety loop), and put that through the stable ring. Tie the rope with a quick-release knot, tucking the end of the rope through the loop, so that the horse cannot pull it undone.

Never tie the horse directly to the stable ring. Should the horse take fright and pull back, he may break the headcollar, pull the ring out, or slip up. If the weather is suitable, and the horse quiet, he can be groomed outside, but the yard gate must be closed. Tie him up to a suitable ring as before. Never tie him to a single post or tree, as he may move round, tangle himself up and have an accident. Never tie a horse to a gate: if he becomes unsettled, he could lift the gate off its hinges and panic.

- If the weather is warm, **remove the horse's rugs**. In cold weather the rugs should be left on while you are attending to the horse's feet; a blanket should be placed over his loins while you are grooming his fore

part, and over his shoulders and back while you are working on his hindquarters.

- **To pick out the feet**, take the hoof pick and pick them out into a skip. Use the hoof pick in a downwards direction towards the ground. Start at the side of the frog, taking care not to damage the softer parts on either side of the frog and the cleft. Make sure the foot is really clean. Check for good condition and that the shoe is secure. With horses who are used to having their feet picked out, it saves time to pick them all out from the near side. The off fore is picked up from behind the near fore, and the off hind from behind the near hind.

- It is usual to **wash the feet** on return from exercise. If the horse is in his stable, hold each foot over a bucket of water and scrub clean using an old dandy brush. Take care not to wet the foot above the bulbs of the heels.

- **To remove dry mud and sweat** from the horse's body, use the rubber curry comb, or a handful of hay or straw, or the cactus cloth. If the horse is sensitive, use a dry water brush. Sticky patches of sweat can be sponged clean with warm water and dried with a towel.

- **To clean the coat**, take the body brush in the left hand and the metal curry comb in the right hand. Start on the near side just behind the ears, push the mane over and thoroughly clean the crest, then work over the body. Stand well away from the horse, so that the full strength of the arm can be used to clean the coat. Brush in the direction of the hair, using both circular and straight movements of the brush to penetrate through to the skin. The hand holding the metal curry comb should rest on the horse's body, with the curry comb strap over the back of the hand. Clean the body brush with the curry comb as necessary. Clean the curry comb by knocking it on the ground outside the door, so that the dust falls out. Sweep up the dust when the grooming is finished.

- **To brush the legs,** put the curry comb on the ground

away from the horse, and with the sharp side down. Use the body brush in whichever hand is most convenient and hold the horse's leg with the other. Crouch – do not kneel or sit – and keep your face away from the horse's knee.

When brushing the hind legs, if the horse is restless, hold his tail and hock with one hand and brush with the other. This prevents the horse from flexing his hock in preparation to kick, and also gives warning of any intention to do so. Stand near to the horse and keep a firm hold on his leg.

- **To brush the mane**, start at the poll, take a few hairs at a time and brush out thoroughly with the body brush. A short-pronged plastic curry comb can be used for this, but care must be taken not to break the hairs of the mane.

- **To brush the head**, untie the horse, brush his face, then slip the headcollar down and buckle it around his neck. Steady the horse's head with one hand, and use the body brush gently in the other to clean the head. Pay particular attention to the lower jaw and gullet and around the ears. Take care not to knock or bruise the bony parts. It is important to be gentle, as horses can become head-shy if they are roughly handled. When you have finished, put on the headcollar again and tie up the horse.

- **To bang or wisp**, take the leather pad, wisp or folded stabled rubber and dampen it slightly. Stand back from the horse and vigorously bang it down on the large muscular areas of the horse on the neck, shoulders, quarters and second thigh. Avoid the loins and bony parts. Use either hand, steadying the horse with the other. Introduce this banging action gently to a young horse, gradually increasing the pressure as he becomes accustomed to it. If a good rhythm is achieved you should see the muscle begin to contract in anticipation of the banging action.

- **To sponge**, wash out the sponge and untie the horse, leaving the rope through the string. Gently but firmly clean the eyes, muzzle and nose, rinsing the sponge as required. Tie up the horse, rinse the sponge or take a second sponge, lift up the tail and clean around and under the dock. If the horse is restless, stand to one side, not immediately behind him.

- **To brush the tail**, stand to the side of the horse, take the tail (approximately at the end of the dock) in one hand, and the body brush in the other. Allow a few hairs at a time to escape from the hand holding the tail, and gently brush these out, taking care not to break the hairs. With thin-tailed horses, it is best to use the fingers, not a brush. Finish by brushing out the top of the tail, dampening it down with a water brush and putting on a tail bandage. Dirty tails should be washed and then brushed out. It is advisable to brush tails infrequently to avoid damage and loss of tail hairs. **Never** try to brush out a muddy tail. It **must** be washed.

- **To use the stable rubber**, fold it into a pad, slightly dampen it and wipe the horse's body over to remove any remaining dust.

- To **lay the mane**, tidy it with a damp water brush and stroke the hair into place.

- **Brush the rug**s and then shake the blanket well away from the horse.

- **Oil the feet**.

- **Put on a tail bandage** to lay the tail if the horse has a pulled tail. (Leave on for half to one hour.)

- If in the stable, **untie the horse**, leaving the headcollar on or removing it, according to the occasion and personal preference.

EXTRA GROOMING

Mane Washing

- Choose a warm day.

- Collect warm water, horse shampoo, a large sponge, a towel and a sweat scraper.

- Put on a headcollar. To avoid staining his head, it

may be necessary to fit a grey horse with a clean nylon headcollar or white rope halter. Do not tie the horse up.

- Wet the mane with the sponge. Starting with the forelock, thoroughly rub in the shampoo, keeping it well clear of the eyes. Continue down the mane.

- Rinse with warm water until all the soap has been removed. Remove surplus water from the neck with a sweat scraper, and rub the ears dry with a towel.

- Brush the mane out with a clean body brush or a plastic curry comb.

Tail Washing

- Collect the washing equipment.

- Put on a headcollar and tie the horse up. If he is restless, ask an assistant to hold him, and to pick up a front leg if necessary.

- Proceed as for the mane, but soak the tail in a bucket of warm water. Take care, because when the water reaches the dock the horse may become disturbed.

- To dry the tail, stand to the side of the horse, hold the tail at the end of the dock and swish the bottom of the tail around in a circle.

- Brush out gently with a clean body brush.

Bathing the Horse

Bathing is a more extensive operation than washing or sponging down after work. It is carried out if there is a non-infectious skin problem, or as a quick measure to ensure a clean horse. Grey horses may have to be washed regularly, as they are difficult to get or keep clean. However, washing should never be a substitute for good grooming.

It is important for the horse to be thoroughly and quickly dried after bathing, so never wash him on a cold day. Lungeing him with a cavesson and boots on, or walking him out with a clean exercise blanket under

the saddle are the best methods of keeping him warm and drying him off.

If possible, enlist the help of an assistant, and proceed as for mane washing. Wash the head first and then dry it with a towel. Tie up the horse and continue the process of washing, making sure that all the shampoo is thoroughly rinsed out. In warm weather, if the horse accepts it, a hose can be used. The legs should be given a final rinse with cold water. Use towels to dry off the horse, or a brisk rub with straw (though not on greys, as the straw may stain the coat). The legs must be bandaged either with Gamgee and stable bandages, or thatched with straw or hay to help them dry.

Bathing removes oil from the skin, and the horse's coat often looks a little dull afterwards. A good strapping will restore the bloom.

Washing White Legs

White legs should not be shampooed in cold weather. Warm water and soap remove the protective oils from the skin, and mud fever or cracked heels may result. However, barrier cream applied to dry, clean legs and heels can assist in preventing this. If there is urgent need, the legs may be individually washed down with cold water, and dried with stable bandages applied over hay or straw.

Sponging Down

Sponging down is used on a sweating horse after exercise, or at a competition.

Method

- If he is still blowing, lead the horse about until he is calm, and his breathing is normal. The girths should be eased, but the saddle left on. Alternatively, the saddle may be removed and the back slapped to restore circulation and then covered with a towel. In cold weather a sweat sheet and light rugs should be put on.

- Pick out the feet. Wash them if necessary.

- Remove the saddle and bridle and put on a headcollar. If an assistant is available, ask him to hold the horse.

- Wash the horse down, using a large sponge and plain water. In hot weather use cold water or, if the horse will accept it, the yard hose. In cold weather, use lukewarm water and sponge only the very sweaty areas. Never use warm or hot water, as it is more likely to give the horse a chill.

- Remove surplus water with a sweat scraper.

- Put on a sweat sheet, rug, roller and breastplate, or alternatively straw beneath an inside-out rug, plus a roller and breastplate. If a hot day, walk dry without rugs.

- Sponge round the base of the ears and face. Dry with a towel.

- Walk the horse about until he is dry. Ensure that the ears are dry and warm. In very cold weather put on stable bandages over hay or straw (thatching).

- After 30 minutes, check the rugs. If the horse is dry, groom him and put on his usual rugs.

Introducing the Horse to Being Hosed

In warm weather hosing is a quick and efficient method of cleaning and cooling off a horse. However, at first most horses are nervous of being hosed down. When first introducing hosing to the horse it is helpful to have a second person to hold him. Start gently with little pressure, and apply the resultant trickle around the front feet. Gradually work up the leg. At no time must the horse be frightened. As he begins to accept the sensation, gradually increase the pressure. When horses become used to the water they usually enjoy it, but some never accept it, and where they are concerned it is best to abandon the process.

Washing the Sheath

Some geldings require regular washing of the sheath.

The signs that this is needed are a strong smell and nodules of greasy dirt collecting in and around the sheath. An assistant may be required to hold the horse and possibly to hold up a front leg. The equipment needed is a bucket of warm water, mild soap, a sponge and rubber gloves. Soak the sheath and thoroughly clean out as much as the horse will allow. Then dry it with a towel. Apply baby oil as a lubricant.

WASHING THE GROOMING KIT

All items should be washed weekly with detergent. Dip the bristles, but do not wet the backs of brushes more than necessary. Rinse thoroughly. Dry the brushes by standing them on their sides in a dry atmosphere, but not near direct heat. Leather-backed brushes should be oiled. Stable rubbers should be washed twice a week, or as necessary. As a safeguard against infection, it is best that each horse has his own grooming kit.

MAKING A WISP

Shake out a length of meadow hay on the ground and dampen it. With the help of an assistant, twist the hay into a rope, make two loops at one end and twist the remaining rope in and out of the loops. Tuck the ends away firmly, then thump the wisp hard. A well-made wisp is firm, secure and small enough to hold in the hand. It should last several weeks.

BUYING A GROOMING KIT

As a rule, the more expensive brushes last longer and are easier and more efficient to use. The bristles tend to fall out of cheaper, machine-made brushes. Grooming equipment can be bought in various sizes. For efficient use, it should fit the hand of the user.

Body Brushes
These should be of soft, natural bristle. Leather backs are preferable to wood, but both are suitable.

Dandy Brushes

These are wooden backed, and the bristles are of a much tougher texture than those of a body brush. Some dandy brushes are made of coloured nylon tufts. Some horses are allergic to nylon, and these brushes clog more easily. Natural bristle is preferable.

Water Brushes

These should be wooden backed, with long, soft, natural bristles.

GROOMING THE GRASS-KEPT HORSE OR PONY

In Winter

• Tie the horse up as before.

• Wash and pick out the feet. Check for any nails or stones which might have been picked up, and for cracked heels.

• When the horse is dry, remove the mud from his coat with the dandy brush or plastic curry comb. The latter is more efficient. Pay particular attention to the saddle and girth areas.

• Clean the head carefully, using body brush or hand.

• Sweat marks should have been washed off after work, before the horse was turned out.

• Wet mud on the legs, knees and hocks may be washed off with cold water and the legs left to dry. Do not use warm water and soap, as this will open the pores of the skin, making the legs more vulnerable to cracked heels and mud fever. It is preferable to leave wet mud until it is dry and then brush it off.

• Clean the dandy brush by brushing it against the sharp surface of a wall or door, well away from the horse. The plastic curry comb may be cleaned by knocking it against a similar surface.

• Sponge the eyes, nose, muzzle and dock area.

• Clean mane with dandy brush or plastic curry comb.

• If the tail is muddy, wash it then brush it out. If it is dry and clean, brush it out with the body brush, never the plastic curry comb. The hairs on a horse's tail, which are easily broken off by rough brushing, take up to two years to grow. Thin-tailed horses and ponies should always have their tails brushed out with a body brush, or simply with the fingers.

• The feet, when dry, can be oiled if required.

In Summer

When grooming for a special occasion, for instance a show or rally, the procedure is similar to that for grooming a stabled horse, but with less emphasis on cleaning down to the skin with the body brush.

GROOMING MACHINES

There are several different types of grooming machines on the market, such as:

• The **vacuum type**, with a suction head, which can be used for removing dirt/mud or loose coat in the spring.

• The **revolving brush type**, which can be recommended for the large yard, when there is insufficient time for daily strapping.

The more modern machines are light and come with a waist belt that fastens round the waist of the operator. This makes them easier to handle than older machines. They have a dust bag attached, which collects all the loose hair and dust.

The stable manager should ensure that staff in the yard do not use the machines as an excuse for inadequate daily care.

A muddy, grass-kept horse or pony required for a special occasion can be effectively cleaned up by the suction-type fixture. This removes mud and dirt, but leaves much of the grease in the coat. Make sure that the coat is quite dry before using the machine.

Because of labour costs, in many racing yards daily strapping has been discontinued. The horses are brushed over and tidied each day. After fast work they are sponged down. They are usually done over with a grooming machine every four to seven days. It has been found that they remain quite healthy and are considerably better tempered in the stable as a result of not being subjected to intense grooming and strapping.

Safety Rules

The operator should wear an overall, headscarf or cap, and rubber boots, and the machine should be plugged into a circuit-breaker. Great care must be taken not to catch the mane or tail in the machine. The tail should be roughly plaited, and then bandaged.

Horses who are difficult to clip often accept a grooming machine and then become easier to manage during clipping. Grooming machines are excellent for cleaning the coat, and the vibratory action of the brush has a stimulating effect on the skin, which is generally much enjoyed by the horse. Care must be taken to use only soft brushes on a thin-coated horse.

The machine should not be used more frequently than once every three days, and the day after hunting or any extra exertion. If it is used too often, the skin may become tender and sore.

11

CLIPPING, TRIMMING, PULLING AND PLAITING

CLIPPING

Clipping is the removal of the horse's coat by machine. A horse changes his coat twice a year. From mid-September to mid-October he grows a thick winter coat, and between February and April he loses his winter coat and grows a lighter summer one. The old coat may become rough and dull during the period of change, particularly in the autumn. The exact timing varies between different types of horses and ponies, and is also affected by weather and temperature.

The growth of coat also varies for a stabled horse; the temperature of the stable and the number of rugs worn are the major influences. A well-bred horse, stabled and rugged throughout the year, grows a very light winter coat, and if he is not in hard, fast work, he may not require clipping.

Grass-kept horses and ponies are left unclipped to give them protection from the weather. If they are in steady work, they benefit from a semi-trace clip, i.e. the removal of the coat on the shoulder and underneath the neck and belly, in October and early November. This makes work easier and avoids any heavy sweating, which can occur if there is a warm spell of weather in October. Ponies protected by a New Zealand rug can be regularly trace clipped up until January. The reasons for clipping a stabled horse are:

- To prevent distress when he is in hard, fast work. This is applicable in winter when the coat is thick, and sometimes even in spring and summer if the horse is

expected to work at maximum effort, for example in three-day eventing or endurance riding.

- To help keep him in good condition. Sweating leads to weight loss.

- To make drying off after work quicker.

- To save time and labour over cleaning.

- To prevent and/or to control skin disease.

The first clip takes place in late September/early October. The weather is often very warm, and a working horse benefits from the removal of his coat. Overweight horses brought in after a summer at grass may have their clipping delayed to encourage sweating and a consequent loss of flesh.

Well-bred horses may require clipping only every four weeks; heavy-coated animals may need clipping every two or three weeks, if they are to look tidy and smart.

A horse who is to be shown should not be clipped after the end of January, as late clips spoil the summer coat.

Show Horses

The owner must decide whether or not to clip show horses. It depends on the individual horse, his type of coat and whether his finer points need to be shown or concealed. If the decision is to clip, in all cases the horse is given a full clip. Coarse blades may be used to give a more natural look.

Show Ponies

Ponies who are slow to establish their summer coat, or who have a rough coat, need to be clipped, particularly if they are being shown early in the season.

Competition Horses

Competition horses involved in demanding sports (three-day events, endurance riding, show jumping and polo) may be clipped immediately prior to competition.

TYPES OF CLIP AND THEIR USE

Full Clip

The entire coat is removed, including the legs and saddle patch. A triangle is left at the top of the tail. The clip may be used on coarse-coated hunters. It makes them easier to clean and quicker to dry off, particularly the legs. It is often used only for the first and second clips. Thereafter the horse does not grow such a heavy coat, and a normal hunter clip can be used.

If a coarser blade is used for the saddle patch and legs, some protective hair will be left. Many consider that even with well-bred horses, if the long hair is left on the legs, it makes them slower to dry, more difficult to keep clean and therefore more liable to mud fever.

Full Clip Except for Saddle Patch

The saddle patch is left on to ease saddle pressure. Thin-skinned Thoroughbred horses are often best left with the saddle patch on.

Hunter Clip

The legs (to just below the elbow and stifle joint) and the saddle patch are left unclipped. Hair on the legs gives protection from the cold, and from thorns, mud fever and cracked heels. The saddle patch gives some protection from saddle friction.

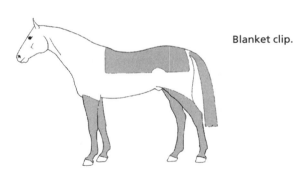

Blanket clip.

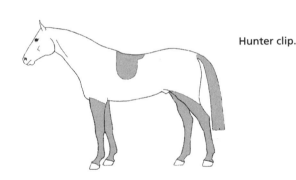

Hunter clip.

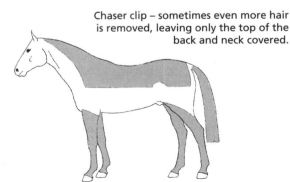

Chaser clip – sometimes even more hair is removed, leaving only the top of the back and neck covered.

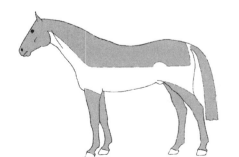

Trace clip.

Chaser Clip

Hair is removed from the head and lower two-thirds of the neck and body. The coat is left on from behind the ear and over the top part of the body to the tail.

Alternative to the Chaser Clip

Hair is removed below a line starting at the stifle joint, along to the bottom of the saddle flap, up the neck to the base of the ears and down the side of the face.

Blanket Clip

Hair is removed from the head, neck, shoulder and belly and a narrow strip on either side of the dock, thus giving the shape of an exercise sheet.

The above three clips may be used for:

- Horses brought into work during the winter who require walking exercise in cold weather.

- Young horses in their first season's hunting.

- Horses who are cold-backed or who easily catch chills.

- Horses competing in winter in show jumping or dressage who may spend time standing around between classes.

Trace Clip

Hair is removed in a strip from the shoulder, belly and thighs, and often from the under part of the neck. This clip is used for:

- Horses who resent having their heads clipped. For them, the line of the clip is taken well up the shoulder, belly and thigh.

- Horses in light work.

- Horses and ponies turned out in New Zealand rugs.

- Young horses, as a first clip to familiarise them with clipping.

PREPARATION FOR CLIPPING

- A warm, still day is best. Start as early in the day as possible.

- Good lighting is essential. A utility box with a non-slip rubber floor is ideal. If this is not available, a spare box may be used, with the bed swept back and a minimal amount of bedding left down as this makes it easier to pick up the clipped hair. Horses should not be clipped on a bare floor, because of the risk of slipping and the danger from electricity if the floor becomes wet.

- A haynet can be tied up in the box to keep the horse occupied.

- A strong box or straw bale should be available for standing on when clipping the horse's head.

- The horse should be clean and dry. A dirty coat clogs the blades, causes overheating and strains the machine. The blades cannot cut a damp coat. If a horse becomes upset and breaks out in a sweat, it is best to stop clipping until he has settled and dried off. A grooming machine can be a great boon, both to ensure a clean horse and to get the animal used to the noise and vibration of a machine. If a horse is known to be difficult, ensure that experienced help is available, and that sufficient time has been allowed to avoid any hurrying.

- After clipping, the horse can be brushed over with a dampened body brush, which will help to remove any surface scurf. An extra blanket will be needed for warmth.

- The person handling the clippers should wear rubber boots or rubber-soled shoes as a safeguard against an electric shock. Overalls and a headscarf or cap protect clothes and hair from clipped hair. A circuit-breaker should always be used.

- With a difficult horse, the holder and clipper are at risk. They may have their feet trodden or stamped on, or receive a glancing blow from the horse's head.

They should wear strong boots and hard hats. The assistant should also wear gloves, which will make the rope easier to hold. He must never place himself in front of the horse in case the animal should strike out. When required, the assistant should hold up the horse's front foot, standing facing the shoulder and holding the front of the foot in the palm of the hand with the fingers. The rope can then be held in the other hand. (See page 62.)

ELECTRIC CLIPPING MACHINES

There are two popular types of mains electric clipping machines. There are also smaller battery-driven machines suitable for trimming, clipping heads and for use on thin-skinned, nervous or difficult horses.

Electric Hand Clippers
The motor and clipping head are contained in one machine, with a cable to a mains power plug and circuit-breaker. They are small enough to hold in the hand. They are an efficient and excellent for the small yard or private owner clipping clean, stabled horses.

Hanging or Heavy-Duty Type Clippers
These are suitable for the large commercial yard where many horses of all types have to be clipped. The machine itself is separate from the clipping head, and is connected by a flexible driving shaft. The machine is heavy and must be suspended from a beam or wall bracket. The actual clipping head is light, but the flexible shaft makes it less easy to manoeuvre, and there is more vibration than with a hand machine. It does give a very clean cut. If possible, such a machine should be kept in a loosebox or utility box reserved for clipping. Grooming machines may also be kept in this box. It will be easier to move the machine about if it is suspended on overhead runners.

Clipping with a heavy-duty machine will be easier if the driving shaft is kept as extended as possible. It requires regular oiling and greasing.

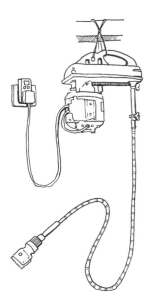

Heavy-duty clipping machine suspended from a beam and fitted with a circuit-breaker.

Battery Machine
Battery machines are becoming more widely available and efficient. There are two general types – one standard size; the other smaller and suitable for clipping faces etc.

The standard-size battery clippers usually have a power pack which is attached to the user's back and is independent of any mains cable. This enables much greater freedom of movement and is safer. The clipping head is light and easily handled. As with all clipping machines, it is essential not to overheat the machine by clipping for too long at a time. The blades must be kept clean and oiled.

Electrical Connections
Those machines which are not self-earthed require a three-wire cable correctly connected to a three-pin plug to ensure safe earthing and must be connected to a circuit-breaker. The correct cable and working instructions are supplied by the manufacturers. The instruction booklet should be kept with the machine, and the maker's instructions followed.

Self-earthed machines should not be used from a

light socket as this is not safe. Should the cable be pulled, either the light socket or the wiring will give way.

The Blades

The clipping blades must be sharp. Blunt blades are uncomfortable for the horse, give an uneven clip and strain and overheat the machine. A spare pair of new or resharpened blades should be available before clipping begins.

Blades quickly become blunt when used on dirty horses and worked hard by those with limited experience of clipping. Blades can be resharpened during the summer.

Blades are packed as a pair. The bottom plate is flat, and is fitted first. The top plate is placed on top of it. They are secured together by a flat-topped bolt, a tension spring and a thumb screw adjustment. Correct tension is essential. If it is too loose, the machine races and does not cut. If it is too tight, the machine is strained and overheats.

When the machine is switched on, light machine oil should be run through the blades and into the oil holes on top of the machine. The machine should be wiped clean before clipping starts. While clipping, the blades must be regularly cleaned. Brush them with a soft brush and then run the machine through a container of blade wash to remove dirt and hair.

Should the machine heat up, stop clipping, clean it and allow it to cool. It may be necessary to fit fresh blades.

When the clipping is finished, remove and clean the blades, then clean and oil the machine, following the maker's instructions. The thumb-screw and tension spring and bolt should be put back on the machine. Wrap the blades in oiled cloth, and store carefully. It is all too easy to drop the blades and chip them. A light knock is sufficient to put them off true, and thus give a strained and uneven clip.

After clipping, always check the cable for wear. At least once a year have the machine checked professionally for safety and have the motor serviced and overhauled bi-annually.

Electric clippers, and in particular hand machines, are often ill used and prone to strain. They can easily seize up in the middle of a clip. It is sensible to have a replacement machine available in case this occurs.

HOW TO CLIP

- If a mains hand clipper is being used, first assemble the machine. Carry out a safety check, including an inspection for any signs of wear of the cable and the plug. Take the machine to the clipping box. Ensure that there is machine oil, a container of blade wash, a soft brush, spare blades and chalk or saddle soap for marking lines. All this equipment should be placed safely out of the way on a shelf outside the box.

- Plug in the machine, and check that it is working correctly. Switch it off, and put it in a safe place. Ensure that the cable can be suspended over a hook, so that when you are clipping it comes directly to the hand, and does not lie on the floor where the horse can tread on it. Check, too, that the horse cannot reach it with his teeth.

- Bring the horse into a well-lit box and tie him up. If it is warm enough, remove his rugs, placing them well out of the way. If it is cold and draughty, place a blanket over his loins and back. Pick up the machine and switch it on. Allow the horse to become used to the noise before starting to clip.

- Switch off the machine and untie the horse, leaving the rope through the safety loop. Hold the end of the rope in the left hand, switch on the machine and then place the left hand on the horse's neck. Start on the near side, and clip the shoulder, running the machine against the direction of the hair. The line taken depends on which clip is required. Keep the machine flat. Do not push it; guide it and allow the machine to do the work. At all times clip against the growth of hair.

- When the shoulder, neck and chest on the near side have been finished, the horse may be tied up. The front part of the belly is then done, the blanket moved forward and the quarters, back and belly completed. The off side is then clipped. The machine may be held in either hand and changed if required.

- Ticklish or difficult parts, such as the head, underneath of the belly, loins and in between the hind legs can be left to last, when an assistant may be required. When clipping around the elbow and chest, a front leg should be held up to help stretch the skin and avoid nicking it. Similarly, hold the skin taut when doing the stifle joint and flank area.

- When clipping a front leg of a fidgety horse, the other front leg may be held up by an assistant to prevent the horse snatching up his leg and possibly breaking the clipping blades.

- It is important to check that the machine is not over-heating. This is bad for the machine, and makes the horse fidgety and resentful.

For a Saddle Patch

Draw an outline with chalk or saddle soap around the horse's own saddle. Make sure the outline is even and regular. Clip into the hair and lift the blades to achieve a tidy line.

For a Blanket or Trace Clip

Ensure an accurate and matching clip on both sides by measuring over the withers and back, and putting lines on the horse's coat. The top of the front and hind legs should be clipped to the same height and angle. The clip should extend over the elbow to assist cleaning.

To Clip up the Neck

Run the blade as close as possible to the mane hair. It is preferable to leave a narrow edge of coat, rather than risk cutting into the mane.

Clipping of Mane

A small portion of mane may be removed where the bridle headpiece lies over the poll. This should not exceed 2.5cm (1in.). A small section may also be

removed at the top of the withers. In each case, if too much hair is clipped, it is unsightly. On a fine-maned animal, trimming is often better done with scissors, as it is very easy to remove too much with clippers. Sometimes hair is removed over the top of the withers although this may encourage rubbing and wear from pressure from the rugs.

DIFFICULT HORSES

Horses who resent having their heads or other ticklish parts clipped often accept a twitch on the nose, and will stand without struggling and allow the clip to be finished quickly and efficiently. The twitch should never be put on a horse's ear. For use of the twitch, see Chapter 9, Restraint, page 61.

Some highly-strung horses will not accept a twitch, and become violent if one is used. In such cases it can be dangerous to persevere and attempts to apply a twitch should be abandoned. If it is only a question of tidying up, it may be better simply to leave the horse as he is. If he is treated quietly and calmly, and not upset, he may eventually accept clipping more easily.

Horses may become difficult to clip for a variety of reasons. The most common causes are connected with rough handling and lack of tact, and include:

- Electric shock.

- Over-heated blades.

- Blunt blades.

- Bruising and/or cutting by the machine.

- Rough or impatient handling.

- Clipping for too long a time without rest, especially the first time the horse is clipped.

Remedies

It can be helpful to stand a nervous horse within sight and hearing of another quiet horse who is being clipped.

- Use a grooming machine to accustom him to the noise and vibration. However, some horses will not accept this.

- Before attempting a lengthy clip, trim the horse so that he becomes used to the feel of the machine.

- Music may have a calming effect, or cotton wool in the horse's ears may help.

- The use of a small battery-driven clipper may be less worrying to the horse, as there is less noise and vibration. With such a small machine, it may not be possible to achieve a full clip because of over-heating.

When clipping difficult horses, any rough handling is self-defeating. Personnel involved should be calm, skilled, unhurried and sufficiently strong to hold the horse. It is an advantage if the person clipping is tall, as he can clip the horse's head with greater ease.

Sedation

It may be necessary to sedate a very difficult horse. In some cases, a total anaesthetic is required, when it will be necessary to take the horse to the surgery, or alternatively pay for the veterinary surgeon's waiting time while the animal is clipped. In most cases, if patience and kindness are used, these extreme measures may be avoided.

There can be problems connected with the use of sedatives on horses for the purpose of clipping. For the sedative to work satisfactorily the horse must be calm and unworried when it is administered. It is advisable to discuss each individual case with your own veterinary surgeon. Sedatives in tablet form or by intra-muscular injection administered by your veterinary surgeon can be quick and effective but correct timing is essential for the dose to fulfil the clipping time required.

TRIMMING

In winter, horses and ponies at grass should be left untrimmed, but in summer they may be trimmed in the same way as a stabled horse.

Whiskers

These can be removed to give a smart and tidy appearance. However, it should be remembered that they are the animal's chief means of feel. Many owners will not have them trimmed, as they are of particular use to grass-kept animals.

Ears

The long hair in the ears may be clipped level. To do this the ear is held closed up and the machine is used in a downwards direction from the tip of the ear to the base. The inside hair should not be clipped, because the internal hairs give protection from cold, dirt and infection. Grass-kept animals should not have their ears trimmed.

Cat Hairs

These are long hairs which grow a week or so after clipping, particularly in January and February. They may be removed by clipping lightly in the direction of the hair using a coarse blade.

Heels and Fetlocks

In winter, these may be trimmed if necessary when the horse is clipped. In summer, well-bred horses may be trimmed by hand with a comb and sharp, blunt-ended scissors. The hair is raised by the comb and then trimmed off, so that there are no rough edges, in much the same way that human hair is cut by a good hairdresser. The operation requires practice and skill. Coarse-coated horses who grow heavy feather are more easily trimmed by machine. A coarse lower blade, called a leg plate, may be fitted. This tidies the hair, but does not clip too closely. Some horses grow very fine feather, which may be pulled out by hand.

In Mountain and Moorland show classes, native ponies and Welsh Cobs are left untrimmed. Arabs grow very little hair and are left untrimmed.

Jaw, Withers and Poll Piece

These need to be kept tidy throughout the year. Either the normal clipping machine may be used, or alternatively a lighter machine, similar to a dog clipper. This machine is also useful for trimming legs and ticklish, awkward parts of the horse. It is very light, easy to handle, and is quieter and vibrates less than a large clipping machine. Some types are battery powered. Chin hairs may also be trimmed with blunt-ended scissors. When trimming the poll and wither area, great care is needed to avoid the trimmed areas becoming larger and larger. Well-bred animals are often more neatly trimmed with scissors.

Hogging

This is the removal of the mane by clipping. In former years hogging was very fashionable, but nowadays it is not often seen except on polo ponies and for showing cobs. It saves labour, and is a means of keeping tidy a mane which has been rubbed, for instance because of sweet itch, or caught on wire.

To hog, the horse's head should be held low and stretched, so that a clean cut can be achieved. Use the clippers from the withers towards the poll. Once hogged, a mane may take up to two years to grow, and at first will be coarse and upright.

The hogged mane is clipped once every two to three weeks and cleaned daily with a damp brush.

PULLING THE MANE

This is done to:

- Improve the appearance.
- Thin out a thick mane.
- Encourage the mane to lie flat.
- Shorten a long mane.
- Make the mane easier to plait.

Arrange to pull the mane after exercise, or on a warm day, when the pores of the skin are open. This makes the hairs easier to pull out, causing less discomfort to the horse.

To avoid making the skin very sore, it may be advisable to spread the mane pulling over several days.

Some horses have very thick manes, and these can be the toughest to pull. Well-bred horses usually have thinner manes. Care must be taken in this case that the mane is not made too short.

Do not wash the mane before pulling, as this makes the hair soft and slippery and difficult to pull.

It is customary to encourage the mane to lie on the off side of the horse's neck. The manes of young horses may be trained to lie over by regular brushing, damping down and, when the horse is stabled, by loose plaiting. The manes of older horses are more difficult to retrain. If they lie on the near side and are tidy, they may be left.

The mane should never be shortened by cutting with scissors. This will leave an unsightly and very obvious edge. A very thin mane may be shortened by carefully breaking off the ends of the hair with the fingers. This leaves a natural looking edge. Clippers should not be used to thin a thick mane by cutting the underlying hair.

The manes and tails of all Arabs and Mountain and Moorland ponies, if they are to be shown in breed classes, are never pulled.

Satisfactory mane-thinning combs are now available on the market. These give the mane the appearance of a pulled mane without subjecting the horse to genuine pulling, which some horses strongly resent.

Method

- Thoroughly brush out the mane. A thick, tangled mane requires combing out. Remove the underneath hairs first and then shorten the top hairs. This should allow the mane to lie flat. It is a matter of preference as to whether a mane comb or a tail comb is used. Start at the poll, with the comb in the right hand. Take a few hairs in the left hand, run the comb up them to the roots, and then with a sharp firm pull, ease them out. The forelock is done last.

- When brushed out and dampened down, the mane should be about 10–12cm (4–5ins) long. If it is still unruly, loose plaiting may help.

PLAITING THE MANE

The mane is plaited to:

- Make the horse look smart, for example for showing, competing and hunting, or when the horse is to be seen by prospective purchasers.

- Train the unruly mane of a stabled horse. Plaiting for this purpose is done very loosely and the plaits are left long and not rolled up. They should be re-done each day. Rubber bands may be used. If the horse starts to rub the plaits, they should be undone and the mane should be washed.

Equipment for Plaiting

- Comb (mane or tail).
- Water brush.
- Needle.
- Thread of a colour to match the mane.
- Scissors.
- Bucket of water.
- A metal feed bin or wooden box on which to stand, if necessary.

Thread the needle and place it securely in a coat lapel. Needles are easily mislaid, which can be dangerous. If they are lost the surrounding bed should be removed, the floor swept and fresh bedding put down.

Method

- Start at the poll. Comb out the mane and dampen it down. With the comb, divide the mane evenly into as many plaits as are required. This varies according to personal preference, prevailing fashion, the length of the neck and the thickness of the mane. The appearance of a fat neck can be improved with a smaller number of tight, bunch plaits; a poorly developed neck can be improved by thick plaits which are not pulled too tight. In all cases it is customary to have an uneven number of plaits up the neck.

- Take the first section of hair, divide it into three and

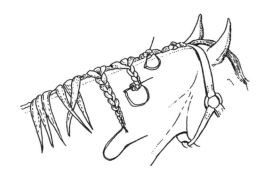

Plaiting a mane.

plait down as far as possible. Take the needle and thread and sew it firmly. Turn the plait under, stitch and then turn up again. Stitch each turn so that it is firm and there are no stray ends. The resultant plait should be button-like, preferably with no thread showing. Proceed down the neck. The two plaits in front of the withers require extra stitching, as there may be friction from the martingale, breastplate and reins. The forelock is done last. In dressage and show jumping the current trend is to leave the forelock unplaited.

- To achieve tidy and firm plaits it is essential to stand at an easy height. Short people may need a bale of straw or a strong box on which to stand.

- If time is short, rubber bands, of a colour which matches the mane, may be used instead of thread. They should never be used where correct turnout is expected. The rubber band is looped several times round the end of the plait. The plait is then rolled up, and the band looped round the whole plait until it is firm and tight. Two bands can be used, one to secure the plaited hair and one to secure the roll.

- It is not advisable to plait manes the night before. Hay and bedding become embedded in the plaits, some hairs break and the result is untidy. Some horses try to rub out the plaits, with even worse results.

- For dressage, white tapes are often wrapped around the completed plait for added effect; this practice was developed in Europe.

PULLING THE TAIL

The tail is pulled to:

- Improve its appearance.

- Show off the horse's quarters.

- Save time when turning horses out for showing, competition, hunting, etc, when the tail must be either pulled or plaited. (It is better not to plait the tail the night before, for the same reasons as for mane plaiting.)

The tails of horses and ponies at grass should not be pulled, as the long hair protects the dock area.

Tail pulling should be done when the horse is warm and the pores of the skin are open. Thoroughly brush out the tail, removing all tangles, but do not wash the top of the tail as this makes it slippery and difficult to pull. The procedure now varies according to the amount of hair to be removed. Well-bred horses should require only a small amount of hair to be removed from each side of the dock for about 15–20cm (6–8ins). The central part should remain untouched so that the natural length of the hair is maintained.

A neatly pulled and banged tail.

Because the pulling causes discomfort and sometimes bleeding, it is preferable to take several days to complete it. The hair is removed by using the fingers, or with the help of a mane or tail comb. The hair is pulled out with a short, sharp tug. Resin on the fingers will give a better purchase. After pulling, the tail should be washed and bandaged.

Horses who grow heavy tails need to have a lot of hair removed. Some sensitive horses object, and with these an assistant is needed to hold up a front leg. If the horse is still difficult, it is advisable to place an upended bale of straw in front of the person pulling. Alternatively, it is sometimes possible to hold the tail over a stable door and so complete the task. A twitch may help to control the horse (see page 62). With very resentful animals, it may be more sensible simply to keep the tail well brushed and bandaged, and to plait it for special occasions.

Once a tail has been pulled, it should be kept tidy and well bandaged and the growing hair should be removed regularly. If the hair of a pulled tail is allowed to grow, it will at first have a bushy and unsightly appearance.

When first learning to pull tails, it is helpful to have an experienced observer on hand. It is easy to pull out too much and to finish with a nearly bald dock. Sometimes tails are shaved in the dock area. This initially a gives neat appearance from a distance, but once shaved it will need constant attention to maintain it. Growing out will involve an untidy, spiky appearance for a long time.

PUTTING UP A TAIL

This is done so that in wet muddy conditions the horse's tail remains comparatively clean and dry. Eventers, polo ponies, and occasionally hunters have their tails put up to help keep them out of the way.

The upper part of the tail is usually pulled. The main portion is firmly plaited down to the end, where it may be either stitched or secured with a rubber band. The bottom half of the tail is doubled up underneath and

A tail tied up with tape.

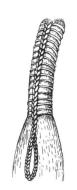

A ridge plait.

secured by stitching, by tapes or by rubber bands. The latter, if too tight, may restrict circulation. Thread, tape and bands should match the tail in colour.

PLAITING THE TAIL

The tail should be washed and then well brushed out. There are two methods of plaiting: one with a flat plait and the other with a ridge plait. The latter requires more skill and practice.

Method
• From the top of the dock, take a few hairs from either side and some from the centre. The centre strands can be tied with thread. These are the three locks of hair that make the plait. Plait them together and then work down the tail, each time taking a few more hairs from the side. No more hair is taken from the centre. Continue plaiting approximately two-thirds of the way down the dock. From then on, continue plaiting without taking any more hair from the sides until the end of the plait is reached. The end of the plait is firmly stitched, and then doubled up underneath and again stitched.

• To achieve the flat plait, the locks of hair are brought inwards and passed one over the top of the other. To make a ridge plait, the locks of hair are passed from underneath, so that a ridge line is made on top. For the first method the fingernails face downwards and for the second they face upwards.

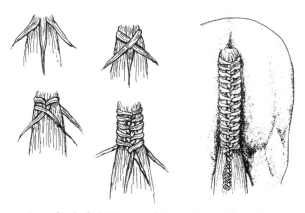

A method of plaiting a tail to produce a flat plait.

12 SHOEING

MODERN SHOEING PRACTICE

Horseshoeing has been an organised craft in Britain since 1356, when the Worshipful Company of Farriers was established in the City of London.

The Farriers Registration Act (1975) states that shoes can only be fitted to horses by persons registered to do so. Registration is confined to persons who have taken and passed the examinations of the Worshipful Company of Farriers (Registered Shoeing Smith until 1977 – now Diploma of the Worshipful Company of Farriers), or persons who at the time of the Act were practising farriery, either on their own or on other persons' horses, with or without reward.

An apprentice training scheme was introduced by the Company in 1960. It is currently administered by the Farriers Registration Council and funded by the Council for Small Industries in Rural Areas. Apprenticeship is for four years, during which the apprentice attends the School of Farriery at Hereford Technical College for twenty-seven weeks, for theoretical and practical instruction.

There are some 2000 persons on the Register of Farriers, including about 300 registered to shoe their own horses. Some farriers operate from a forge to trim and hot shoe horses brought to them, but many others travel around their area. Factory-made shoes are available in many sizes, and are finished to a high standard. They save the craftsman many hours of work at the fire and anvil, and are an acceptable substitute for hand-made shoes. The key to successful shoeing is regular and correct trimming of the hooves to maintain the correct balance, length and angle.

The modern farrier is a skilled craftsman. It is the responsibility of the owner to:

- Make regular appointments for attention to his horse's feet.

- Provide a clean, well-lit place with a level floor and protection from the weather.

- Ensure that the horse is trained to stand quietly for trimming and shoeing.

BASIC STRUCTURE OF THE FOOT

Internal Structure

The bones in the foot are:

- The second phalanx (short pastern bone) – half in the foot; half above the coronet band.

- The third phalanx (pedal bone).

- The navicular bone.

Tendons run down the leg and are attached to the bones, the superficial and deep flexor tendons at the back and the extensor tendons at the front. The former flex the joints, the latter raise and extend the limb.

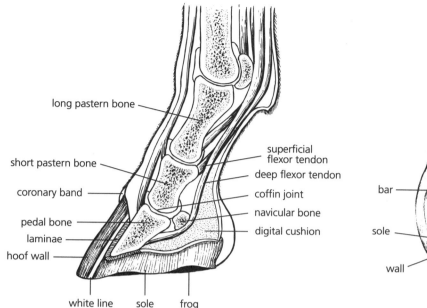

Cross-section through the fetlock, pastern and hoof.

long pastern bone

short pastern bone

coronary band

pedal bone

laminae

hoof wall

white line   sole   frog

superficial flexor tendon

deep flexor tendon

coffin joint

navicular bone

digital cushion

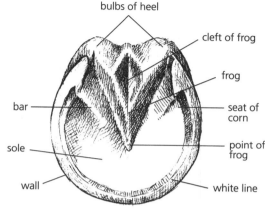

Parts of the foot.

bulbs of heel

cleft of frog

frog

bar

sole

wall

seat of corn

point of frog

white line

Ligaments hold the tendons and bones in place but at the same time allow necessary movement.

The **lateral cartilages** extend from the wings of the pedal bone towards the heels and form elastic shock-absorbers.

The fleshy structures are tough, elastic pads consisting of:

- The **plantar cushion** or **sensitive frog**, which lies at the back of the foot.

- The **coronary band**, which is a strip extending round the coronet and from which the hoof grows.

- The fleshy **sole**, which lies beneath the pedal bone.

The **sensitive laminae** cover the outer surface of the pedal bone, and form a junction between the fleshy and horny parts of the hoof. The fleshy structure and the sensitive laminae supply nutrition to their corresponding parts of the foot.

The foot is well supplied with **blood vessels** and **nerves**.

External Structure

The hoof forms a non-vascular and insensitive covering for the internal parts of the foot. It consists of the wall, the sole and the frog.

The **wall** is that part of the hoof which is visible when the foot is on the ground. It is more or less semi-circular, the front feet being rounder in shape than the hind feet. At the extremity of the heels it bends sharply back to form the **bars.** The area between the wall and the bars is called the 'seat of corn'.

The wall is thicker at the toe than at the heels, and expands slightly at the heel when the foot takes weight. It grows from the coronary band, and takes an average of nine to twelve months before coming into wear at the toe, and about six months at the heel. When working on hard surfaces, the wall cannot grow fast enough to replace the wear and shoeing is necessary.

The **periople** forms a shiny protective layer on the surface of the wall. It prevents undue evaporation; its removal by unnecessary rasping can cause the feet to become brittle owing to excessive loss of moisture. The water content varies from 25% in the wall to 40% in the frog.

The junction between the wall and the sole is marked by a distinct **white line.** This forms an important guide to the farrier as to the thickness of the wall covering the sensitive laminae. It also marks the union between the sole and the wall.

The **horny sole** forms the bottom surface of the hoof. It is softer than the wall, and grows from the fleshy sole. Natural exfoliation maintains a constant thickness to the sole, but its shape and thickness vary between different horses.

The **frog** is of a tough, elastic consistency. It is triangular in shape, and fits into the space between the angle of the heels. The depression in the centre is called the cleft. When the foot comes to the ground, the foot expands, acting both as a shock-absorber and a weight-bearing surface. This expansion, followed by contraction, has an important function in maintaining the circulation of the blood in the foot.

CONFORMATION

Good or Normal Feet

The **front foot** should be rounder at the toe than the hind foot, and viewed from the side should usually make an angle with the ground of between 45° and 50°. Although the angle should be the same as the pastern axis, it may vary between horses. The inside wall is normally slightly steeper than the outside wall. The line when the foot is picked up should be symmetrical around an imaginary central mid-line. The heels should be wide apart and the frog large and well developed, with a definite cleft. This should not extend up between the heels. The sole should be slightly concave. Both front feet should match in shape and size.

The **hind foot** is normally narrower and more upright than the front foot, with the toe less rounded. The angle with the ground should be between 50° and 55°. The sole is more concave, and the frog smaller. The

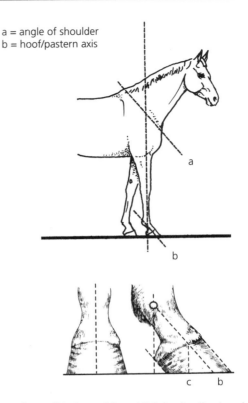

a = angle of shoulder
b = hoof/pastern axis

The well-balanced foot. Weight distribution should be equal in front of and behind c.

hind feet should be a matching pair.

The surface of the wall may have a number of rings. These are normally insignificant in size, and smooth, and merely indicate changes in diet. But they may be larger and ridged, in which case they could indicate a previous illness.

Some horses have unusually high heels. This may be the natural conformation of the feet, or it can be the result of the way in which they are shod. Any excessive lowering of the heel, especially without corresponding cutting back of the toe to maintain the correct foot/pastern axis, can cause problems.

When the horse is standing naturally, all feet should point directly to the front. Any deviation affects both the movement of the leg and the wear of the shoe. Pigeon-toed or 'toed-in' horses show excessive wear of the outer branch of the shoe; those with 'toes out' produce extra wear on the inner branch.

The Hoof/Pastern Axis

It is important to consider the balance of the foot with regard to the relationship between the slope of the pastern and the slope of the hoof wall. To ensure cranio-caudal (forward to backward) balance, the slopes should be similar. It is also important to maintain lateral to medial (side to side) balance, which ensures that the foot lands level on the ground. If the hoof/pastern axis is broken back, for example by excessive length of the toe or by excessive lowering of the heels, undue strain is applied to the flexor tendons and suspensory apparatus of the distal limbs. Any imbalance will furthermore apply undue strain to the bones, joints and ligaments of both limbs.

Poor or Abnormal Feet

FLAT FEET show a decreased angle of the foot with the ground. The heels are low and the sole flat and often thin. The frog compensates for the poor conformation of the rest of the foot, and is usually large and well developed. Such feet are likely to sustain bruising of the sole and if they are not carefully shod, they are susceptible to corns.

BOXY OR UPRIGHT FEET show an increased angle of the foot to the ground. The heels are high, the frog small and the horn texture is often very strong. Many native ponies running unshod on their natural terrain have this type of hard foot, but with a good, well-developed frog.

CLUB OR MULE FEET are a more pronounced version of

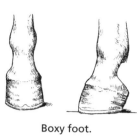

Boxy foot.

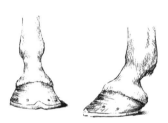

Flat foot with long toe and collapsed heel.

boxy feet, with an even greater angle to the ground.

LONG OR FLESHY FEET can be similar in some aspects to flat feet. The toe is long and cannot be reduced to normal size without risk of injury to the internal fleshy structures.

THIN SOLES are very sensitive soles.

DROPPED SOLES are convex and in acute cases below the surface of the wall. Dropped soles are often a sign of chronic laminitis with rotation or sinking of the pedal bone.

PRONOUNCED RIDGES OR RINGS are a sign of an alteration of the growth rate. They can be caused by past or present disease, such as laminitis, or by incorrect hoof trimming. The rings are widely spaced at the heel and converge at the toe. Those associated with a change in diet run parallel to the coronet and are smooth.

BRITTLE FEET can be partly inherited or may be due to incorrect diet. Careless rasping of the wall – resulting in loss of moisture – or dry weather – resulting in the natural loss of moisture from the hoof without its being replaced – can be contributing factors. Careful balancing of the diet, regular exercise and skilled farriery may improve brittle feet.

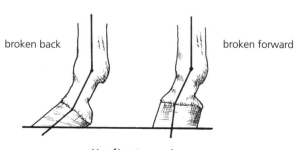

broken back   broken forward

Hoof/pastern axis.

CARE OF THE FEET

A horse turned out on good pasture, or a stabled horse on a balanced diet, should be receiving all the necessary vitamins and minerals for the production of strong horn. However, the metabolism of some horses is such that in spite of a suitable diet, they still have weak and brittle feet. Improvement can then be made by adding biotin, a member of the vitamin B complex, to the diet. Methionine, zinc, or glycerine are also useful additions and may help. Stable hygiene and the type of bed used is also important for the maintenance of strong feet.

Methods of Improving Horn Growth

- Regular attention from the farrier can improve the shape of the foot, and reduce the risk of damage to the wall by a lost shoe. Rasping of the wall must be minimal. The frog and sole should be trimmed, to remove untidy and dead fragments, but not cut back.

- Stimulating ointment rubbed daily into the coronary band encourages the growth of horn but will not improve texture. Care must be taken not to blister the coronet.

- Regular oiling of the wall will prevent undue evaporation, but too much oiling of the feet, particularly of the sole and coronary band, restricts the amount of moisture that the foot is able to absorb, and can be counter-productive, causing brittle feet.

Unshod Horses and Ponies in Work

Horses who are ridden on soft surfaces, e.g. in a school, with little road exercise, can work without shoes. Ponies with naturally hard horn often work without shoes, although some may need shoeing in front, particularly if they work on rough or flinty surfaces.

In a dry season most horses'/ponies' feet are hard, and will stand up to wear, although some horses' feet in dry conditions lose moisture and become brittle. In a wet season the feet may absorb too much moisture, and the sole and horn become soft, thus making shoeing necessary.

When the shoes are first removed, the horse/pony may go a little short until the feet harden. If there is no improvement, then the horn texture is not standing up to work without shoes.

Horses/ponies in work without shoes will on occasions wear the feet down to a degree where they become foot sore. It may then be necessary to put on shoes and allow sufficient new horn to grow before removing them again.

An alternative is to rest the animal until the horn has grown, or to work on soft going only.

Care of unshod feet entails:

- Regular inspection of the feet, including removal of small flints and debris which may cause infection and lameness.

- Every four to six weeks the farrier should dress the feet to maintain their balance. Some animals wear the toe, some the heel area, and some wear more on one side than the other. It is important to level up the foot, and to give the animal a level bearing surface.

Horses at Grass and Not in Work

Animals with good, strong horn growth should not require shoes. The feet should be dressed every 4–6 weeks so that the foot does not grow out of shape.

Some horses have poor horn growth and a tendency to flat soles. This type of animal is better kept shod in front. Shoes should be a lightweight pair of normal shoes. The farrier should replace them every 4–5 weeks.

In a dry year when there is little grass, and on a rough or flinty surface, horses can become foot sore and may need shoeing.

Young Horses

Young horses with foot and/or gait problems may have them considerably improved if their feet are looked after from an early age by a skilled farrier.

PARTS OF THE SHOE

- The **web** refers to the width of the material from which the shoe is made.

- The **branch** is the parts of the shoe from toe to heel.

- The **heel** and **toe** are self-explanatory.

- The **quarter** is the part of the hoof between the toe and the heel.

- The **bearing surface** is the part of the shoe in contact with the foot.

- **Clips** are drawn at the toe of the front shoe, and usually at the quarters of the hind shoe, although some ponies and cobs not used for fast work and jumping may have a toe clip on a hind shoe. The clips should be solid, but not larger than the thickness of the shoe, and rounded rather than diamond shaped. If a shoe becomes loose or dislodged, the horse often treads on the clip, which can cause a serious injury if it is sharp. Clips assist in keeping the shoe in place.

- Permanently **raised heels (calkins** and **wedges)** on the shoes of horses working on hard surfaces dramatically reduce the weight-bearing area of the shoes and therefore apply extreme pressure to the foot at the point where it is raised. They are not normally seen in modern farriery.

- **Studs** or **nails** can be fitted at both heels but should only be used if extra grip is absolutely necessary.

TYPES OF IRON

- **Plain** is heavy, straight-edged, flat iron.

- **Concave iron** has a wider foot-bearing surface than ground-bearing surface. This reduces suction and gives a lighter weight shoe, but it also reduces the ground-bearing surface.

- **Fullering** refers to the groove made in the ground-bearing surface of the shoe. It provides a secure bedding for the nails and a better foothold.

COMMON TYPES OF SHOE

PLAIN-STAMPED SHOES are made of plain iron with nail holes stamped through the full thickness of the web. Both front and hind shoes usually have one toe clip. These shoes are only suitable for horses doing slow work, as they give a less secure foothold, and are also more likely to cause interference, such as brushing or over-reaching.

HUNTER SHOES are suitable for most riding horses and ponies. They are made of fullered and concave iron, with pencilled heels, and have one clip in the front shoes, two in the hind. Dressing the heels of the shoes to the angle of the feet lessens the risk of their being pulled off by a tread. Two clips in the hind shoes allow the shoe to be 'set back', the front edge to be rounded off, and the inner edge bevelled, all of which reduces the risk of an over-reach or of the shoe pulling across the foot when jumping and turning.

To give better grip, hind shoes used to be fitted with a calkin and wedge heel, although it is now more usual to use road studs, if anything at all. It should be remembered that any raising of the heel can give a false angle to the foot, cause side-to-side imbalance and reduce the weight-bearing area of the shoe. Modern farriery favours the balanced foot with minimal interference from introduced studs or heels.

Rolled toes on hind shoes are used on horses who drag their toes. The front of the shoe is turned up to make a thick, wide toe. The two clips can still be drawn if required. Reinforced steel or tungsten can be put in the toe to give longer wear.

On front shoes, rolled toes are used for horses with long toes who stumble, but if the foot is short and upright, they can increase the stumbling. The shoe is set back on the foot, and the toe surface is rounded off.

ALUMINIUM SHOES are lightweight shoes used for 'plating' racehorses. They may also be put on show ponies and hacks. Plating is a term for shoeing horses with lightweight shoes for racing. These 'plates' are usually removed the following day and replaced with normal exercise shoes.

RACING PLATES are lightweight steel or aluminium shoes for racehorses. They are often fitted without clips.

Various REMEDIAL SHOES, for example bar-shoes, egg-bar shoes, heart-bar shoes and raised-heeled shoes, may be recommended in special circumstances; but since these are likely to interfere with the primary requirement of trimming and shoeing – that is, to establish or maintain lateral-to-medial and cranio-caudal balance – specialist advice should always be sought.

SHOEING TOOLS

Forge Tools for Making the Shoe

- The **anvil** should be sufficiently heavy to withstand the blows of a heavy hammer during the making and shaping of a shoe. The beak, or pointed end, provides a surface for shaping the shoe.

- **Fire tongs** are long-handled tongs used to hold hot metal so that it can be turned in the fire.

- **Shoe tongs** have shorter handles and are used to hold a hot shoe on the anvil.

- The **turning hammer** is a heavy hammer used to shape the shoe and to draw the clips.

- The **stamp** is a punch used to make the nail holes. By altering the angle of the stamp, the nail holes are drawn in closer to the inner edge at the toe of the shoe where the horn is thicker. These are called 'coarse holes'. As the wall becomes thinner towards the heels, the nail holes are made near the outer edge and are called 'fine'.

- The **pritchel** is a long steel punch used to finish the nail holes begun by the stamp. It is also used to hold the hot shoe against the foot.

Tools for Preparing the Foot and Nailing the Shoe

- The **shoeing hammer** is a lightweight hammer for driving in the nails. The curved claw at one end is designed to grip and twist off the point of the nail.

- The **buffer** has a blunt chisel end for cutting clenches, and a pointed end for punching out broken nails.

- The **rasp** is a heavy, flat, steel file about 41cm (16ins) in length. On one side it is coarse cut, and on the other fine cut. It is used to make a level surface for the shoe, and also to smooth off any rough edges of horn after the shoe has been fitted.

- The **drawing knife** has a curved blade with a bent-over, very sharp end. It is used for cutting off ragged bits of horn, and to cut out the small piece of wall for the toe clip. It may also be used to cut away any diseased portions of the frog.

- The **searcher knife** has a lighter blade than the drawing knife, and is used to search and open up a punctured foot.

- The **toeing knife** has a short, flat blade. It is used with the shoeing hammer to prepare and trim the foot surface before finishing with the rasp. It is used for the same purpose as hoof parers, but usually on heavy horses.

- **Pincers** are used to lever off the shoe, to remove nails and to assist in clenching up.

- **Hoof-cutting pincers** have overlapping blades (one thick, one thin) and are used to trim overgrown feet.

- The **tripod** is a three-legged iron support, which the farrier may use to support the horse's foot when he is cutting the clenches, or when finishing the foot.

SHOEING

The average horse requires a new set of shoes, or must have his old shoes removed and replaced, every four to six weeks. The time interval depends on the growth rate of the feet of the individual horse, and how he wears his shoes. Horses who are heavy on their feet, and who do a lot of roadwork, may need new shoes every two weeks.

Indications that re-shoeing is necessary:

- A lost or loose shoe.

- The shoes are thin, either all over, or in one part.

- The clenches have risen.

- The shoe or shoes have spread.

- The foot is beginning to grow over the shoe, resulting in the heels of the shoes pressing into the seat of corn.

- The wall of the foot has grown and the foot is unbalanced.

The Process of Shoeing

Before shoeing a new horse the farrier should study the horse standing square in his old shoes, and then see him walked and trotted up, so that he can observe any abnormalities of action. When viewed from in front the long axis of the limb should be over the centre of the shoe. He should also check the old shoes for any uneven wear. This may have been caused by faulty conformation or by unsoundness, but it can also result from faults in the dressing of the foot or the fitting of the shoe. When the old shoes are removed, the horse should be observed standing unshod and walking.

To Remove a Shoe

1. The foot is held between the knees, and the clenches are opened, using the shoeing hammer and the sharp edge of the buffer. Risen clenches are easily cut; tight clenches require more skill. The corner of the buffer can be placed against the clench, but on no account should the buffer be driven into the wall to raise the clench. With a heavy horse, it may be easier to place the foot on a tripod when cutting clenches.

2. With the foot between the knees, the pincers are used to ease the shoe at each heel, and then round the toe. The shoe is removed by gripping it with the pincers at the toe, and pulling it backwards towards the frog. On occasions it is necessary to remove the nails individually. In this case the clenches are cut, the shoe is eased at the heels and the shoe is knocked back with the pincers. This pushes out the heads of the nails, which can then be removed with the pincers.

3. The old shoe should be put out of the way, with the nails hammered flat, or it may be taken to the fire so that the new shoes can be matched up. Any loose nails should immediately be picked up so that the horse cannot tread on them.

Preparation of the Foot

It is necessary to restore the balance of the foot, and to prepare a level bearing for the new shoe. The farrier should look at the shape and angle of the foot when it is on the ground. When viewed from the side, the angle of the foot and pastern should be the same. The foot is then picked up to observe the bearing surface. In most cases it is necessary to remove more horn from the toe than from the heels. In normal circumstances, the foot is reduced with the rasp; first the sides and toes are done and then, if necessary, the heels. If there is excessive growth, it is taken off by using either the hoof cutter or the toeing knife. Any untidy pieces of sole or frog may be cut off, but, apart from this, if they are in healthy condition, neither should be touched. After the final balancing of the foot, the horse should again be observed standing unshod.

Hot Shoeing

The prepared shoe is heated, shaped and then applied to the foot to check for fit and bearing. To avoid excessive burning of the foot, it should be at a dull heat, and should not be left against the foot longer than is necessary to ensure level bearing. It is then taken to the anvil, where any necessary adjustments are made. After a final check on the foot it is cooled in water.

Nailing On

Nails must be of a suitable size (no larger than the wall of the foot can absorb) and must fit into the fullering. The end of the nail is shaped so that when it is driven correctly, it emerges about one-third of the way up the wall. It is usual to have three nails on the inside of the shoe and four on the outside. The shoe is nailed on first at either the toe or the heel, and then on alternate sides. If the foot is broken or weak, it is important that nails should only be driven into sound portions of horn. In this case it may be necessary to drive the nail higher in the wall than is normally desirable, and also to make sure that nail holes in the shoe are stamped, so that weak parts of the wall are avoided.

If a nail is driven too close to the sensitive tissue it causes a nail bind. It must be withdrawn immediately. This mishap is unlikely to happen if the horse has strong, well-shaped feet. It can occur more easily if the horse has broken or brittle feet, when considerable skill is required to fix the shoe firmly. In rare circumstances, the farrier may 'prick' the foot, that is drive a nail into the sensitive tissue. Often, when the nail is withdrawn, it shows traces of blood. The horse may not be lame at the time, but infection is likely to develop in the foot and to cause acute lameness within sixty to seventy-two hours. The wall is of equal thickness from the coronary band to the ground surface and therefore the height of the nail is not likely to cause lameness.

As each nail is hammered home, the point is either immediately twisted off, or bent over and twisted off later. The end of the nail now forms a clench. The closed pincers are held against the clench, and the nail hammered tight. This turns the clench over. The clenches are shaped with the rasp, and then hammered flat against the wall. The rasp is used to smooth off the clenches, and the edge of the wall, thus making it tidy and finishing the shoeing.

Cold Shoeing

Hot shoeing has been preferred to cold, but farriers without a portable forge may have to shoe cold, and as long as it is done correctly it is as good. In fact, more problems can occur with hot shoeing, from over-burning a shoe on the foot, burning a foot unlevel, burning an unlevel shoe on to a foot, etc.

Arrange for the farrier to see the horse, measure the feet and see if there are any particular problems. A skilled farrier who makes his own shoes can shoe cold very satisfactorily. However, in the case of factory-made shoes which arrive complete with nail holes, a less satisfactory fit is obtained. Because the nail holes are always in the same position, when the shoe is put on nails may well have to be driven into weak horn. This results in a less secure shoe.

To Measure a Foot

If it is necessary to give the farrier the approximate size of the shoes, measure the foot from toe to heel and then across the widest part. Any broken or weak parts of the wall should also be mentioned. Alternatively, place the foot on paper or cardboard and draw the outline.

THE NEWLY SHOD FOOT

When the Foot is on the Ground

- The foot should be suitably reduced in length at both toe and heel. The line of the pastern should be continued by the foot until the shoe and the ground surface are reached, i.e. straight hoof/pastern axis.

- The frog should be close to the ground.

- The clenches should be in line, well formed and bedded, and the correct distance up the wall. They should not have been driven into old nail holes.

- The place for the clip should be neatly cut and the clip rounded and well bedded.

- The wall should have been rasped only lightly, and not above the line of the clenches.

- The foot should be level on the ground.

When the Foot is Picked Up

- The shoe should fit the foot. There should be no dumping of the toe (excessive cutting back of the toe). The heels of the shoe should be the correct length, neither too long nor too short.

- The type of shoe and the weight of iron should be suitable both for the work the horse is doing, and for the size and shape of the foot.

- There should be no unnecessary paring of the sole or cutting of the frog.

- The foot should be evenly rasped down and the heels level, and not opened up by cutting the bars.

- No daylight should show between the shoe and the foot, particularly in the heel area.

- The nails should be of a suitable size and should have been well driven home, so that they do not project above the surface of the shoe.

- The correct number of nails should have been used, usually three on the inside and four on the outside.

- The horse should be sound when he is trotted up.

FAULTS IN SHOEING

DUMPING The shoe is set back, and is often too small. The toe of the foot is rasped off to make it fit the shoe. The rasping of the wall exposes the soft underlying layers, which then become brittle and will not hold the nails. The foot itself has a neat appearance, but the fault can be easily observed because the angle of the wall changes as it comes towards the shoe. The foot itself has a smaller bearing surface.

NB: Dumping of the toe should not be confused with rasping the toe to obtain a normal hoof/pastern axis.

OVER-LOWERING OF THE TOES raises the heels and reduces the bearing surface. The angle of the wall on the ground is increased, and the balance of the foot, and therefore of the horse, is upset. Frog pressure is reduced and more weight is placed on the front of the foot. The heels should be one-third of the height of the toe from the coronary band.

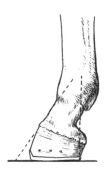

Dumped toe.

OVER-LOWERING OF THE HEELS puts weight on the back of the foot, with consequent extra strain on the tendons and ligaments. The angle of the wall to the ground is decreased, and the resultant long toe can cause stumbling, and the possibility of the development of navicular disease.

Opening the heels by CUTTING AWAY THE BARS of the foot causes a weakening of the wall and of the heels, and results in contraction of the heels and in increased concussion.

OVER-LOWERING OF THE BEARING SURFACE can occur through lack of judgement when a farrier first shoes a horse with flat, open feet. The bearing surface is rasped down too much and the horse becomes footsore. An uneven bearing surface occurs when one side of the foot is rasped away too much. When the foot is picked up and the shoe checked, the uneven surface can be seen. It can also result from a lost shoe, when one side of the foot is worn down or breaks off.

RASPING THE WALLS results in the removal of the periople, and consequent drying out of the foot and brittleness. Ideally, rasping should be minimal, never above the clenches and only sufficient to finish and tidy the foot. However, in the case of long-toed feet, some extra use of the rasp is necessary.

SHOES TOO HEAVY may cause some form of interference, such as brushing, over-reaching, etc. If the horn is not strong, a shoe may come off and often it will take a piece of the wall with it.

NAILS TOO LARGE may split and damage the wall. The resultant loss of holding power may cause problems when the horse is next shod. There is also a greater risk of nail binding. The head of the nail sits above the fullering, and thus receives more wear.

EXCESSIVE BURNING OF THE HORN when applying a hot shoe to the foot. The shoe should be at no more than a dull, red heat, and should be held in contact with the foot only for as long as is necessary. If it is subjected to too much heat, the wall surface is weakened and may well crumble away, leaving a loose shoe. More often, the interior soft tissues of the foot are affected by scalding.

NAIL BIND occurs when a nail is driven too close to the sensitive laminae and causes pressure. When the horse is trotted up after shoeing he goes unlevel or a 'little footy'. The nail must be removed and another nail driven in at a less steep angle. This lameness can occur as long as seven to ten days after shoeing.

NAIL PRICK is when a nail is mistakenly driven into the sensitive laminae. The horse shows pain, and the nail when removed may show traces of blood. Often, the horse trots up sound at the time, and may not go lame for two or three days, by which time pus will have formed in the wound. The shoe will have to be removed, the nail route traced and opened at the base, and the foot poulticed.

STUDS

The use of studs should be considered with extreme caution. They affect the angle at which the shod foot meets the ground, alter the bearing surface and balance of the foot, and increase the wear of the shoe at the toe. These problems can be resolved if the studs are always used in pairs. In this way, the balance and angle of the foot can be maintained.

Studs are made from specially hardened steel, and are used to give a better foothold. Road studs are put in for slippery roads and jump studs for jumping.

Road studs can be hammered in and remain a permanent feature of the shoe. When the horse is re-shod, it is often possible to remove the studs from the old shoes and to use them again. They are usually placed on the outer heel of both hind shoes, but they may be put on the outside quarter of the shoe. For even greater security, they can be put on the front shoes too, although this can cause greater concussion and the risk of tendon strain from steadying too abruptly. There is also a risk of treads (see page 217) if studs are placed on the inside heel.

The screw-in type road stud can be used if it is necessary to replace it with a larger stud for jumping.

Jump studs come in several shapes and sizes. Generally, large square studs are used for deep mud, and smaller pointed studs for hard ground. The actual choice of stud is very much one of individual opinion, and must relate to ground conditions at the time.

On a narrow web shoe, it may be necessary to widen the shoe at the heel to give room for the stud holes, which are made when the shoe is hot. To ensure correct fitting, and because studs can vary, it is important to show the farrier the studs you plan to use.

If the stud holes are not filled by road studs or plugs, they should be greased and packed with cotton wool. This ensures that they are kept clean and facilitates screwing in of studs at a competition.

Studs in a horse's shoe can result in an extra risk of injury to both horse and rider from a strike or kick. For this reason, studs are not permitted in racing. Their use for cross-country competitions must depend on the going and on local conditions. For dressage, again it depends on the conditions of the day. Studs can give more security to the horse on corners and in transitions, but on hard going they may shorten his stride and emphasise any unlevelness. The horse must be allowed to get used to working with studs.

Equipment Required for Fitting Studs

• Nail for removing cotton wool.

• Metal tap for cleaning out and reviving the thread of the screw hole.

- Spanner for removing and screwing in studs.

- Fresh oiled wool for placing in the holes when the studs are taken out.

- Screwdriver to fix or remove plugs, if used.

Drive-in Plugs

These can be more effective than studs on a slippery road surface and in snow and/or ice. They are fitted round the shoe, and therefore give an even bearing and do not upset the balance of the foot.

PADS

Leather or synthetic pads can be used, temporarily to protect an injured sole, or on a permanent basis to minimise concussion and bruising in the case of horses with thin soles.

Before a pad is fitted, the sole and frog should be treated with Stockholm tar and the sole padded with Gamgee or a similar material.

The synthetic pad is cheaper than leather, but because it is more likely to make the sole and frog sweat, it makes it more difficult to keep these structures healthy. A synthetic pad shaped to the shoe, and continued across the base of the frog, can ease concussion. The bar over the frog can alter the balance of the foot and care should be taken over this. Leather pads are also affected by conditions – when wet, they compress between the shoe and the foot, causing clenches to rise; when dry, they can become rock hard. A more effective protection for bruised or thin soles, and one which minimises concussion, is a rubber or plastic cushion, which fits under the shoe but leaves the frog area clear.

Wedge-shaped pads can be used to ease the pressure on the tendons and to reduce concussion. In snow or freezing conditions, pads fitted under the shoes help to prevent snow balling in the feet.

13 BEDDING

Bedding is material that is put on the floor of a loosebox or stall to:

- Encourage the horse to lie down and rest.

- Enable him to do so in comfort, without risk of injury.

- Keep him warm and minimise draughts.

- Encourage him to stale.

- Reduce jar to his legs from standing and moving around on a hard surface.

- Prevent him slipping up when moving about the stable.

Bedding is also used in a trailer or horsebox when a horse is travelling to cover the floor and to give a softer and less slippery surface.

TYPES OF BEDDING

Straw

Straw is still a widely used bedding.

WHEAT STRAW Modern farming methods have resulted in wheat being grown with a much shorter stalk. When baled it is often heavily compacted and brittle, affecting its durability and reducing its value as bedding. Good wheat straw makes excellent bedding but it is not easily available.

BARLEY STRAW is usually longer, of better quality and a brighter colour than wheat straw. Modern combine harvesting removes the awns, which used to irritate the horse's skin. Barley straw can cause problems if the horse is prone to eat it to excess.

OAT STRAW is palatable and more expensive. It quickly becomes saturated, which makes it the least suitable straw for bedding.

If it is of good quality, straw of any type is likely to be eaten by the horse. In a similar way to hay, the straw and dust will be a host to fungal spores. When inhaled by the horse these can cause, or re-activate, allergic coughing.

The eating of straw can cause common horses to become gross and can affect the fitness of horses in fast work. Non-toxic disinfectant sprayed on the bed helps to discourage the eating of straw.

ADVANTAGES
- The bed has a clean, bright appearance.
- Straw manure can be disposed of more easily than other types.
- In a good harvest year it can be cheap.

DISADVANTAGES
- It can be eaten by the horse and is likely to cause allergic coughing.
- The manure is heavy.
- In a bad harvest year it can be very expensive.

Straw Deep Litter

ADVANTAGES

- It is economical. Once the bed has been put down (plenty of straw must be used) the daily requirement will be less than for a conventional straw bed.
- It is warm, and more secure for foals and young stock, reducing the risk of injury.
- It provides a solid bed. The horse cannot get through to the floor when rolling, and there is less risk of injury.
- It is labour-saving on a daily basis.
- It is suitable for use where there is an uneven floor or poor drainage.
- When mucking out there is less disturbance of the bed, so that fewer fungal spores and less dust are released into the atmosphere.

DISADVANTAGES

- It can be unsightly and unhygienic if it is badly managed. If the bed starts to smell, remove it, disinfect the floor and put down a new bed.
- It is not suitable for restless or very dirty horses as it will be impossible to keep them clean.
- Extra care must be taken to ensure that the horse's feet remain healthy.
- The eventual cleaning out is very heavy work and requires extra help. A tractor and trailer can be useful. It should be done when the height of the bed becomes inconvenient, which can take from one to three months.

Shavings and Sawdust

These may be collected locally, or compressed baled shavings may be bought from merchants or by direct delivery from manufacturers. Shavings are generally preferable to sawdust as they are less dusty.

ADVANTAGES

- They provide a spore-free bed.
- They will not be eaten by the horse, therefore his fitness and weight will not be affected.
- The bed is lighter to work with than straw.

- There will be less smell on clothes.
- Grey horses will be easier to keep clean.
- In some areas shavings and sawdust can be collected free.
- Baled shavings are conveniently packed in polythene bags for easy storage and are comparatively dust free.
- They may be used with advantage on an uneven floor where there is no drainage.

DISADVANTAGES

- Some samples may be dusty. They may require damping down when the bed is first laid.
- They cannot be used for horses with open wounds.
- Manure may be difficult to dispose of, although it can be burnt.
- Shavings, if sharp, may irritate the heels of some horses.
- Rugs cannot be put down in a corner of the box as in a straw bed.
- Care must be taken when putting down any articles, including grooming equipment.
- If the bed round the walls is left undisturbed and allowed to compact and to heat, there is a risk of the wooden walls rotting. To avoid this, the bottom 60cm (2ft) of the wall should be treated with black bitumen paint.

Shredded Paper

Shredded newspaper or waste paper from offices can be bought by the bale. It is also possible to buy a shredder and to cut paper on the yard.

ADVANTAGES

- It provides a totally dust- and spore-free bed, so it is excellent for horses who are allergic to straw and dust.
- Being packed in polythene bags, it is easy to store.
- It is easily disposed of because it can be burnt when dry, or put on to gardens when rotted. It breaks down more quickly than other types of bedding.

DISADVANTAGES

- It is unattractive to look at.

- It needs very careful management.
- Horses with white legs may be allergic to printer's ink.
- Contractors may be unwilling to collect the manure.
- It can be more expensive than other types of bedding.
- Problems may arise in keeping the yard and muck heap tidy.

Peat Moss

This is seldom used nowadays.

ADVANTAGES
- It can be used on gardens when rotted down.
- It is suitable for horses with fungal-spore allergies, or for confirmed bed-eaters.

DISADVANTAGES
- It is a difficult bed to manage and can be dusty when put down.
- The dark colour means that wet parts of the bed are not easily visible.
- Horses become easily stained.
- It cannot be burned.

Hemp

Hemp is a natural fibre derived from the flax plant. It is a popular bedding with many attributes. Comparable in price to good quality shavings, it establishes well as a deep-litter type bedding. It is highly absorbent, dust free and unpalatable. Once settled it is easy to maintain by regular removal of the droppings, removal of wet areas, approximately weekly, and regular topping up with fresh bedding.

MANAGEMENT OF BEDDING

Putting Down the Bed

Whatever the type of bedding used, the bed should be between 15cm (6ins) and 30cm (12ins) deep. A good bed will make a horse feel better and work better. Banks round the sides to a height of 30cm (12ins) increase warmth, lessen draughts and give some protection should a horse become cast. For added warmth the bed should if possible be taken up to the door, particularly in winter.

RUBBER FLOORING is sometimes used to reduce the amount of bedding required. After the initial cost of the rubber flooring savings can be made on labour as less bedding can be utilised. Rubber flooring used to the exclusion of any additional bedding is unattractive and smelly. Rugs and bandages become stained and tainted and the horse may be reluctant to lie down regularly.

Taking the Bed Up

In many stables straw beds (if it is not deep litter) are taken up during the day, the floor allowed to dry and a thin layer of straw spread on the floor.

ADVANTAGE
- If there is a drying wind the floor will dry and so, too, will some of the bedding.

DISADVANTAGES
- There is a danger of a horse slipping on the floor.
- There is a danger of a horse lying down on an insufficient bed and injuring himself.
- On a damp day the floor will not dry.
- The horse may be reluctant to stale if bedding is sparse.
- Bedding the horse down in the morning saves time and labour in the evening, when time is always short.
- A good bed in the day adds greatly to the comfort and well-being of a horse.

The only way to economise on bedding without detriment to the horse is to check on the amount of usable bedding which is put on the muck heap in the morning. Any usable bedding should be kept; clean bedding should be mixed with it and spread over the top. It is all too easy to throw away clean bedding which has dung on top of it instead of shaking it clear.

This is not to say that horses should have dirty or wet beds. The principle is to provide good beds but not to waste bedding. Regular skipping out to remove droppings keeps the bed tidy and well maintained.

Equipment

LONG-HANDLED TWO-PRONGED FORK, usually known as a pitchfork or hay fork. The handle should be of a length appropriate for the height and build of the groom. These forks are not suitable for picking up manure and wet straw. They should be used for shaking up the straw bales and for shaking up the bed when it is being re-laid. In careless hands they can be dangerous if used when the horse is in the box. It is easy to misjudge the weight or depth of a forkful of straw and to nick the horse on the leg. A wise precaution is to have the ends of the prongs blunted.

FOUR-PRONGED FORK, short or long handle. A tall groom may find the long-handled type more comfortable as it will necessitate less bending, but the short-handled fork is easier to manage and to control. The ends should not be sharp. The fork is used for mucking out straw bedding and spreading other types.

MULTI-PRONGED OR CLOSE-PRONGED LIGHTWEIGHT FORK This is used for collecting dung particles from a shavings, sawdust or similar bed.

SHOVEL, of a suitable weight for the groom. Very light shovels rapidly rust from contact with urine.

BROOMS Small nylon brooms are easy to handle and are hard wearing. The wider, heavier brooms are more suitable for large areas of yard sweeping. Small, soft brooms with long handles are useful for sweeping walls and for removing cobwebs.

Yard brooms, 60–90cm (2–3ft) wide, are excellent for yard sweeping but soon wear if used in the stable. Brooms are made of birch or nylon tufts. Both materials give good wear. The handles must be of a suitable length and size for the user.

SKIP This is used for collecting the droppings from the bed; the droppings are either scooped or forked into it. Plastic washing skips are cheap, suitable and easy to keep clean.

WIRE RAKE This is used for raking shavings beds. The type used for raking moss out of a lawn ('Springbok') is suitable.

RUBBER GLOVES may be used to skip out a box by hand, particularly when the bed is of shavings or sawdust.

WHEELBARROWS Lightweight wheelbarrows are not suitable, as they quickly rust through and do not stand up to the work required of them. Two-wheeled barrows are expensive but they are easy to load, very stable and easy to push. Single-wheeled barrows are suitable providing they are made of a heavy metal, but they are hard to push and easily turn over. Wooden wheelbarrows are heavy and not recommended.

Before using, it is worthwhile giving the inside of all wheelbarrows a second coat of paint. This gives them some added protection – an undercoat of rust-resistant paint gives even better protection.

Split polypropylene **MUCK SACKS** with rope handles may be used instead of wheelbarrows for conveying stable manure to the muck heap. They are easier to fill than wheelbarrows, but heavy to carry. When swinging them on to the back it is easy to strain a muscle. They are easily washed clean.

Split jute sacks were formerly used but these are now difficult to obtain. They tend to become saturated and unpleasant to handle.

All stable tools should be washed regularly – at least once a week. This improves their appearance, makes brooms more efficient to use, and makes shovels last longer. Wheelbarrows profit from being washed out daily and then stood up so that the water drains out. If left in contact with urine and manure, metal will quickly rust.

When not in use, all stable tools should be hung up in a dry shed. This lengthens their useful life and means that in frosty weather grooms will not have to contend with ice on the handles. Stable tools, particularly forks and rakes with long handles, can be a hazard if left lying about for either staff or horses to trip over or stand on. Tidiness and care in the use of stable tools is a must.

Preparation of the Box Before Putting Down a New Bed

EQUIPMENT
- Long-handled pronged fork.
- Shovel.
- Wheelbarrow.
- Hard broom for sweeping up and scrubbing walls.
- Soft broom for sweeping walls and ceilings.
- Large sponge for washing windows.
- Leather for polishing windows, or window-cleaning equipment.
- Water bucket.
- Disinfectant.
- Well-built step-ladder for reaching ceilings and windows.

METHOD
- Take the horse out of the box.
- Remove old bedding and water.
- Dust walls, ceilings and windows.
- Clean out and wash fixed mangers. Cover them with polythene bags to keep them clean.
- Wash and clean windows.
- Wash and scrub floor. Leave to dry.
- Put down bed.
- Check stable for dust, and re-do if necessary.
- Clean and re-fill water buckets or water bowl.
- Check fixed manger. Remove polythene cover and clean again if necessary.

Putting Down the Bed

Choose the required type of bedding and put it down to a suitable depth appropriate to the type and with banks around the walls. For deep litter seal any drains and allow the deep litter to gradually develop.

Daily Programme

MORNING
- Muck out.
- Set fair the bed.
- Sweep up the stable area and yard.

MIDDAY
- Skip the box out (the box should be skipped out throughout the day). NB: The horse should be tied up whilst the box is skipped out.

EVENING
- Skip the box out.
- Add fresh bedding if required.
- Set the box fair.
- Sweep the yard and leave everything tidy for the night. (If labour is available, the yard can also be swept at midday.)

Mucking Out

METHOD
- Collect equipment.
- Put a headcollar on the horse.

EITHER
Tie the horse up in his box. The disadvantages are that he is then exposed to dust and fungal spores if straw bedding is used, and to dust if shavings, sawdust or peat moss are used. It is quicker to muck out a box if the horse is not in it, although this is not always practical in a large busy yard.

OR
Tie the horse up outside the box. This should only be done, when outside looseboxes are used, if the horse is used to it and if the weather is dry. It is not advisable on a windy or very cold day. The yard gate should be closed. With inside boxes it is convenient to have the horse out of the box. However, when he is tied in the

passageway he causes an obstruction and may be a hazard if other, neighbouring horses are also tied up outside their boxes. It is likely that he will still be within inhaling distance of fungal spores and dust caused by the shaking up of a straw bale.

OR

Tie the horse up in a spare loosebox. This is the preferable arrangement, though it should not be considered if there is any virus infection or other problem in the yard.

All stabling should be kept as dust-free as possible. Walls should be dusted down regularly.

Whatever bedding is used, mucking out is a time-consuming task. Any new ideas or systems which can shorten the process of both mucking out and sweeping the yard, without loss of efficiency, must be of benefit to both staff and management.

Labour-Saving Ideas

- Deep litter

- The use of a tractor and trailer.

- The putting of manure directly into transportable skips which are collected, when full, by a contractor.

- An automatic sweeper.

- Barn-type stabling.

- Split sacks or polypropylene muck sacks with rope handles, which are easier to fill than barrows, though they can be heavy to move.

Avoiding Dust Allergies

During the process of mucking out and shaking up a straw bed, there is a very considerable increase in the amount of fungal spores and dust in the atmosphere. This is not good for any horse but has a disastrous effect on those animals who suffer from allergic coughing and/or COPD. If possible, whenever straw is to be shaken up, the horse should be out of the box. With interior barn stabling, consideration should be given to horses in adjacent boxes.

DAILY CARE

- Skip out the bed regularly during the day – the horse should be tied up – using the skip.

EVENING STABLES (between 4pm and 5.30pm)

- Tie the horse up.
- Skip out the bed.
- Shake up and re-arrange the bed, adding fresh bedding if required .
- Empty and refill water buckets.
- Untie the horse and remove headcollar.
- Empty the skip either into a wheelbarrow or straight on to the muck heap.

NB: It is time-saving and convenient to hang a skip on the outside wall of a stable so that it is always available. Except with deep-litter beds, the floor should be washed weekly with disinfectant, preferably when the horse can be out of the stable for an hour.

THE MUCK HEAP

Siting and construction are covered in Chapter 48, page 332.

Building the Muck Heap

The muck heap requires daily care. If the manure is thrown up, it must be built in steps with a flat top. This construction allows the muck heap to absorb rain, which assists in the rotting-down process. The sides must be kept vertical and well-raked down and the surrounding area swept up and kept clean.

If it can be arranged for the approach path to be higher than the muck heap, wheelbarrows may be emptied by tipping them, which saves time and effort. It is also possible to have a pit dug for the same purpose, though this type of muck heap may be difficult to empty by fork-lift lorries.

Whatever the type of muck heap, the actual building and raking must still be done each day. An untidy muck heap is a disgrace to a stable yard and reflects badly on both management and staff.

Disposal of Manure

Under EU regulations the burning of waste bedding is prohibited.

It is essential to try to make satisfactory arrangements for the disposal of manure. Regular collections may be undertaken by contract with firms supplying market gardens and mushroom growers, although these are increasingly difficult to arrange satisfactorily.

Local farmers may be willing to spread stable manure on their fields, but ensure that they do collect it regularly, not just once a year.

Large yards, which can guarantee a regular supply, have less problem making permanent arrangements, but it can be difficult for the small yard, which only stables horses in winter. The financial aspect is of little importance compared with regular clearing.

In the past, some firms would only accept wheat-straw manure. However, as straw-only manure has become less readily available, it has become easier to get rid of the mixed muck heap, including paper bedding.

It is not recommended that horse manure be spread on fields belonging to the yard, unless these are to be ploughed. There is an increased danger of worm infestation and such fields are likely to become 'horse sick', an ever-present problem in equestrian establishments.

With the increased cost of inorganic fertilisers and the trend towards 'organic' farming, well-rotted horse manure can be sought after by large farms. If a heap is accumulated away from the yard it can be allowed to decompose well for twelve to eighteen months and then be of value as 'organic fertiliser'.

Disposal of Paper Bedding Manure

Paper bedding manure breaks down quickly and may be spread within a few weeks or used as mushroom compost. It should not be put on to a grass surface to form a riding track because in wet weather the track will become slippery and dangerous.

14 RECOGNISING GOOD HEALTH AND CARING FOR A SICK HORSE

Condition is the bodily state of the horse. An experienced manager with a trained eye will be able to recognise the condition of all the horses in his care, and will be quick to observe any change.

SIGNS OF GOOD HEALTH

- A light, alert expression.

- Clear eyes. The membrane under the lids should be pale salmon-pink in colour.

- No discharge from the eyes or nose. Discharge from eyes in summer could indicate irritation by flies.

- Mobile ears.

- A glossy coat which lies flat.

- Loose skin, which moves easily over the underlying bones.

- The horse stands with all four feet evenly on the ground or, if relaxed, one hind leg resting only.

- No visible sign of sweating when at rest, except in very hot weather or in humid conditions.

- Cool limbs, with no unusual swellings.

- Fairly opaque and colourless or pale yellow urine, which should be passed several times a day.

- Droppings that are free from any offensive smell and

are passed several times a day. They may be of various colour and consistency, according to the diet.

- Normal appetite and water consumption.

- A temperature of 38°C (100.5°F). This can vary by up to half a degree night and morning (usually lower in morning).

- A pulse rate at rest of 36 to 42 beats a minute.

- A respiration rate at rest of 8 to 12 inhalations a minute.

GRADES OF CONDITION

The Very Poor Horse

The animal is emaciated, with spine, ribs, tail, head and hip bones prominent. The bone structure of the backbone, withers, shoulders and neck are clearly visible. There is no fatty tissue under the skin, which feels tight over the bones and is not easily moved. The muscles of the shoulders, loins and quarters are underdeveloped and flaccid. The outline over the top of the quarters is hollow and falling away. The animal is often dehydrated, which is indicated by the slow recovery to normal of a fold of skin when pinched.

Horses in very poor condition are likely to be infected with worms, but such starved animals should be wormed with care (see Chapter 15, Internal Parasites). Veterinary advice is essential. They are also likely to be infested with lice, which cause irritation, and bare patches of skin may be visible. Such animals require easily digested food in small quantities at frequent intervals. The quantity of food should be increased only gradually. Riding horses and ponies should not be worked until their condition has improved. This may take many weeks.

The Thin Horse

A thin horse is one with little fat between muscle and skin. The backbone and ribs are easily felt and visible. The skin is tight, but less so than in the very poor horse. The neck may be hard, but lacks substance. If such a horse is in work, there may be some muscle development, but if exercise is reduced, muscles will waste and the loins and quarters will fall in. There is a risk of saddle sores developing.

In winter the appearance of ponies at grass can often be deceiving. They grow long, thick coats, which may help to give an appearance of plumpness. They should be closely examined, particularly in the neck and backbone area, to make sure that they are in an acceptable condition.

A pony in good bodily condition usually grows a dense but not long winter coat. His summer coat is likely to come through earlier than that of a pony in poor condition. The latter often retains patches of long winter coat until well into the summer.

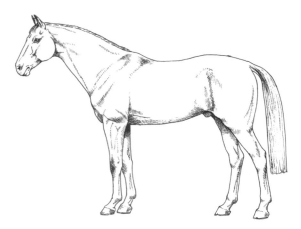

A horse in good condition.

A horse in poor condition.

The Fit Horse

The backbone is not visible, and the flesh on either side is level. The ribs are barely distinguished, though they can just be felt. The shoulders and neck blend smoothly into the body. The neck, when felt, is hard and has substance. The skin over the ribs is loose and easily moved. The muscles are well developed and firm to the touch.

Four categories of fitness may be recognised:

- **The ridden show animal**. Fairly well muscled up, but often overfat. Ribs not visible and can only just be felt.

- **The show jumper/dressage horse**. Well muscled, but may carry some fat. Ribs cannot be seen, but can be felt.

- **The event horse**. Very well muscled, but should not be carrying much excess weight. Ribs can be seen and felt. A three-day event horse will be much leaner than a one-day event horse.

- **The racehorse**. Very well muscled but with no surplus fat. Muscles are rounded and can be clearly seen and felt, including above the bony points, especially over the loins and croup. There is little fat in the spaces between muscles. This horse, to an uneducated eye, may appear thin.

The Plump Horse

Fat deposits can be felt over and between the ribs and along the withers, shoulders and neck. The crest feels solid. Food must be restricted and no fast work undertaken. This condition is typical of horses, and particularly ponies, turned out on good grazing in the summer.

The Fat Horse

The fat horse has fat deposits on either side of his backbone. His ribs cannot be felt. There is fat in the areas around the withers, behind the shoulders, along the inner thighs and around the tail and head. The crest feels thick and solid.

The Very Fat Horse

The very fat horse has an obvious crease down his back. His ribs are covered with a layer of fat and cannot be felt. There is bulging fat around his tail, head, withers, shoulders and neck. His flank is filled with fat, and his crest very thick and solid.

In the last two cases there is an urgent need to restrict severely the horse's food. The extra weight is a health hazard to heart, wind, legs and particularly joints. Laminitis may be imminent, if not already apparent. Animals in such condition may be said to be equally as neglected as those who are allowed to become too thin. Exercise is advisable, in conjunction with strict control of diet.

SICK NURSING AND THE CONTROL OF INFECTION

At all times the attendant must be observant and use common sense to ensure the horse's welfare. Close attention must be paid to comfort, cleanliness and diet. The veterinary surgeon's instructions should be written down and carefully followed.

Stable

A very sick horse may well settle and be more peaceful if he is put in a loosebox in a quiet or separate part of the yard. However, some horses become upset if they are required to change boxes, and are better left in their own environment where they have company.

Horses with eye trouble should have light restricted and smaller windows, and the top-door openings should be covered. Essential ventilation must be maintained.

Warmth

In cold weather, or in some forms of illness, warmth is essential, which should be achieved by using extra lightweight clothing, rather than by restricting ventilation. There must be an ample supply of fresh air; the atmosphere in the box must never feel stuffy. Outside looseboxes may have the top door closed, providing there is another adequate form of draught-proof ventilation. Stable bandages, warm, lightweight blankets and, on occasions, a hood, all help to combat the cold. An infra-red lamp fixed in the ceiling adds to the horse's comfort. This can usually be run from a suitable light fixture if no permanent socket is available.

In warm weather there must be an ample supply of fresh air but no draughts. Light rugs may be put on, but care must be taken that the horse does not become overheated, as excessive warmth can be damaging. A sweat rug may be used during the day, and a night rug put on when the temperature falls in the evening.

Bedding

An ample bed, with banked-up walls, should be provided. It should be kept clean and regularly skipped out. A weak or injured horse's leg may become entangled in deep straw, so this type of bedding should be well flattened and not shaken up unnecessarily. If shavings or paper are used, the problem is avoided and there is also less disturbance and dust when the bed is mucked out. Shavings or sawdust are not suitable when a wound has to be left uncovered, or during foaling.

Grooming

In the case of very sick horses, grooming should be restricted. The eyes, nose, sheath and dock area should be sponged clean, and the horse may perhaps appreciate a light body brush or hand rubbing over his head and body. Stable bandages should be removed twice a day, and the legs hand rubbed towards the heart. The feet should always be carefully picked out. A clean cotton sheet worn under the rugs can make a horse more comfortable.

In the case of infectious or contagious diseases, grooming should be discontinued, as brushing can only assist the spread of the infection.

In cases of lameness, or in convalescence, a thorough grooming can be beneficial and is usually enjoyed by the horse.

Hygiene

In cases of infectious or contagious disease, stable tools, grooming equipment and food utensils should all be kept separate. The attendant should wear a special overall, headgear and rubber gloves when looking after the horse. A bucket of strong disinfectant should be kept outside the door for washing boots. Jeyes Fluid is suitable.

Water

A constant supply of clean, fresh water must be available. In all cases where the horse's water intake has to be monitored, automatic watering bowls should be disconnected and water supplied by bucket. All water containers must be cleaned out and filled with water several times a day. This is particularly important where there are nasal discharges.

Diet

In all cases a low concentrate diet will be necessary. If possible it should be based on the one which the horse is used to, with the forage level being increased and the concentrate level decreased. It should include nutritious and easily digestible ingredients. Soaked sugar beet, breadmeal with chaff or molassed chaff, plus hay or vacuum-packed forage would provide an open yet balanced diet. Meadow hay is easier to chew and to digest than seed hay, and horses with respiratory diseases are best fed from the ground, with the feed damped and the hay soaked. Fresh-cut grass makes a suitable and appetising ration. Apples, carrots or other roots may also tempt the appetite.

It is essential to appreciate the preferences and needs of the particular animal. All food should be freshly prepared and offered in small quantities. In some cases, hand feeding may be necessary. Uneaten food should be removed after half an hour and discarded.

Feed bowls and mangers should be washed out after every feed.

For special cases the veterinary surgeon may suggest a suitable diet.

Administration of Medicines

If medicines are mixed in feeds or added to the water, they may put the horse off his food or discourage him from drinking. It is better to mix them with water and administer them by syringe into the mouth, or mix them with treacle and put this on the tongue (see Chapter 9, Handling the Horse).

ROUTINE CARE

The horse's **temperature, pulse** and **respiration** should be taken morning and night and noted on a chart. This chart should also contain information as to the state of the dung and urine, how often they are passed, and also details of water consumption, food eaten and the horse's appetite.

In epidemic situations it is sensible to chart the temperatures of all horses in the yard morning and night.

As stated earlier, the normal temperature of a horse is 38°C (100.5°F), although there can be slight variations from horse to horse.

To take the temperature

The horse should be held by an assistant. Sharp horses should have a front leg held up. The thermometer should be the type with a short, stubby end. Shake it until it registers below 37°C (99.5°F). Grease it with Vaseline and, standing to the side of the horse's quarters, insert two-thirds of it into the rectum. Leave it for half a minute, or as directed. The thermometer should be firmly held, as it can be drawn into the rectum and be difficult to recover. After taking it out and noting the reading, shake the thermometer down, wash it in cool disinfectant, dry it and replace it in its case. Modern digital thermometers obviate the need for shaking down before or after use and are quicker and easier to read.

To take the pulse

The normal pulse of a horse at rest is 36 to 42 beats to the minute. The pulse can be taken by a light feel of the artery as it crosses the horse's jaw bone. It is a skill which needs practice. It is helpful to have a watch with a second hand and a large, easily seen face.

Respiratory rate

The normal respiration of a horse at rest is 8 to 12 inhalations per minute. This can be observed by watching the movement of the horse's flank.

CONTROL OF INFECTION AND CONTAGION

It must be accepted that in many cases it is not possible to prevent the spread of airborne germs, or to stop other horses from becoming infected. The offending animal may be infectious for several days before showing any signs of illness. In any case, an isolation box is unlikely to be of use unless it is at least 400m (1/4 mile) downwind from the yard. Birds and flies also carry germs.

Contagious conditions, for example ringworm, are passed by contact (direct – touch, or indirect – e.g. via a brush), and can be successfully isolated with strict care. Infectious conditions are airborne and it is almost impossible to be sure of successful isolation.

Isolation boxes should be used for housing animals new to the yard, particularly those bought from sales or any who have travelled long distances.

PRECAUTIONS

- The person attending the horse should not be in contact with any other horses.

- The veterinary surgeon should see other patients in the yard before seeing the infectious horse.

- Overall washing facilities and paper towels should be provided.

- A door grille prevents any passer-by from handling the horse and helps to prevent infection by contact.

- A warning notice should be pinned to the door.

- A supply of paper towels and cotton wool should be available for dealing with nasal discharge.

- Used material should be burned. Rubber gloves should be disinfected after use, or plastic disposable gloves used.

- In the case of contagious skin diseases, for example, ringworm, the most likely points of infection and contact should be carefully watched on all other horses, for example the girth, saddle and bridle areas and where the rider's leg rests. Early signs of the disease can thus be noted and treatment given before it spreads. All riders should disinfect their boots before riding another horse.

- Horses with ringworm may be rideable. They should have their tack kept separate, and this should be disinfected each day. If numnahs and nylon tack are used, disinfection is easier and leather saddlery will not be harmed. Attendants and riders should bath each day, and if possible wash their hair. Special soap can be obtained. Particular attention should be paid to any place where there is friction from clothing, for instance the wrists, neck, etc.

CONVALESCENCE

A horse who has had a serious illness may need several months convalescence before being fit to work. Exercise may start with five minutes walking out and progress gradually as the horse gains strength. When work is resumed, it should progress in very slow stages, so that at no time is the horse put under strain.

The veterinary surgeon will advise as to a suitable programme.

The horse must be kept warm without being over-heated. Hill work of any sort should be avoided. Walking on long reins can give more control and less strain than lungeing. Work on a circle can put undue pressure on slack muscles.

CLEANING AND DISINFECTING STABLES AND EQUIPMENT

- Competition and racing yards usually employ professional cleaning services, and the buildings concerned can be steam cleaned. This is an expensive process, and the private owner or the small yard is more likely to do the work personally.

- To remove grease and dirt, use washing soda and bleach. Used together these are also an effective disinfectant, and kill fungus. A strong disinfectant, such as Jeyes Fluid or Lysol, also kills fungus.

- The Jockey Club recommends Halamid. The Ministry of Agriculture recommends creoline derivatives (synthetic Phonolee disinfectant, e.g. Hycolin).

- All bedding should be removed and burned.

- The roof or ceiling should be washed down with a jet hose to remove all dust. The walls, doors (inside and out), windows, mangers and doors should be scrubbed clean with washing soda and bleach. When dry, unpainted woodwork can be creosoted.

- Brick or block walls may be Snowcemmed or emulsion painted.

- Clothing and grooming kit should be soaked in disinfectant (not bleach or soda) for six hours, then washed in detergent and carefully rinsed.

- Stable tools, buckets and portable mangers should be washed with strong washing soda and bleach.

- Leather saddlery which has been in contact with a fungal infection (for example, ringworm) should first be cleaned with washing soda and water, then with a strong disinfectant. It can then be oiled. Soda is bad for leather, but a daily application of glycerine soap should restore it. Non-leather girths, numnahs and nylon tack should be soaked in strong disinfectant (Jeyes Fluid or Lysol) for six hours and then washed.

POULTICES

Poultices may be used warm or cold, and can be applied anywhere on the horse where they can successfully be kept in place.

Warm Poultices
- Soften the tissues, so that pus can escape.
- Increase the blood supply to the injured area, and thus assist healing.
- Soothe bruising.
- Should be put on at a heat just bearable to sustained contact with the back of the hand.

Cold Poultices
- Reduce inflammation.
- Constrict blood vessels and arrest internal haemorrhage.

As a basic principle, cold poultices are for immediate use to ease inflammation; warm poultices are for foot infections, deep puncture wounds and to help the healing process once the original inflammation has subsided. Poultices are usually left on for twelve hours and then renewed. The poulticing of wounds on a joint, particularly the knee, should be done only on veterinary advice.

Types of Poultice

- **Kaolin** is a natural clay with medicinal properties. It may be applied warm or cold. It can be used on open wounds, but is better encased in muslin. It is used for bruising, strains and all foot injuries. It is economic to use when several days poulticing is required.

- **Animalintex** is a prepared medicinal lint. It is applied warm or cold. It is used where there is an open wound, or for a punctured foot. It has less of a clogging effect than kaolin. It is more expensive than kaolin.

- **Glycerine and Epsom Salts** mixed to a paste may be applied cold, and are of particular value for coronet abscesses and swollen cannons.

- **Bran and Epsom Salts** 227g (½lb) bran and 57–85g (2–3oz) of Epsom salts are mixed with hot water to form a stiff dough which is used for foot problems. The horse may be tempted to pull off and eat the bran poultice. A poultice boot usually prevents this.

Leg Poultices

KAOLIN If applied warm, kaolin has to be heated in its tin or in a small container in a saucepan of boiling water. The lid should be eased so that steam can escape. The water must not be so deep that it can overflow into the tin. To test for temperature, stir well and then test a small portion on the back of the hand to ensure that it is not too hot. An over-hot poultice will blister the horse's skin. Kaolin may also be heated by spreading it approximately 6mm (¼in.) thick on a suitable sized piece of cotton or linen. This is then put on a flat container and placed either in an oven or under a grill on a low heat. It heats very quickly, and care should be taken that it does not dry out.

If kaolin is heated in a tin, the poultice is then spooned out on to a suitable sized piece of plastic, tin foil or brown paper. Test it again for heat and then place it on the injury; cover with thick Gamgee and secure with a stable or crepe bandage. This must be firm, but not tight. If the kaolin has previously been spread on cloth, this is placed on the injury, covered with plastic and Gamgee and bandaged as above. The purpose of the plastic is to keep in the heat, exclude air and retain moisture, thus allowing the poultice to work.

ANIMALINTEX A suitable sized piece of lint is cut off and placed on a plate or other clean surface. The lint is then saturated with boiling water. The surplus is removed either by placing another plate of equal size over it and squeezing, or by placing it in a clean cloth and squeezing. Test the lint for heat on the back of the hand and apply directly to the injury, dressing side down. The lint is covered and kept in place in a similar way to kaolin (the plastic covering is essential). It can also be used as a dry poultice, or dampened with water if applied cold (see instructions on packet).

GLYCERINE AND EPSOM SALTS Place the paste over the wound, cover with plastic, and bandage, as for kaolin.

Foot Poultices

All surfaces of the foot should be scrubbed clean.

USING AN EQUI BOOT The selected poultice is placed over the area of the foot to be treated. Plastic is put on top and the Equi Boot then put on and secured to the foot. If the boot fits well, it is sometimes possible to allow the animal to be turned out.

USING A POULTICE BOOT The poultice is put on as above. Bran or Epsom salts or yeast may be used instead. Place the poultice over the sole of the foot, cover with plastic, place the foot in the poultice boot and fasten securely. The leg should be protected with Gamgee and a stable bandage. The heel of the foot, if not the seat of injury, should be protected by a layer of thick grease.

USING A THICK PLASTIC BAG AND/OR SACK The selected poultice is either placed on the sole of the foot or, if it is bran based, is placed in the bottom of a strong plastic bag. The foot is then placed in the bag, which may be covered with sacking. This is then wrapped round the fetlock and leg. The leg should be protected with Gamgee. The sack and/or bag can be tied round the pastern with an old crepe bandage or string. A stockinette bandage is then put on to secure the Gamgee and the top of the sack or plastic bag. This is

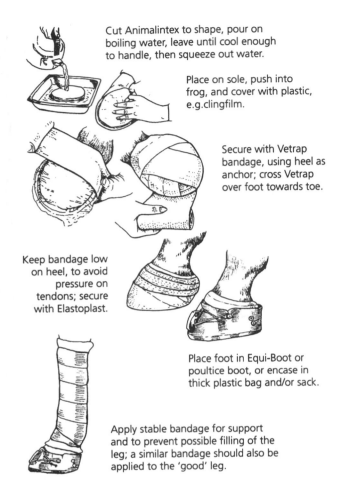

Cut Animalintex to shape, pour on boiling water, leave until cool enough to handle, then squeeze out water.

Place on sole, push into frog, and cover with plastic, e.g.clingfilm.

Secure with Vetrap bandage, using heel as anchor; cross Vetrap over foot towards toe.

Keep bandage low on heel, to avoid pressure on tendons; secure with Elastoplast.

Place foot in Equi-Boot or poultice boot, or encase in thick plastic bag and/or sack.

Apply stable bandage for support and to prevent possible filling of the leg; a similar bandage should also be applied to the 'good' leg.

Poulticing a foot using Animalintex.

made more secure by putting on a stable bandage. With heavyweight horses it may be necessary to reinforce the bag or sack. The pressure of weight quickly wears through a thin layer of sacking or plastic.

Cold Kaolin Poultices

These should be chilled in a refrigerator before use. They can be applied with advantage to all four legs of a horse after any type of severe work. The kaolin is put on the leg by hand, then covered with brown paper and thick Gamgee. A secure, but not tight, stable bandage is then put on. Various forms of prepared bandage for use as a cold poultice are now available.

Inflamed Backs
Inflammation under the saddle can be reduced by putting cold kaolin on the area and covering it with plastic. This can be held in place by Elastoplast.

NOTE: Both warm and cold kaolin eventually return to blood heat.

KNEE AND HOCK BANDAGES

Apply the dressing or poultice as directed, and cover all surfaces of the joint with thick Gamgee. With the knee, it is the bony projection (pisiform bone) at the back of the knee joint which may become susceptible to friction, particularly when the joint is moved. This area must be kept free from bandage pressure.

The bandage, preferably crepe or 'Vetwrap' type, is started above or below the joint with two secure rounds. It is then put into a figure-of-eight around the joint, leaving the back of the knee or point of the hock free. The bandage can finish above or below the joint but should end with one more round to complete and assure security. A stable bandage should be placed on the leg below the joint to prevent the upper bandage slipping down.

Putting on knee and hock bandages requires consider able skill and experience. It is advisable to practise beforehand, so that should the necessity arise the bandaging itself will cause no problem.

Horses often attempt to chew at a knee covering, especially over a wound (see Chapters 9 and 16).

Various patented forms of leg support and protection are now available, some of which can usefully be used to retain poultices and dressings, particularly on awkward areas such as the knee or hock (e.g. Tubigrip or Vetrap bandages).

NB: In all cases of leg or foot injury, thick Gamgee and a support bandage should always be placed on the opposite leg to prevent filling and to ease strain.

PRESSURE BANDAGES

This type of bandage is used after a suspected tendon injury. The bandage helps to limit any swelling caused by the injury. Such swelling interferes with the healing process. The bandage, preferably crepe, should be put on over a thick layer of evenly applied Gamgee. It should be removed each morning and night, and the leg should be given a brisk massage by hand, the hands working in an upward direction towards the heart. Veterinary advice should be sought before long term treatment with pressure bandages.

TUBBING

This is useful for puncture wounds in the foot. It draws out pus without softening the horn.

METHOD
- Scrub the foot clean. Grease the heel with Vaseline.

- Use a clean container of heavy-duty plastic or rubber. The handle should be removed.

- Dissolve 227g (8oz) of Epsom salts in hot water.

- Place the Epsom salts and water in the container and reduce to blood heat.

- Place the horse's foot in the container and soak for ten minutes. Top up with water if required.

- Repeat three or four times a day.

FOMENTING

This is a method of applying heat to areas which cannot be poulticed.

METHOD
- Dissolve 220g (8oz) of Epsom salts in hot water.

- Place the Epsom salts and water in a bucket and reduce to blood temperature.

- Soak an old towel in the water, wring it out and apply to the affected area.

- Repeat every minute and continue for fifteen to twenty minutes. Top up with hot water as required.

COLD HOSING

This treatment is appropriate for the immediate relief of bruising, for instance following a kick, or a bang when jumping. It is also appropriate for cleansing a wound and assisting in restricting blood flow to arrest bleeding.

METHOD
- Ask an assistant to hold the horse, who may need to be bridled for extra control.

- Run the water gently over the foot at minimum pressure.

- Gradually work up the leg to the bruised area or wound, and allow a slight increase of water pressure.

- For bruising continue the treatment for ten minutes, and repeat four or five times during the day.

- An alternative, if available, is to stand the horse in a running stream or in the sea.

Fomenting a forearm.

Pressure bandages, if practicable, should be applied between hosing sessions.

COLD BANDAGES

Patented forms of cold pack are now available. Makers' instructions should be followed. These packs can be held in place on the leg with a light bandage. Gamgee, or similar, should be placed underneath the bandage.

An **ice-pack** can be made by crushing ice cubes in a cloth and placing them in a plastic bag. Cover the leg with Gamgee, or similar, and put the ice bag on the leg and secure with Gamgee or a light bandage. Purpose made 'ice' preparations are widely available. Vetrap bandages are light, versatile and effective, though expensive.

Great care should be taken with any type of frozen/iced application that 'burning' does not create greater problems than the curing effect may have.

15 INTERNAL PARASITES

'Worms' are the common name given to an important group of internal parasites of the horse. All horses and ponies, except newborn foals, may suffer from a worm burden.

Worms cause anaemia, debility, unthriftiness and colic. This means that horses are not able to work up to their true potential, food is wasted and young horses do not grow properly. Almost 90% of all colic cases can be traceable to worm infestation. In some cases they lead to death.

A proper routine to control worms is, therefore, an essential part of any horse-care programme. Money spent on wormers (anthelmintics) is excellent insurance for cheerful, healthy horses and ponies, who can enjoy their work and give of their best.

Symptoms of Worms

The following symptoms should be cause for concern, although some may occur for other reasons:

• The horse eats well but puts on no condition.

• He has a large belly but very little condition along his top line.

• His skin feels dry and inelastic. His coat is harsh and 'staring'.

• He is anaemic. This can be confirmed by veterinary investigation.

• He has little energy and tires quickly.

• His droppings are loose and smelly; sometimes there may be diarrhoea.

• He has recurring attacks of colic.

Any, or a combination of these symptoms, warrants veterinary investigation.

NB: It is possible for horses and ponies to have a serious worm burden without visible outward signs.

Worm Egg Count

Worm infestation can be assessed by counting the number of eggs per gram of faeces. If any eggs are counted this indicates that there are adult worms of some species 'passing'/laying eggs. These eggs can be hatched out in a laboratory and the larvae allowed to develop to be identified by a specialist.

The number of eggs per gram can vary from as few as fifty to tens of thousands. A high count would indicate a large number of adult females. A low count simply tells the story on that day and the number could rise rapidly within a few weeks.

The number of eggs per gram is usually required by the veterinarian either to investigate a current problem and/or to help plan a control system.

Horse owners should accept the fact that all horses at grass have some worms. There is no 100% effective way of eradicating worms, only reducing numbers to a manageable level.

Life Cycle

Under conditions of moisture and temperature in the sward, eggs can hatch in a few days or lie dormant for many months. The optimum time for hatching and development is spring and early autumn. Hatching larvae are called first stage larvae and, subject to moisture and temperature, these moult to become second and third stage. The interval between stages varies with the time of year.

There is a large loss of eggs and larvae under normal conditions, particularly in excessively dry or cold weather. Eggs and larvae in stages one and two cannot develop further if eaten; they are digested. Stage three cannot feed, and unless consumed within a few days, they will die. They take up position in droplets of moisture on grass and these droplets travel up the blade to become more readily eaten. Once consumed, irrespective of species, the third-stage larvae penetrate the mucosal lining of the gut, where they moult again to become fourth stage. After a variable period within the gut lining they leave to live on the surface of the bowel, and there become adult males and egg-laying females.

There are variations once larval penetration into the gut has taken place.

Strongylus Vulgaris (Large Redworms)

Due to the development of the drug Ivermectin, which kills migrating larvae, the incidence of worm infestation has greatly reduced. The worm larvae pass through the bowel wall into the blood vessels and travel along them until they reach the junction of the gut artery and the aorta. Here they may sit for up to three months before returning to the bowel to become adult egg-laying worms. The waiting larvae can cause a thrombus (blood clot), which may partially or completely block the blood supply to the large intestine, impairing the gut's ability to digest food, and in the worst scenario causing death.

SYMPTOMS
Loss of condition may be visible if the horse has a heavy infestation of redworm. Intermittent colic for no apparent reason is the most common symptom.

CONTROL
Ivermectin, dosed every 8–10 weeks.

The other species of large redworm (**Strongyle Edentatus**) migrates through the peritoneal cavity setting up streaks of peritonitis with adhesions which could entrap parts of the small intestine. The migration of Edentatus is not so long as Vulgaris. Symptoms and control are as for Strongylus Vulgaris.

Cyathostomes (Small Redworms)

There are forty species of small redworm, not all appearing in any one horse. Although they do not leave the gut they can become encysted within the mucosa and remain in that dormant situation for weeks or months. In this state they are much less easy to kill and their numbers increase. It is the sudden release en masse of these encysted worms into the lumen that causes acute enteritis, often with diarrhoea, but more frequently the pathological effect is one of abdominal shock.

SYMPTOMS
Poor condition and sometimes very acute weight loss accompanied by the onset of diarrhoea. This may indicate a life-threatening situation.

CONTROL
Treatment as for large Strongyles.

Parascaris Equorum (Roundworms)

These are most common in foals and yearlings. The predilection site is the small intestine. The eggs are deposited in stables and pasture. They can live for many years and are resistant to most chemical disinfectants. The eggs are taken in by the mouth during suckling from a contaminated teat, from udder skin or when grazing. The eggs hatch in the intestines; the larvae penetrate the gut wall and enter the lymph stream, travelling through the liver and then via the blood to the heart and the lungs, then into the airways to the throat and so back to the small intestine.

SYMPTOMS

Unthriftiness, emaciation, dry staring coat, pot belly. Sometimes there is an unexplained respiratory illness and coughing.

CONTROL

All common anthelmintic drugs are effective. Foals should be treated from six weeks of age every four weeks with a wormer safe for foals.

Gasterophils Species (Bots)

The predilection site of these larvae of horse bot flies is the stomach. The female fly lays eggs on the hairs of the horse's legs. They cause irritation and, when licked by the horse, hatch and migrate to the stomach. They are passed out in the dung as mature larvae which pupate and develop into flies during the summer.

SYMPTOMS

Very unlikely to show specific symptoms unless present in huge numbers, when colic could result.

CONTROL

Ivermectin should be given in one annual dose. If eggs are seen on the legs during the summer, they can be inactivated by washing vigorously with warm water containing an insecticide; alternatively they may be shaved off.

Strongyloides Westeri (Threadworms)

The predilection site is the small intestine. These parasites can be free-living, especially in warm, humid conditions such as deep-litter bed. They are of importance in foals. Infection can occur by two different routes. First, and most common, is via the dam's milk. The second is via penetration of the skin.

SYMPTOMS

There is persistent scouring in the foal after the foal heat. The mare is simply a carrier of infection and shows no clinical signs.

CONTROL

The stables should be kept clean and dry. With very young infected foals, veterinary help should be sought. Thiabendazole or Ivermectin are the usual drugs used. It is difficult, if not impossible, to control the larvae in the mare and therefore the resultant infection in the foal.

Oxyyuris Equi (Pinworms)

The predilection site is the large intestine. The eggs containing larvae are taken in by mouth. These hatch and the worms develop in the wall of the large intestine before returning to the lumen of the gut. When mature, the female moves to the anus to lay her eggs.

SYMPTOMS

Due to the irritation caused by the egg-laying females, the horse rubs the top of the tail and sometimes breaks the skin. Similar signs are seen in sweet itch and louse infestation.

CONTROL

The treatment for other worms also deals with pinworms. The eggs should be washed off from under the tail with a disposable cloth.

Dictyocaulus Arnfeldi (Lungworms)

The predilection site is the major airways of the lungs. Lungworms are particularly common in donkeys, who are the main carriers of infection, although they may show no symptoms. The larvae are taken in from pasture and migrate through the lymphatic and blood system to the lungs. Mature worms produce larvae which then pass up the throat, are swallowed and hence excreted in the dung to develop to the infective stage on pasture. In horses over a year old the worms remain small and immature, and do not produce larvae, which makes diagnosis from the dung impossible. This is not so in foals and donkeys, where infection can frequently be diagnosed by finding larvae in the dung.

SYMPTOMS

These only show in horses over one year old. A horse develops a persistent, chronic cough. Often, diagnosis is retrospective, after a favourable response to treatment with an anthelmintic effective against lungworms.

CONTROL

Donkeys should be routinely treated in the spring. A single dose of Ivermectin, or twice the normal dose of Mebendazole for five days running, has been found effective. Treatment of horses may prove difficult, and veterinary advice should be sought.

Tapeworms

There is evidence that horses can suffer problems from tapeworms although clinical symptoms are relatively uncommon. A double dose of Pyrantel should be given in September.

WORM CONTROL

The worming of all horses, ponies and donkeys, especially those living at grass, should be carried out on a regular basis. In accordance with veterinary advice. It is a false economy to cut back on worming, especially with young animals.

Some parasites can build up a resistance to the group of Benzimidazole drugs. If a Benzimidazle compound is used then worming should be carried out every 4 weeks; if Pyrantel is used then every 6–8 weeks is advisable; with Ivermectin, every 8–10 weeks should suffice.If the wormer being used does not control bots (see Methods of Dosing, below), Ivermectin should be given once during the winter (see bots, page 117). Ivermectin is lethal to large and small strongyles and will kill bots. There is evidence that it will control blood-sucking lice. Two doses, eight weeks apart, between November and January is worthwhile.

Foals are particularly susceptible to worm damage, as they have little immunity. This damage, apart from stunting their growth or even killing them, may affect

them throughout their working lives. They should be dosed every month from six weeks onwards with an anthelmintic safe for foals and effective against *Parascaris equorum*.

As grass is the commonest method of access for worms, the importance of clean grazing cannot be over-emphasised. Rotation of pasture, cross-grazing, preferably with cattle but also with sheep, especially during the spring and summer, helps reduce its population of infective larvae on the grass.

The daily picking up of droppings in small paddocks greatly reduces the chances of infestation and increases the amount of palatable grass.

Resting the fields for several months or taking a crop of hay from them also reduces infestation.

All new horses should be routinely dosed and kept in for 72 hours before being turned out and preferably turned out onto 'clean', rested pasture.

Even one untreated horse can contaminate the pasture and cause reinfection of his companions.

At present, nothing is known which can be put on grassland to kill the eggs or larvae.

Methods of Dosing

Powder (as such or in suspension) – added to feed
Granules – added to feed
Pellets – added to feed
Paste – in a dosing syringe

Active Principles of Anthelmintics

It is very important to read the name of the active principle as well as the trade name, so that the correct anthelmintic is used.

Benzimidazole Group

The advantages of this group of drugs are that they are of a broad spectrum, that is, they deal with many species of worms, and they are very safe. However, resistance to the whole group may be caused by using one of these drugs exclusively over a long period. Their use must be monitored with faecal egg counts, measured before and after worming. If the egg count does not noticeably drop then this compound should be abandoned and an alternative type of drug should be used.

The present range includes:
- Thiabendazole
- Mebendazole
- Fenbendazole
- Oxfendazole
- Oxibendazole
- Febantel

NB: This group of drugs is not effective against bots.

Ivermectin

This is a broad-spectrum drug, which also kills bots. It is unrelated to the Benzimidazole group.

Pyrantel

This is a broad-spectrum drug which, like Ivermectin, has the advantage that it is not related to the Benzimidazole group. It has no activity against bots.

Frequency of Dosing

The interval between treatments recommended for routine parasite control programmes is four to eight weeks. For this period the dung will be free of eggs so that further pasture contamination is prevented. There are so many variations in the effect worms have on horses, in conjunction with environment, age, grazing, numbers, etc. Not only should suspected parasite infestation be under the guidance of a veterinary surgeon but a programme of control should be planned between veterinarian and yard owner.

Adult horses which are well managed and fed particularly on pasture where there is a faecal clearing programme, rarely suffer from what should be a comparatively small adult worm burden. The aim of routine treatment is to reduce the number of eggs passed out throughout the year and thus reduce the number of larvae which mature on the pasture. Especially nowadays with the small redworms, it is the level of consumption of their larvae which can prove dangerous to stock, in particular young horses.

16 STABLE VICES AND PROBLEM BEHAVIOUR

Most stable vices may be classed as nervous habits. They are acquired by the horse as a result of stress or boredom, or by imitation. The tendency may be inherited, that is it may be a question of temperament, but it can often be avoided through good stable management and firm but considerate handling of the horse. The following measures help to relieve stress and to avoid boredom:

- For feeding, grooming, exercise and stable routines a regular timetable should be adhered to.

- Sufficient bulk food should be provided to occupy the horse when he is in his loosebox. Many nervous habits start when horses in training have their hay reduced.

- Extra care must be taken with a horse who is on complete nuts and reduced bulk. A lump of rock salt should be put in his loosebox. To prevent chewing of wooden stabling, place a length of tree branch in the stable and allow the horse to bark it.

- Since most horses like music, transistor radios played quietly are acceptable items in the stables.

- Because travelling can cause stress, whenever possible long journeys should be avoided. Never drive too fast, or travel a horse in a noisy trailer.

- Avoid subjecting the horse to undue pressure during training, by either the trainer or the rider asking too much.

VICES

Weaving

This is a nervous habit most common among well-bred horses. The animal rocks himself to and fro, the head swinging with the movement. Confirmed cases lift each foot in turn as the forehand is swayed from side to side. As a result, extra strain is placed on the tendons of the front legs, and lameness can occur. Established weavers give themselves little rest, and often lose condition. The habit can be catching, being quickly imitated by some other horses.

Animals sometimes continue the habit even when they are tied up in a loosebox or stabled in a stall. In the latter case, if they are turned round and put on pillar chains (chains or ropes attached to the rear side pillars of the stalls and then hooked to the headcollar), they may become less bored and so lose the habit.

Some horses weave when they are first brought up from the field; others only when they become fit and are competing or racing. Some start weaving when they are stabled on their own, others when they are in a line of boxes. There are no fixed rules or indications; each animal reacts in his own particular way.

Remedies
- Try to find out the cause, and rectify it. A confirmed weaver is incurable, but if a young horse is caught in time, the habit may be checked.

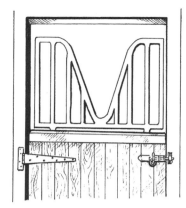

Anti-weaving bars fitted on a stable door.

- Fix an anti-weaving grid to the top of the lower door of the loosebox. This allows the horse to put out his head, but prevents him moving from side to side. Some yards have these grids fitted on all doors to discourage horses from starting the habit.
- Alternatively, fix a grille to the top of the lower door to prevent him putting his head out. But bear in mind that although this prevents a horse weaving over the door, boredom may still lead him to continue the habit while standing back in his box.
- Take a length of wood, slip the top half through the handle of a rubber dustbin lid and bolt the bottom half firmly to the door. This cheap, easy-to-make device allows the horse to put his head out over the door, but prevents him weaving.

Crib Biting and Windsucking

The original cause is usually boredom, though both habits can also be developed by one horse watching another. In crib biting, the horse seizes hold of any projecting edge or object with his teeth, arches his neck and gulps down air. In doing so, he makes a distinctive grunt. When windsucking, the horse does not seize anything with his teeth, but gulps air in a similar manner. A crib biter placed in a loosebox with no suitable surfaces to seize often turns to windsucking.

Crib biters often continue the habit when turned out to grass. The top of fence posts and gates make convenient surfaces. Other mature horses turned out in the same field rarely develop the habit, but young animals may easily do so.

There appears to be no infallible remedy for these two habits. Numerous preventive measures can be tried. In some cases they may work, but in others they are only successful until the horse learns to outwit the prevention. If caught in the early stages and if no suitable surface is made available, a crib biter sometimes loses the desire and hence the habit.

To prevent crib biting, the loosebox should have no projecting surfaces. A metal grille fitted to the door allows the horse to look out but protects the top of the door from his teeth. It must be sufficiently fine that he cannot seize hold of it.

Should a suitable box not be available, a muzzle is an effective deterrent. The horse should wear it at all times in the stable, except of course when he is being watered or fed or having his hay.

There are various patterns of anti-crib biting straps, of which the Meyers pattern is probably the best. It is made of vulcanite and has a leather strap which runs over the poll and fastens to the headcollar. The vulcanite or metal section has a V-shaped part, which fits tightly up into the gullet, and prevents the horse arching his neck (see diagram overleaf). The Matthew Harvey Grinder's Bit is also reputed to stop horses crib biting. When it is removed they do not resume the habit.

A flute bit may be used, fixed to the headcollar. This has a perforated, hollow mouthpiece that disperses the

Windsucking.   Crib biting.

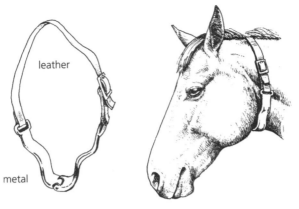

Anti-crib biting strap.

gulp of air and prevents it being sucked in.

In recent years, a surgical operation on the neck was at first thought to be effective, but this has since proved less successful.

Weaving, crib biting and windsucking are all stable vices. They must be declared at the time of sale, and should be noted on the veterinary certificate. If this information is knowingly withheld, the sale can be declared null and void.

PROBLEM BEHAVIOUR

Kicking the Stable Walls

This sometimes starts when mares become irritated by the presence of a neighbouring horse. It can also be a sign of ovary trouble in a mare. Moving the horse to a different box may help. Some horses do it only when they are stabled on their own, others only when they are stabled in company.

Rats or mice in the stable can also start the habit, or it may be a sign of frustration. The noise seems to be addictive.

Remedies
• Line the box with rubber matting.
• Hang a sack of straw from the roof, so that it rests just behind the horse's hindquarters.

• Hang a bunch of gorse or a broom head in the same position as a sack of straw – although the horse may eat it.
• Fit kicking hobbles – though these may cause other problems and need to be used only by an experienced person as a last resort.

Try to arrange for the horse to spend several hours in the field each day.

Rug Tearing

Remedies
The skin and mane should be checked for skin infection and/or parasites. If present, these should be dealt with. This habit is often caused by boredom, and the provision of toys, such as a ball in a haynet or plastic bottles hanging on pieces of string, often helps. A bib of strong leather fastened to the back and side squares of the headcollar, so that it lies behind and below the chin, checks most offenders. Those who persist can be muzzled. A bar muzzle, which allows the horse to pick at hay, is the best type.

Banging the Stable Door

This is an irritating habit, which can quickly spread round a yard if it is not checked. It is usually started by

Bib fitted to headcollar to prevent rug tearing.

horses impatient for feed, particularly if some are fed and others are kept waiting. However, it can become addictive, and horses will bang doors with a front foot or knee as a nervous habit. It can cause bruising and concussion.

Remedies

- Fix a grille over the door, so that the horse is kept back and is unable to reach the door with his foot or knee.
- Fix an iron bar or a chain across the doorway, so that the door can be left open in working hours. This is usually successful, as long as the arrangements are secure.
- If it is essential for the door to be kept closed, set a bar at an angle across the doorway. This will also keep the horse back from the door.
- Because many horses appear to enjoy the noise, fix thick matting or lining on the door. This will muffle the noise and may discourage the horse from persisting with the habit.

Gnawing Wood

Remedies

- Apply creosote, or some other strong-flavoured mixture, to wooden stabling.
- Cap all accessible surfaces in the stable with metal.
- Check the fibre and mineral content of the horse's food because in some cases it can be a diet problem.
- A strand of electric fencing inside post and rails will prevent the horse reaching the fence to chew.

Eating Droppings

This is usually considered to start as a diet problem, but it can become an addiction.

Remedies

- Check the mineral and vitamin content of the horse's diet and correct any deficiencies.
- Put a forkful of turf, with its roots and earth, in the box.

- Worm the horse regularly. Seek veterinary advice on a programme of worming and diet supplementation.

Eating Straw Bedding

Most horses bedded on straw eat a certain amount of it, but a greedy horse eats so much that it affects his condition and wind.

Remedies

- Bed the horse on shavings, sawdust or paper.
- If straw bedding has to be used, spray it with non-toxic disinfectant.

Pawing the Ground and Stamping

The attraction of this habit appears to be the noise which it creates, but it can also be a sign of internal discomfort and/or pain. Heavy, solid rubber matting put down at the front of the loosebox by the door can help to check it. The concrete flooring underneath will sweat, so it should be swept daily and, when possible, exposed to the air.

Biting

Horses who bite when being handled and groomed can be controlled by a muzzle, or by being tied up to two rings across the corner of the box. The cause may be rough grooming – some horses are very sensitive – poor handling when young, or the feeding of tit-bits.

Tail Rubbing

A horse may rub his tail, either because of the presence of whip worms in the area of the anus and rectum or, in summer, because of sweet itch caused by midges.

Remedies

- If the tail rubbing is caused by worms, the horse should be dosed with an appropriate wormer. Keep the area around and under the tail clean, and the tail itself regularly shampooed.

- If the cause is sweet itch, a special lotion should be applied. Seek veterinary assistance.

Refusal to Lie Down

This puts extra and unnecessary strain on the legs. It is usually caused by one of the following:

- Insufficient bedding.
- Getting cast.
- Slipping when getting up.
- Worry in a new stable, especially if there are no other horses.

Remedies

- Put the horse on a deep shavings bed or, if straw is used, adopt the deep-litter system.
- Provide a pony or donkey in a neighbouring field or loosebox for company. This often helps.

Once the horse has overcome his nervousness and has lain down, providing the above precautions are taken, the problem should not recur.

Tearing Bandages

A muzzle, cradle or crossed side-reins are the most effective remedies for this. In the latter the horse will not have freedom to lie down at will. The muzzle must be removed for feed, water and hay.

Pulling Back When Tied Up

This habit arises from fright. It usually starts because the horse, when first tied up in a stable, possibly with too short a rope, pulled back and frightened himself. Therefore, always attach a safety loop of string to the ring and tie the rope to the string, never to the ring itself. Unfortunately, once learned, this habit is difficult to cure, and many remedies have attendant risks.

Careful handling in the stable may eventually effect a cure. When grooming, try threading the rope through the ring, but not actually tying it. The problem often arises when a horse is being tied up in a trailer if he is

allowed to pull back before the back strap is done up, or before the back of the trailer is closed.

Halter and Headcollar Slipping

This habit is often acquired as a result of pulling back, tilting the head and learning that the halter or headcollar slips off. Ponies and cobs can be checked by being tied up with thick neck straps instead of a headcollar. This is not wise with young or highly strung horses, as they may pull back, lose their footing, and give themselves a fright and, often, an injury. Such horses should never be left on their own when they are tied up. If they are not allowed to persist in the habit, they may in time forget it.

PROBLEMS IN THE STABLE

Choking

Choking results when there is a blockage of the tube down which food passes from the mouth to the stomach. The obstruction can usually be felt by examining the left and lower side of the neck.

Possible Causes

- An apple or a large piece of root fed whole and not chopped.
- A greedy horse bolting his food.
- A tired horse who is suffering from mild dehydration and whose mouth is not salivating.
- Feeding certain foodstuffs dry, when they should have been soaked.

The symptoms are salivation and frequent attempts at swallowing, causing a gurgling sound. The head is either drawn in tight to the chest, or stretched out towards the ground. The neck muscles are tense. There may be a discharge from the nose. The symptoms are alarming.

Treatment

- Do not give water.
- Do not try to drench the horse.

- Smear a lump of butter on the tongue. It may help to ease the obstruction.
- Lightly massage the left side of the jugular groove.

Stay with the horse. If he can relax the spasm, the obstruction may move itself. Should the condition not be relieved within half an hour, seek veterinary help.

Horses who are known to bolt their food, particularly cubes, should have their feed mixed with bran or chaff. An alternative is to soak the cubes before feeding, and then feed them mixed with soaked sugar beet.

A Cast Horse

If a horse rolls over too close to the wall of his loosebox and has insufficient room in which to right himself, he may become cast. Many horses become very frightened and/or violent, and by kicking and struggling do themselves injury. Other more placid horses lie quietly until help arrives.

Horses often become cast at night. The tell-tale signs in the morning are a disturbed bed and scratch marks on the wall. This shows that horses, given time, often right themselves without help. If a horse has been cast, he must be closely examined for injury, and then walked out and trotted up in hand.

Horses may roll in their box at any time, but are

more likely to do so:

- When clean bedding has been added.
- When first put back in after work.
- When they feel the onset of an attack of colic.

The sight of a cast horse is alarming. As he tries to right himself, his thrashing hooves put a helper at risk. The task of righting a cast horse should not be attempted by inexperienced persons.

Anti-cast ridges or grooves, which help a cast horse to right himself, can be placed on loosebox walls. His hind feet will catch on the ledge, and he can then push himself over.

Methods of Getting a Horse Up

One person on his own may well be able to right a pony or lightweight animal under 15hh. Considerable strength is required to right a large animal, and two or three people are required, particularly if the horse is very violent.

One person on his own restrains the horse by sitting or kneeling on the horse's head. Make sure the animal's nostril is free. A cloth placed beneath the head will help prevent damage to the eye on the underside. Reassure the horse with voice and hand, and consider the situation. Sometimes the horse will remain quiet and in a few minutes is able to right himself. If he does not, stand up and release his head. Remove any ridges of bedding immediately behind his back and shoulders. Then take hold of his tail and pull his quarters round. The horse will slightly pivot on his middle and shoulders, which should give him room to extend his legs and get up without trouble.

Two or three people: Collect one or two strong ropes or webbing lunge lines. One assistant sits on the horse's head as before, and the second removes the bedding from behind the horse. Most experienced people fasten the ropes round the front and hind legs nearest the wall using a slip knot. The first assistant releases the horse's head and is then ready to help pull. By pulling on the ropes the horse can be rolled over. He is then able to stand up. Release the ropes as soon as possible.

Dealing with a cast horse.

WATERING AND FEEDING

CHAPTER 17 WATERING

THE NECESSITY OF WATER

Water is essential for life and health. Horses can stay alive without food for a few weeks but without water they would probably die after five to six days.

Water makes up approximately 65% of an adult horse's body weight and almost 80% of a foal's body weight. Every cellular activity requires water. It is present in all body fluids, which are in three compartments: blood, intercellular fluid (between cells) and intracellular fluid (inside the cells).

Water is essential for:

DIGESTION – as saliva, which helps swallowing; water also provides a fluid medium for the food to pass along the digestive tract and provides the basis for digestive juices.

BLOOD – the fluid containing blood cells and nutrients etc. which circulates round the body and also carries waste from the tissues.

LYMPH – to drain tissue and help maintain the right balance of body fluid. It is important for defence against disease.

URINE – to excrete the waste products, and as a vehicle to regulate levels of sodium, potassium and other electrolytes as determined by kidney function.

FAECES – to supply fluid to aid excretion.

BODY – to regulate the body temperature by transferring excess heat to the surface.

SKIN – to get rid of surplus heat as sweat.

EYES AND NOSTRILS – in the form of tears and mucus as a lubricant.

JOINTS – in the form of oil as a lubricant.

MILK – to make 91% of the milk of lactating mares.

Stabled horses eat conserved food, with a water content averaging about 14%. It is therefore necessary to have water always available so that they can make up this deficiency. Some horses will drink between eating mouthfuls of their concentrate food. Most will drink after eating concentrates and roughage, especially the latter.

A stabled horse on dry food requires 20–40 litres (5–10 gallons) of water a day depending on the work he is doing and the weather. A lactating mare with a foal at foot requires considerably more.

An 8% deficiency of water, however lost from the horse's body, causes sub-acute illness. A loss of 15% causes dehydration and possibly death, especially if associated with heat exhaustion.

(See also page 251 on care in extreme heat.)

The Principles of Watering

Horses should have clean fresh water available at all

times. If this is not possible, for example after exercise or when travelling, they should be offered some water as soon as possible and before they are fed. Horses that have free and constant access to water rarely drink very much at one time.

Care should be taken that horses do not drink excessively before fast work. For racing, when maximum effort is required for only a short time, water may be removed before a race but never for longer than 2 hours (see also watering competition horses, page 131).

Purity of Water

Horses are extremely fussy about water. If it is tainted in any way they will go thirsty rather than drink it. Even a change of water may cause rejection.

Any suspicion of dung in the water will prevent a horse drinking. Water may be contaminated by ammonia (NH_3) absorbed overnight from a soiled stable. Fresh water should therefore always be available in the morning. It is sometimes recommended that medicines are put in the drinking water. This can be unwise as the horse will usually refuse to drink it. Water that has stood for any length of time in plastic containers may be rejected. Sometimes the addition of a little glucose or molasses will persuade horses to drink.

'Hard' water is probably preferable to 'soft' but the important point is that it is clean and fresh. The main storage tank for the stable yard should be regularly checked, and a dust- and vermin-proof cover used.

If horses are seen to play and fuss with their water this is likely to be due to contamination, but could be a sign of a sub-acute colic attack.

SYSTEMS OF WATERING FOR STABLED HORSES

Automatic Water Bowls

ADVANTAGES
- Water always available.
- Labour-saving. May be essential in a large yard.

DISADVANTAGES
- Can cause physical injury unless boxed in.
- Liable to destruction from horse activity.
- Liable to freeze in cold weather or malfunction.
- Unable to monitor amount of water being drunk unless each individual bowl is metered.
- Not suitable for mares and foals.
- Difficult to clean.
- On a communal gravity-fill system supplying a number of stables, the risk of infection passing from horse to horse would be greater.

Water bowls must be large enough for the horse to get his muzzle in with ease. They should be of the self-filling kind with a small ball-cock, and preferably with a plug to facilitate emptying and cleaning. The inlet pipe must be well insulated to prevent freezing and should have a stopcock.

Gravity-filling bowls are sometimes installed but can prove troublesome. Also, bowls which have a plate for the animal to press with its nose are not suitable for horses.

Water bowls must be emptied and cleaned frequently. In big yards they must be inspected daily by a single competent person.

Buckets

ADVANTAGE
- Water consumption can be monitored.

DISADVANTAGES
- Labour-intensive.
- Can be knocked over.
- Horse may be without water at times.

Buckets should be large (14 litre/3 gallon for preference) and of heavy-duty material. They are usually made of PVC.

Buckets should be emptied and refilled twice daily. They should be topped up at midday, or more often. Cleaning should be done with plain water and a brush kept specially for the purpose. If light colours are chosen for the buckets it will be easier to see if they have been properly cleaned.

Buckets should be put in a corner of the box with the handle turned away from the horse. They should be well banked up with bedding to prevent them from being knocked over. They should not be put by the door but preferably near the drainage outlet.

Buckets with handles are not suitable for foals.

WATERING COMPETITION HORSES

See also Section 5, Specialist Care of the Competition Horse.

Event Horses

Providing the horses have had free access to water they will not need a long drink before the competition starts. This will particularly apply at a three-day event where the horses are stabled nearby, but at a one-day event, if the horses have travelled some distance, they should be offered water immediately on arrival. Half a bucket of water drunk 20 minutes before starting work will do no harm because the dressage phase requires no fast, strenuous exertion. Event horses should be allowed a small amount of water between each phase to maintain their body fluids. More may be given if the gaps between phases are very long.

At a three-day event most veterinary surgeons now consider that horses should be offered a few mouthfuls of water – 1 litre (2 pints) – in the box before the cross-country. If it is possible to arrange, a few mouthfuls should be given at the end of phase B, the steeplechase. This is especially important in warm, humid conditions.

After the cross-country is completed, and as soon as the horse's breathing is recovering steadily and he shows no signs of distress, he may be given 2¼ litres (½ gallon) of water every 15 minutes until he is satisfied. Electrolytes may be given, but plain water should also be offered. It is important that the horse drinks and if he appears distressed, veterinary advice should be sought immediately. (See Signs of Dehydration, page 132.)

Endurance Horses

Much veterinary research has been done on endurance horses. It has been found that the best method of avoiding dehydration is to allow the horses ad lib water whenever possible during the ride. If this has been done the horse should be allowed 2¼ litres (½ gallon) every 5 minutes after finishing the ride. If water has not been available on the ride 2¼ litres (½ gallon) every 15 minutes may be allowed. Electrolytes may be added but the horse must drink, so plain water should also be available.

Hunters

Hunters should be allowed unrestricted access to water before hunting.

If the journey home in the box or trailer is longer than 30 minutes, the horses should be offered water before they start travelling. On arrival home, if they are cool and dry and do not appear distressed they should be allowed water ad lib.

If they are hot they should be allowed 2¼ litres (½ gallon) of water every 15 minutes. If they are distressed, veterinary help should be sought. (See Dehydration below.)

DEHYDRATION

Dehydration results when more water and salts are lost from the body than are taken in and the horse has not enough fluid in his body to maintain normal physiological conditions. It is usually associated with fast, energetic work but there are many other situations in which dehydration can occur.

Possible Causes of Dehydration

- Lack of available water or failure to drink.
- Sweating – prolonged and marked.
- Diarrhoea – especially in foals.
- Excessive urination when not accompanied by a compensatory thirst.
- Haemorrhage.
- Heat exhaustion.

Dehydration may result in:

- Reduced performance.
- Muscle damage.
- Colic.
- Reduced kidney function.
- Laminitis.
- Azoturia (rhabdomyolysis).
- Coma.
- Death.

Signs of Dehydration

- Skin loses its pliability (see Pinch Test, below).
- Lack of inclination to graze or eat.
- Listlessness.
- Loss of normal colour from the membranes of gums and eyes and reduced membrane refill time.
- Muscles quiver.
- Pulse becomes small.
- 'Thumps' (see below).
- Thick, patchy sweat.
- Panting (only if in association with heat exhaustion).

Any horse showing the above symptoms after severe exertion should have veterinary attention immediately. Dehydration can be confirmed by blood-testing.

Pinch Test

A fold of skin on the neck is picked up between finger and thumb. If when released it does not return to normal in 5 seconds, the horse is dehydrated to some extent.

'Thumps' (synchronous diaphragmatic flutter)

The diaphragm contracts in the same rhythm as the heart beat so the horse's flanks synchronise with the pulse. The contractions may even be heard. Veterinary advice should be sought immediately.

Causes of Dehydration

Lack of Water

This may be a chronic, ongoing, low-grade dehydration where animals are not provided with a sufficient supply of fresh, clean water over a long period of time. Poor performance will follow and is one cause of the horse's fitness being 'over the top'. It is likely to occur in the following situations:

- Ponies at grass, particularly those on tethers, who are watered from a bucket or other inadequate supply.
- Stabled horses who are only offered water at set times.
- Stabled horses whose water bowls are not cleaned regularly.
- Stabled horses whose buckets are not cleaned, emptied and refilled regularly and thereafter kept full. The available water may be tainted.
- Water being withheld deliberately.

It is not unknown for dealers, and riders, especially in dressage, show jumping and showing, to withhold water so that the horse becomes partially dehydrated and therefore quieter. This is iniquitous and can result in permanent damage. If this happens at a three-day event, even if the horse is given water ad lib after the dressage, he will start the cross-country below par.

Sweating

Man and horse are the only animals that regulate their temperature by sweating. When the horse sweats, he will lose valuable electrolytes; the most important being sodium, chloride and potassium.

Excess sweating is usually caused by long-term strenuous work such as long-distance riding. It is the duration of sweating which is important but temperature and humidity are important factors.

Other causes of sweating include:

- Travelling.
- Illness, especially obstructive colics.
- Stress.
- Fear.
- Pain.
- Excitement.
- Airless hot stable with high humidity prevents evaporation and may lead to heat stress.

If these cause dehydration, water and electrolytes (essential salts) are necessary for recovery. In severe cases veterinary assistance is indicated.

Travelling

Travelling, even short distances, can be a stressful experience for the horse. Some horses, however accustomed they are to travelling, will refuse to eat while in the vehicle and will not drink even when stationary. These horses must be treated with great care as there is a danger of dehydration and colic is more likely when they start eating at the end of the journey. During any breaks in travelling, watering is a top priority and, if possible, some grazing.

Horses can also suffer dehydration due to excessive sweating on the journey. This is particularly likely with young horses unaccustomed to travelling. It is not always appreciated how much they have sweated as the air blowing through the vehicle may dry them off and hide the extent of the sweating.

Diarrhoea

Diarrhoea will cause dehydration. It is especially serious in the case of foals. Acute diarrhoea will require replacement therapy urgently. Horses that are ill or are suffering from the effects of worm infestation causing chronic diarrhoea should have electrolyte salts added to their diets in addition to the anthelmintic treatment given under veterinary supervision.

Excessive Urination

This may be the result of over-prolonged dosing with a diuretic. Veterinary advice should be sought. However, the horse with polyuria (excessive urination) will usually develop a compensatory polydypsia (increased thirst).

Electrolytes

These are a mixture of body salts which are added to the water or feed after strenuous exercise. It is pointless giving them a long time before a competition, but to get horses used to the taste in water they may be offered daily after work for a week or so.

With a commercial preparation it is important to check that the proportions are correct. If glucose is incorporated, it is there to increase absorption and not as a source of energy.

Electrolytes as powders added to the feed or in an oral paste in a syringe (similar to a worm dose) are an easier way of guaranteeing that the horse takes in the dose. Veterinary advice regarding electrolyte supplementation is worthwhile.

If electrolytes are given after a competition, plain water should also be offered. It is extremely important to get the horse to drink. In cases of extreme exhaustion the veterinary surgeon may give the horse fluid intravenously as in this case oral rehydration can be dangerous.

18 FOOD AND DIGESTION

THE NECESSITY OF FOOD

Food is needed by the horse for:

- Maintenance of life.
- Growth and/or work.
- Repair of tissues.

It is the source of energy. Energy is needed to sustain life, and for the production of all body tissues (e.g. muscle and blood). Energy is needed for the maintenance of muscle activity, digestion, excretion, reproduction and all body processes.

DIGESTION

In this chapter the emphasis is on those particular features of the horse's system which are practically related to a better understanding of the rules which govern the art and the science of feeding. Knowledge of these features can be considered to be a basic requirement of responsible horsemanship at any level. Such responsibility is directed towards the skills concerned with feeding, which is one aspect of the welfare of the horse.

The aims of feeding are to:

- Maintain health.
- Maximise performance.
- Minimise nutritional mistakes.
- Reduce the risks of malnutrition in all its aspects.

The more advanced the athletic performance of the horse, the greater the attention needed to detailed application of the skills. Although background knowledge, as distinct from this essential knowledge, is not necessarily required, it is assumed that stable managers so involved will wish to increase their theory. The 'what', 'when', and 'how' are more efficiently put into practice if the 'why' is also understood.

Any theoretical knowledge gained should be primarily directed towards the practical ability to:

- Appreciate when all is well, and that the horse is at ease.

- To recognise quickly when all is not well, and the horse is at 'dis-ease', as judged by the clues produced by the structures and functionings of the system(s) involved.

The 'feeding of the horse is in the eye of the master'!

The art of feeding horses has been practised and taught for over 3000 years. The scientific understanding is much younger, since most research into the subject has been carried out during the last fifty years. It must not be forgotten, however, that the horse has not basically changed in many thousands of years. Its instincts and behaviour as a nomadic, gregarious herbivore, its appetite and its digestive system, have not altered since it became *Equus caballus*, an established plains dweller living by grazing and browsing.

It is still constructed and 'programmed' to survive by eating vegetation, and although the vegetation to which it has access has changed, it has done so only to a degree. These changes first occurred as the various 'types' of horse spread out into Europe and Asia to settle in different climatically controlled areas:

- Mountain and moorland.
- Lush lowland pastures.
- Steppes and prairies.
- Deserts.

Early domestication within these broad locations still depended upon 'natural' eating.

Farming practices in those countries where seasonal changes in grass growth occurred (temperate zones) eventually produced methods of conserving herbage for over-winter feeding, followed by the development of harvested indigenous grains.

More recent centuries recognised the need for energy intakes on a daily basis in excess of that which hay and unimproved grain could supply, and horsemen learnt how to concentrate the ration by using progressively more improved cereal crops of denser, therefore richer, starch content so that they met the energy requirements within the limitations of the horse's daily capacity. Thus began 'unnatural' feeding practices.

Modern feeding may seem to be a complicated business of permutations of many indigenous and foreign materials but, with few exceptions, these are all vegetations and their by-products and they are basically similar to the horse's 'natural' foods but with marked differences in starch content and mineral ratios. It is these which make such rations 'unnatural'.

The Characteristics of Plants – seen from a 'horse feed' aspect

Vegetation converts the energy of the sun on leaf chlorophyll (via photosynthesis) into plant energy for:

- Growth. To this end the metabolic processes produce compounds of carbon, hydrogen, oxygen, nitrogen, etc. as carbohydrates, proteins, lipids and vitamins

and, along with minerals from the soil, these are the solubles and suspensions in the fluid or sap; and a compound carbohydrate, cellulose, which gives shape and strength to the plant.

- The development of the reproductive system, the flower and ultimately the seed, both of which replicate the constituents of the parts; except that the seed requires a concentrated form of nutrient to nourish the eventual germination. This is starch, another form of carbohydrate.

All these are potential sources of energy for the grazing horse. It can be said that plants store energy and that horses expend energy (in growth, repair and work).

The most important component is **cellulose.** Under natural feeding this carbohydrate forms by far the greatest part of the horse's staple diet **but** its energy value can be released only by microbial fermentation within specialised areas of the digestive system: the caecum and large intestine. It cannot be digested in the ordinary omnivorous way, although the other constituents can. Some of these solubles are also used to nourish the micro-organisms.

When cellulose is broken down it produces large quantities of volatile fatty acids which are absorbed into the system and metabolised into energy sources and stores.

The Micro-organisms

There are several species always present in the digestive tract. The relative proportions vary with the predominant pasture strains. Some are adapted for fermenting lush grass cellulose, others for the stemmier types and yet others for the husks of seeds. Yet again other strains are concerned with those undigested solubles and starch that reach the large intestine.

The Horse and its Herbivorous Diet

Grass, even when fresh, succulent and high in water content, is still a fibre-rich plant containing comparatively small amounts of solubles. The horse

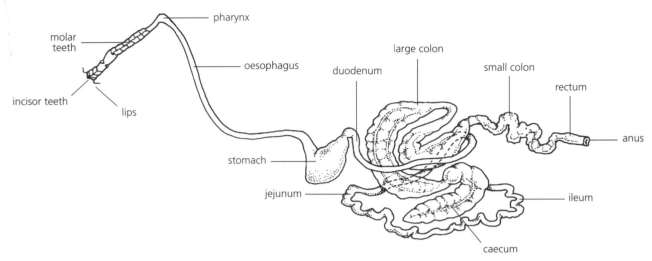

molar teeth

incisor teeth

lips

pharynx

oesophagus

duodenum

large colon

small colon

rectum

anus

stomach

jejunum

ileum

caecum

The digestive system, simplified.

must ingest large quantities: on average, approximately 21/2% of his own body weight as dry matter intake. He eats for the greater part of a 24-hour day and for long periods at a time. Such is called 'trickle feeding'.

An inherited instinct is the desire not only to eat but to eat foodstuffs which have the stimulus to instigate **mastication.** The horse 'feels the need to chew' and the fibrous nature of even wet grass guarantees this satisfaction.

Mastication takes time, requires powerful jaw muscles to move the lower jaw across the upper, and long jaws to accommodate large molar teeth. These are set back in the jaws away from the incisor teeth, and the tushes if present, the spaces between being called the bars. This anterior (rostral) area allows the food, picked by the **lips** and cut off by the **incisor teeth** to be collected by the **tongue** and propelled backwards to between tongue and **mouth roof** and the molar arcades. (Space in the lower jaw for the bit is fortuitous!)

Characteristically the grazing (eating) horse continually eats and chews. As it does so, mastication stimulates the flow of copious **saliva** from the parotid gland and this begins the initial digestive process. In addition to acting as a lubricant the watery saliva carries certain electrolytes and antacids. As the chewed food passes into the **throat** (pharynx) it is moulded into

elongated boluses before entering the **gullet** (oesophagus) en route to the **stomach** wherein digestion proceeds. The cellulose particles pass through that organ and the small intestine and in no more than 6 hours enters the **blind gut** (caecum) and the **large intestine** (large, compound and small, simple colon) before the undigested material is voided after some 40 hours or so.

The stomach never quite empties but does intermittently fill according to the grazing pattern. The large intestine and caecum never empty. There is always a constant topping up and an ongoing fermentation with audible gut sounds.

The stomach is relatively small and so is liable to overfill, especially when feeds are lush and suddenly available after periods of scarcity. Its construction is such that burping (eructation) and vomiting cannot occur.

Mastication serves other important functions besides stimulating saliva:

- It breaks up the stem, leaf, flower and seed.
- It liberates the sap with its solubles.

Incidentally, a horse does not salivate at the sight, smell, taste or 'sound' of foodstuffs.

Under 'natural' grazing the horse may show herbage variety preferences but all the nutrients in each and every grass, legume or plant are ingested and utilised. There is, of course, a residual indigestible component.

Changes in nutrient ratios related to growth rates of the herbage are minor or slow in onset, associated more with the seasons than with weather variables. The micro-organism colonies, once established, therefore remain fairly steady and they suffer no 'dis-ease', as can happen with sudden and dramatic feed intake changes. Consequently gut function, subsequent metabolism, and ultimate life and work remain in a 'steady state'.

Stabling, winter feeding and feeding for high performance require quite marked changes for the horse's digestive system. Grass is replaced by hay and allied products with its higher dry matter, stemmier consistency, i.e. more cellulose and less solubles, now collectively called **roughage.**

The energy input is obtained from increasing use of grains – cereal seeds – which are rich in starch, densely packed, hence the name **concentrates.**

It is any change to the dietary regime which demands good stable management.

Hay, which as a general rule should not be less than 50% and never less than 30% by weight of the total

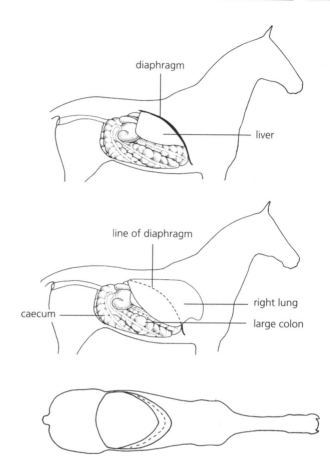

Relative positions of digestive organs and lungs. Lungs project backwards over either side of the 'dome' of the diaphragm, whose contraction and relaxation causes inspiration and expiration.

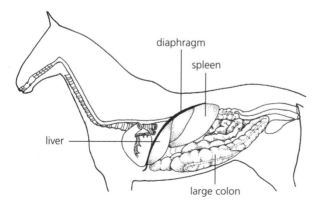

Abdominal contents, simplified. Note the greater volume occupied by the abdominal contents relative to heart and lungs (see also illustrations above right) and the relative, much larger volume of large intestine (large colon) and caecum compared to the stomach and small intestines (small colon).

ration, is eaten slowly but it is impossible not to have periods within the 24 hours when eating and chewing is not required. Corn is eaten quickly. It should be mixed, 'opened', with chopped roughage, bran or soaked sugar beet to slow up ingestion and encourage better chewing.

The effect of mistakes in these changes are dealt with elsewhere but always:

- Offer hay after grain whenever possible. Alternatively offer hay half an hour before the concentrate feed to stimulate the digestive system, e.g. in early morning.

- Offer the largest amount of hay after the evening feed.

- Change from grass to hay slowly over 7–10 days.

- Establish on hay before introducing grain.

- Offer no more than 500g grain (roughly 1lb) for a 500kg (1100lbs) horse on day one and increase daily by no more than 200g (½lb) daily till maximum is reached. Remember that the work which requires concentrates progresses slowly anyway.

- Never put more than 2kg (4½lbs) of feed in the manger at one time.

- Feeds should be divided into at least two per day and more as feed is increased.

- Make sure that there is always plenty of water.

ROUTINE HORSEMANSHIP

Once settled on to a pasture, a horse will graze for up to 16 out of 24 hours. This will be broken up into variable sessions, more at night in warm weather but usually the longest periods in daylight. It is not known what regulates the horse's appetite.

Horses tend to walk forward as they graze. Occasionally they'll lift their head and neck to check out the environment, but continue to chew. Most will stop grazing and come to, or at least heed, a call from the owner/handler.

At the end of a grazing session various activities will follow:

- Close contact with a 'friend' for mutual grooming.

- Exercise with other horses.

- Movement to excretory areas, but horses rarely void urine and faeces at the same time.

- Rest: standing, lying down.

Deviations from known behaviour should attract attention. Interpretation and assessment is described in other manuals but beware the horse that stands 'away from herd', refuses to stand up on stimulation, or shows no interest in his preferred tit-bit or plucked grass. It is fortuitous if defaecation is observed during any visit to inspect the horse. The sudden appearance of a soiled tail and bottom in the grazing animal should be investigated.

In the stable, evidence of a horse 'off food' is seen in the manger and haynet, if feed and hay are untouched or thrown on to the floor. A badly disturbed bed and marked alteration in the number of droppings, especially if down to nil, and changes in their colour, consistency and smell are all important clues. Partially chewed, dropped out wads of hay, grass etc. may indicate an acute mouth problem.

Copious salivation, especially down the nostrils, with intermittent marked swallowing attempts, indicate oesophageal choke which requires urgent veterinary attention.

IN THE STABLE

It takes a horse about 40 minutes to consume 1kg (roughly 2lbs) hay; straight hard feeds take about 10 minutes for 1kg (2lbs). When openers, especially chaff, are added, this time will be doubled or trebled.

A horse will defaecate eight to twelve times in 24 hours. (During working hours there is no excuse for more than one 'pat' in the bedding at any one time, assuming that the stable is regularly skipped out.)

The horse can be seen to eat, seen and heard to masticate; and seen to swallow (the bolus passes down the left jugular groove at intervals during eating and chewing). He can be heard to produce intestinal rumbles (borborygmi); heard and smelt to have passed wind (flatus); and seen and heard to pass faeces.

FAECES

Faeces are the undigested residue of the feedstuffs consumed. Even the dry-looking droppings of the horse contain a percentage of water, at least 15% by weight.

The small colon and rectal wall secrete a water-based mucus to lubricate the faeces' passage but the bulk of the water content is that left after reabsorption of the water in the large colon.

The water found in the digestive system consists of:

• that drunk;
• that extracted from the foodstuffs; and
• that secreted by the many glands:
  salivary;
  biliary;
  pancreatic;
  gastro-intestinal wall mucus and digestive enzymes.

During the passage along the digestive tract there is also a two-way (secretion/absorption) movement of water across the gut wall to transport:

• the digested constituents **inwards;** and

• the **exchange** of electrolytes involved primarily in determining the pH level of the different regions of the tract, from stomach to small colon.

This considerable quantity of water, greater than that in the blood and in the extracellular fluids, has to be reclaimed.

The residue of water is a reflection of:

• The water content of the original food eaten, e.g. spring grass gives wetter (softer) faeces.

• The 'health' of the digestive system.

It is recognition of **unusual** consistencies of faeces which is important in horsemanship. Beyond this is the appreciation that the rations fed (and eaten) according to the rules have maintained the desired condition in the light of work, temperament and weather.

To maximise this end result regular attention should be given to dental care: the 3500 jaw movements per 1kg (2lbs) food masticated soon wears the molars to produce sharp points in both upper and lower jaws in stable-fed animals.

CHAPTER

19   THE RULES OF FEEDING

The rules of feeding are based on scientific fact and it is important for the horsemaster to know and understand them:

1. Feed according to the horse's condition, work, size, temperament and age.
2. Feed only good quality foodstuffs.
3. Feed sufficient roughage (fibre).
4. Feed little and often.
5. Make no sudden changes in the type of food fed.
6. Keep to the same times of feeding every day.
7. Feed something succulent every day.
8. Do not work immediately after a full feed.
9. Water should be freely available at all times. Where this is not possible, water should be given before feeding.
10. Keep all watering and feeding utensils clean.
11. Know the weight of the volume of food in the stable scoop or bucket so that weight is the basic measurement.

The reasons for these rules are as follows:

1. Feed According to the Horse's Condition, Work, Size, Temperament and Age

It is important that the rations are suitable for the horse in question bearing in mind:

- Present condition.
- The individual's metabolism, i.e. difficult to keep in condition, or puts on weight too easily.

- The workload.
- The ability of the rider.
- Whether extra food is needed for growth.
- His general behaviour.

This subject is discussed in more detail in Chapter 21.

2. Feed Only Good Quality Foodstuffs

It is a false economy to feed poor quality food. Poorly harvested hay or bad oats can cause many problems and these include colic and COPD (chronic obstructive pulmonary disease). Performance will be affected and fussy feeders will reject food.

For competition horses, the higher the level at which they compete, the greater the care required in assessing the quality and overall composition of the ration. In terms of performance the ratios of the major nutrients and that of minerals, trace elements and vitamins become critical.

3. Feed Sufficient Roughage

The horse requires plenty of roughage in his diet in order to digest his food properly. Even fit competition horses should never have less than 30% by weight of the ration made up of bulk food. This is mostly hay or its equivalents, but also includes chaff.

4. Feed Little and Often

As already stated, the horse is a trickle feeder. If the stomach is overloaded with food, some will be pushed into the small intestine without being properly digested. Not only will this be a waste of food but it may also cause colic due to fermentation.

The small size of the stomach (see digestive system) means that not more than 2kg (4½lbs) of concentrated food should be given at any one time and part of this should be chaff or soaked sugar beet.

It follows that the total daily concentrate ration must be divided into separate feeds to adhere to this 'not more than 2kg (4½lbs) at a time' rule and the feeds spread as evenly as possible over a working day – a long day may be necessary.

5. Make no Sudden Changes in the Type of Food Fed

Because the bacteria (micro-organisms) in the horse's gut are 'food specific' it is important that any new foodstuff is introduced gradually. Therefore, as an example, horses who do not receive a small amount of bran each day should not suddenly be given a bran mash. A new supply of hay should be gradually mixed with the old supply so the change is not sudden.

When a new horse enters the yard enquiries should be made as to his previous diet. If possible, it should be adhered to and only changed slowly.

When horses travel and compete it is important that all the rations they require are taken with them. A new supply of hay or a different brand of cubes must not suddenly be introduced.

6. Keep to the Same Times of Feeding Every Day

The horse is a creature of habit and will do much better on a regular routine.

Make the timetable suit the yard routine. It is no good arranging for a late night feed unless it can be given regularly every night.

For the one-horse owner who is at work all day the feeds should be given early and late, with the horse having plenty of hay to keep his digestion working.

Regular feeding times should also apply to horses living out otherwise they are apt to spend much time waiting at the gate.

7. Feed Something Succulent Every Day

Ideally all horses should have some grazing each day. To the horse green grass is the acme of enjoyment. Grazing old pasture will give him the chance to find different grasses and valuable herbs. Carrots, turnips, swedes etc. are appreciated by the horse, as are apples.

Carrots are very high in carotene and a good source of vitamin A for horses that are stabled all the time.

Succulents will help to tempt the appetite of the poor feeder and will provide a natural source of minerals and vitamins. Many vitamins are present in fresh, green food. In the spring, dandelions, both leaves and roots, are much relished by stabled horses.

Some yards now feed barley grass. This is hydroponically grown in a purpose-built cultivator.

8. Do not Work Immediately After a Full Feed

It takes approximately 20 minutes for a horse to eat a full feed and at least 1½ hours to process it through the stomach and small intestine.

Blood is involved in the digestion of food. If there is a simultaneous work effort with muscle activity, the digestion will be impaired. A full stomach will also restrict room for lung expansion, which could cause strain if hard work was expected.

It follows that if it is necessary to exercise early in the day, either the first feed should be after work or it should be very small. Feed should be given early enough to allow time for digestion. This is especially important for competition horses or hunters whose workload is extra demanding.

This rule also applies to the bulk food. Whilst bulk food takes a short time to pass through the stomach and small intestine, it can dwell in the caecum and large intestine for up to 72 hours.

9. Water Should be Freely Available at All Times

This subject is covered in detail in Chapter 17.

10. Keep all Watering and Feeding Utensils Clean

Horses are very fussy feeders with an acute sense of smell. All water buckets and feeding utensils should be kept scrupulously clean; all stale food should be thrown away. Food must be kept free of contamination from mice, rats and other vermin. Some horses will starve themselves rather than eat tainted food.

Warmth and humidity enable micro-organisms to flourish and manufacture toxins. Regular emptying and cleaning of all food containers is essential.

11. Know the Weight of the Food in a Scoop or Bucket

Weigh the empty scoop/bucket, then fill it with each food and re-weigh it. Subtract the weight of the empty scoop/bucket to find out the weight of the food. Feeds can then be measured out in terms of scoops/buckets rather in kilos or pounds.

FOOD – COMPOSITION AND TYPES

COMPOSITION OF FOOD

Broadly speaking, there are five categories into which the horse's food can be divided. They are:

- Carbohydrate – which includes starch, sugar and cellulose (fibre).
- Protein.
- Fat/oil.
- Minerals, including trace elements.
- Vitamins.

Carbohydrate

Carbohydrate is a food substance found in vegetable tissue comprising starch, sugar and cellulose (fibre). In the feeding of horses carbohydrates are important and should be the main source of energy for the horse (two-thirds of the weight of the concentrate feed ration should be carbohydrate).

- STARCH is the major energy store in plants.

- SUGARS are the simplest form of carbohydrate.

- CELLULOSE (fibre) is insoluble carbohydrate.

Crude fibre contains cellulose, and lignin. The horse can digest cellulose fibre but not lignin.

Fibre is an essential part of the horse's diet: without it he cannot digest his food properly. It stimulates the contractions of the gut. At least 30% of his rations must be fibre.

The main sources of fibre for the horse are grass, hay and straw, which are known as bulk food.

Protein

Protein is a complex organic compound containing nitrogen. Different proteins are formed from varying combinations of twenty-two amino-acids and their derivatives.

Protein is a builder; it is necessary for growth and replacing wastage. It is made up of chains of amino-acids linked together. Some of these amino-acids can be synthesised (made) by the horse. Others are made by the micro-organisms in the gut. However, others known as essential amino-acids must be included in the diet. An ordinary hay/cereal ration is low in these essential amino-acids, so for growing or working horses, additions should be made to the ration.

Lysine, methionine and tryptophan are three of the essential amino acids, which are not available in 'straight' grain feeds.

Foods that are high in lysine are:

- Peas and beans, including soya-bean meal.
- Dried milk pellets (not a suitable source for adult horses).
- Fish meal.
- Lucerne/alfalfa.

Not all high protein foods have easily digestible protein. Reputable compounders making up a mix usually ensure that the protein is digestible.

Some nutritionists maintain that the total percentage of digestible protein in the ration should not be over 10% except for lactating mares, foals and young stock.

The protein levels will naturally alter with the changes in the quality of cereal fed. There is no evidence that athletic horses require high protein supplements although individual National Hunt and three-day event horses might. In fact it is thought that the overfeeding of protein to competition horses can be detrimental to their performance.

Fats and Oil

All foods will contain some fat/oil as an integral fraction of their composition (i.e. their analysis).

Fat and oil are important for the growth and maintenance of cell membranes, the metabolism of cholesterol and the formation of prostaglandins. They are also used by the horse as a source of energy but, as far as is known, only after metabolism from body fat deposits. Such fat deposits play a role in control of body temperature.

Traditional rations are low in oil and many feeders now add corn oil to the diet.

Dietary fat, in the form of corn oil, is digestible and can reduce the requirements for cereal carbohydrate (starch). Recent studies show that it also lessens the decline in blood glucose during endurance rides and seems to accelerate the recovery of heart and respiration rates during the first 10 minutes of rest.

Since corn oil, along with other oils and fats of plant and animal origin, provides twice as much energy as does starch per g/oz fed it is a useful part-substitute for cereals in endurance and three-day event horses on high rations.

Minerals (including trace elements)

These are essential elements of the diet but are required in very small quantities. They are inorganic substances.

Macro-minerals

The major minerals are as follows (their abbreviations are in brackets):

- Calcium (Ca)
- Phosphorus (P)
- Potassium (K)
- Sodium (Na)
- Chlorine (Cl)
- Magnesium (Mg)

Micro-minerals

These are the trace elements of importance to horses:

- Zinc (Zn)
- Manganese (Mn)
- Iron (Fe)
- Fluorine (F)
- Iodine (I)
- Selenium (Se)
- Cobalt (Co)
- Copper (Cu)
- Sulphur (S)

Minerals (the usual name for macro-minerals)

Wild horses roaming over large, open areas were able to choose variable herbage and were 'naturally' certain of a reasonably correct intake of minerals in terms of range and quantity. Properly balanced feeding and good pasture management are therefore essential for the domesticated horse's well-being. Supplementation may well be required.

The feral or wild horse, particularly when adult, is not athletic in competition or work terms. The need for high-energy rations is considerably less. Cereals as such will not be eaten and it is cereals which unbalance rations *vis à vis* minerals. Both minerals and vitamins interact with one another, so an imbalance can affect the correct functioning of many body processes.

Adding minerals to the horse's diet should always be carried out with great care. Many of them can be very harmful if given in excess and 'guestimates' of what should be fed may well produce serious imbalances.

Calcium and common salt may have to be added to

the diet, especially for performance horses and young stock, but magnesium, manganese, copper, fluorine, iodine and cobalt are usually present in sufficient quantities. Sometimes potassium, zinc, iron, sulphur and selenium may be deficient.

Some experts consider that iron may be needed in high performance horses.

Calcium (Ca) and Phosphorus (P)

These are the commonest minerals in the horse's body and it is particularly important that they are correctly balanced and are supplied in the correct proportions.

Calcium should be present in the diet in a ratio of 2:1, i.e. 2 parts calcium to 1 part phosphorus. Most cereals have a very poor calcium to phosphorus ratio, so seek advice from a reputable feed merchant or from your veterinary surgeon.

Bran inhibits calcium uptake. A calcium supplement, possibly in the form of ground limestone, will be needed if much grain or bran is fed. Calcium-rich foods such as lucerne/alfalfa and sugar beet are the best way to introduce calcium into the diet. The body can more efficiently utilise calcium within a food substance rather than as a supplement.

Potassium (K)

Potassium is needed for body fluid regulation. It is likely to be deficient in water-extracted foods but is high in lush pastures. Well-saved forages are higher in potassium than cereals. Supplementation may be necessary when horses are working hard on a high cereal diet but care is required. Molassed sugar beet is a good source of potassium.

Sodium (Na) and Chlorine (Cl)

Sodium and chlorine are also needed for body fluid regulation and are likely to be lost in sweat. The best supplementation is probably common salt.

Iron (Fe)

Essential in the formation of haemoglobin, the means by which oxygen is carried by the blood. Lack of iron causes anaemia.

Magnesium (Mg)

Magnesium is an important element in the blood and in most cellular activities. Its major interaction is with calcium, phosphorus and vitamin D.

Soya-bean meal, good permanent grazing, well made hay and other forages are all good sources of magnesium but most cereals, especially maize, are poor.

Problems such as poor performance syndrome and obscure muscle stiffness, have been shown to be associated with imbalances of these major minerals but not necessarily in the rations. Individual animals seem to be unable to make full or correct use of what is present.

Veterinary advice based on laboratory findings is necessary before attempting to supplement the diet.

Selenium (Se)

When horses graze pasture which is low in selenium there may be problems and diets should be supplemented with selenium and vitamin E. A deficiency of either selenium or vitamin E can cause muscle disease in young horses. As with most trace elements an excess can be toxic.

Cobalt (Co)

Cobalt may be used in cases of anaemia. It is a component of vitamin B12. A deficiency may cause liver dysfunction or poor appetite.

Copper (Cu)

There are parts of the British Isles where copper deficiency occurs. Copper is necessary for the formation of the blood pigment haemoglobin. A deficiency may cause anaemia or reduced growth.

Macro-minerals play important parts not only in tissue structure but also in metabolic processes and some in the electrolyte balances of intracellular and intercellular fluids.

Trace Elements (minor minerals)

All these play important parts in enzyme action.

Vitamins

Like all mammals, horses need vitamins, although in very small amounts. Most of these will already be present in the diet of the correctly fed horse, although supplementation may be necessary under some circumstances. Little research has actually been done into the amounts required by the horse although the signs of deficiency are known.

The main vitamins are: A, C, D, E, K and the B complex. They can be divided into two groups: fat soluble and water soluble.

- A, D, E and K are all fat soluble and can be stored in the body.

- B complex and C are water soluble and cannot be stored.

Fat-Soluble Vitamins

Vitamin A (Retinol)
May be referred to as Beta-carotene.

The horse obtains vitamin A naturally from green herbage which contains carotene. This is converted into vitamin A. There is some carotene present in newly made hay and other preserved grass but this deteriorates considerably with age.

A summer out on good grass will provide enough vitamin A, which can be stored in the horse's liver, to last through the winter.

Horses living in throughout the year and not receiving any green food may benefit from the addition of cod liver oil to their rations, although the amount must be carefully controlled.

Succulents such as carrots, apples etc. also supply some vitamin A.

Broodmares and young stock, particularly newly weaned foals, may also require supplementation.

Retinol occurs in animal fat although a synthetic substitute is made.

Lack of vitamin A can cause anorexia, poor growth, night blindness, inflammation of the cornea, infertility, respiratory disorders. Excess vitamin A is toxic.

Vitamin D (Calciferol)
The horse can obtain vitamin D naturally from sunlight on the skin. However, this is limited by: (a) the strength and amount of sunlight; and (b) the amount of time the horse is outside.

The amount of sunlight in the British Isles in winter is probably of little value but the vitamin can be stored by the horse.

A horse living out during the summer will probably accumulate enough to last him well into the winter but thereafter he may need supplementation. A horse that lives in all the time, or is turned out only occasionally, or who is out in a New Zealand rug, will require vitamin D added to his diet. Cod liver oil is a natural source. Synthetic vitamin D can be very potent and needs adding with great care as too much is toxic.

Vitamin D is necessary for the metabolism of calcium and phosphorus Lack of vitamin D can cause bone and joint problems, as can excess vitamin D.

Vitamin E (Tocopherol)
Fresh food such as grains, seeds and herbage are all natural sources of vitamin E. Its levels in stored foodstuffs daily falls as it is 'used' by the foliage and grains as a protection antioxidant.

More vitamin E will be needed when selenium is deficient and in an oil-rich diet. Late winter supplementation with vitamin E and selenium may be necessary. Synthetic vitamin E deteriorates and all compound food and supplements should be date-checked when buying new supplies. Wheatgerm oil is a good but expensive natural supplement.

Vitamin K
Vitamin K is synthesised by gut micro-organisms. It is also present in leafy matter.

Vitamin K is a blood-clotting mechanism and supplementation is normally unnecessary.

Water-Soluble Vitamins
The water-soluble vitamins include:

- B1 – thiamine
- B2 – riboflavin

- B6 – pyridoxine
- B12 – cobalamin
- niacin (nicotinic acid)
- pantothenic acid
- folic acid
- biotin
- choline
- C – ascorbic acid

Most of these vitamins are either present in high quality food, especially green food, or are synthesised by the gut bacteria. Horses on good quality rations should not need supplementation but horses suffering from disruption of the gut bacteria due to sudden changes of food or antibiotic treatment will need additional supplementation. The important vitamins in this group are thiamine, folic acid and biotin.

THIAMINE is important for horses in hard work as it assists with the breakdown of lactic acid.

FOLIC ACID is high in good grazing, especially that with a high legume content. Stabled horses may be deficient, when a form of anaemia may develop. This shows after approximately three months of stabled feeding.

BIOTIN is thought to be of value for horses with poor hoof formation.

TYPES OF FOOD

The modern horse's food can be divided into four groups:

- Roughage (bulk food).
- Concentrates.
- Succulents.
- Additives and supplements.

ROUGHAGE (BULK FOOD)

Grass

Grass is the most natural food for horses. If the quality and quantity are adequate grass is a complete maintenance food from April to September. This is why, if not otherwise organised by man, mares have their foals in late spring.

As the season progresses the energy value falls. In May it may be as high as 12MJ per kg of dry matter, falling in July to 9MJ (MJ = megajoule – a measure of digestible energy). Sometimes the value may rise again in the autumn. If saved at the optimum time it will supply suitable amounts of fibre, protein and minerals for the resting horse.

Grass is conserved by various means so that it is available as a basic ration throughout the year. The earlier in the spring it is cut, the higher the protein value. It can be saved in the following forms:

- Hay.
- Silage.
- Vacuum packed (as haylage).
- Dried and cubed.

Hay

The most common way of saving grass as hay is to cut it, allow it to part dry in the field, bale it and bring it under cover.

Hay that is grown for sale is usually setting seed and therefore depreciating in food value. It is cut at this stage because the bulk will be larger, the grass more stemmy, and easier to 'save'.

The way the hay is made is of almost more importance than anything else. If it is rained on, this washes out valuable nutrients. It should be dry when it is baled otherwise mould spores will generate inside the bales. Hay baled too wet may also cause the bales to overheat and some stacks can even catch fire.

Judging the value of hay by looking at it is almost impossible. Knowing its 'history' does help. In a large establishment, particularly one with competition horses, it is worth the expense of having the hay analysed. For small establishments or the single horse owner, the cost of a full analysis would be uneconomic, so in this case it must be judged by the following criteria:

- It should be a good colour. Hay varies from greeny/grey to pale fawn but the greener it is, the better.

- It should smell pleasant, not musty, nor of vermin.

- It should shake out well and not stick together.

- It should be free from dust and disintegrating leaf, especially clover.

- It should not have any trace of damp or mould. To test this, it is essential to check the centre of the bale.

- It should have a good proportion of flowering heads.

- It should be free from weeds such as docks, thistles, nettles, bracken and, most particularly, ragwort. Ragwort, when dried, becomes palatable to horses but it is **very** poisonous, causing liver damage which will eventually result in death.

New Hay

New hay can be harmful to horses. This is because grass made into hay continues to mature chemically in the bale. Although the maturing process uses up some of the sugars, it converts the unwelcome nitrates into amino-acids.

The basic sward from which hay is made is rich in sugars and nitrates. New hay is still maturing and horses fed on it are likely to develop digestive problems. These can result in colic, laminitis and other metabolic disturbances. Filled legs are a typical result.

Hay should not be fed until two to three months after it has been made. It should then be introduced gradually by mixing it with the 'old' ration in increasing amounts proportionally over three to four days.

Types of Hay

Hay can be made from grass or lucerne. It may be specially grown or taken from permanent pasture. Grass hay is usually divided into two types:

- Seed or hard hay.

- Meadow or soft hay.

Seed Hay

Seed hay is made from grass that has been sown specifically as a crop. The grass is usually left down for only one to five years before being ploughed in. It consists of only a few varieties of grass or, in some cases, one variety only. The grasses commonly used are rye grass, timothy or cocksfoot. Some red clover may be added.

Seed hay feels hard to the touch and looks quite 'stalky', hence the alternative name 'hard' hay.

ADVANTAGES

- The grass used for the hay will be of a variety that is nutritionally valuable.
- Because it has been grown for re-sale, it is likely to have been made with more care.
- It is unlikely to have weeds in it.

DISADVANTAGES

- Because only one or two types of grass are included, the horse may be limited in mineral intake.
- It is expensive.
- It is less palatable.
- It is often mistakenly considered to be of greater nutritional value than meadow hay. In fact it contains more fibre.

Meadow Hay

Meadow hay is made from fields that are down to permanent grass and as a result contain many more types of grass and in addition herbs and possibly weeds as well.

The grasses are usually much finer and shorter in the stalk; thus it is sometimes known as 'soft' hay.

ADVANTAGES

- It is made up of many different types of plant, so the horse has access to a wide range of minerals.
- It is less expensive.
- It is more palatable.
- It is usually less fibrous than seed hay and of greater nutritional value.

DISADVANTAGES

- The optimum time for cutting is difficult to determine as the grasses flower at different times.
- Weeds, such as docks and thistles, may be incorporated. Worse still, there may have been ragwort in the field.

Clover Hay

Clover hay is seldom seen in Britain as it is hard to make and inclined to break up and become dusty. If used, it should be fed with care as it is high in protein.

Lucerne/Alfalfa

Lucerne, or alfalfa, is made from lucerne. It is very high in protein and is ideal for young, growing horses, especially if they are working. Many racehorse trainers use it for the two- and three-year-olds in training. It is usually very expensive but less concentrate food will be needed when lucerne is fed. It can, however, be low in phosphates so may need supplementing.

Other Roughage Foods

Silage

Silage is made from young cut, but not dried, grass and either put into a clamp with the air excluded, or sealed in large, airtight bags (known as big bales). It is often made with additives of different kinds. It is important to check that any additives used are suitable for horses. Deaths from botulism after eating big-bale silage have been recorded in horses. Bacterial contamination is more of a risk if the silage has been cut very close to the soil.

Vacuum-Packed Grass (Haylage)

This is a fairly recent introduction. It comes in small, sealed plastic bags, and to differentiate between hay and silage, the vacuum-packed contents are known as haylage, although there are many trade names. If it is used sensibly, horses seem to do well on it. Haylage is of particular value for horses with respiratory problems. It should not be used for two months after cutting. It is made on the same principle as silage, although it is wilted for one or more days by excluding the air from cut grass. It is important, therefore, to make sure that the bags are not punctured. Mice and rats chewing the plastic can be a problem.

Haylage is usually higher in protein than hay so a horse's concentrate ration should be reduced. The manufacturer should state the protein level. Haylage may have molasses added.

Due to the higher water content in both silage and haylage it is important that pound for pound the weight of forage compared to hay is not reduced or the fibre intake for the horse would be affected. If the horse thrives on haylage then the concentrates could be reduced accordingly.

Silage or haylage whether in big or small sealed polythene bales must be carefully stored and once opened the contents should be used in two to three days. Particularly in hotter weather, once opened the haylage/silage will begin to deteriorate in quality as the presence of air allows breakdown and chemical changes to start again.

Dried Grass or Lucerne Cubes

These are made from grass or lucerne which has been cut at intervals during the growing season, usually when it reaches about 20cm (8ins) high. It is taken straight to a drier and then made into cubes.

It cannot really be classed as a bulk food as it is high in concentrates, both carbohydrate and protein. It is, however, a valuable food and can be mixed with soaked sugar beet to make a mash. The cubes are best soaked before being fed.

Oat or Barley Straw

This can be a useful bulk food in certain circumstances. Its value is often equal to poor hay and, in a bad hay-making year, it is better to feed clean, dry straw than musty, mouldy hay. In this case, however, the horse will need more concentrates as straw's energy value is low.

Straw can be very useful for ponies, and for horses turned out on rich, succulent grazing. Made into chaff, it can be added to the feed to prevent bolting and to give bulk.

Barley straw used not to be acceptable, as the awns, which are very irritating, were left on the straw. Now combine-harvesting removes the whole top of the stem, leaving only the straw behind.

Chaff

Chaff, or 'chop', is chopped hay or straw. It can be bought ready chopped or cut at home in a chaff-cutter. A chaff-cutting machine, which is operated by electricity or by hand, consists of a long trough with knives at one end, and the hay or straw is fed through the cutters to emerge in pieces about 5cm (2ins) long. Proper safety precautions must be taken to prevent accidents.

Only clean, dry hay or straw should be used; oat or barley straw is better than wheat. Beware old, dusty and mouldy material.

Not more than a week's supply should be chaffed at any one time. It should be stored in sacks in a dry place as it easily picks up damp from the atmosphere.

Chaff mixed with molasses can be bought, but it is an expensive method of adding bulk to the food. It is, however, useful for horses on a 'slimming' diet as it gives 'something in the manger' and therefore prevents the horse from getting bored or jealous of its companions. Vacuum-packed haylage is dust- and mould-free and can be made into chaff for horses with respiratory problems, but its feed value must be recognised. The same applies to chopped, dried lucerne or grass which is also now obtainable. One of the popular trade names for lucerne is 'Alfa A'. One of its main attributes is the readily available source of calcium within it.

CONCENTRATES

Seeds contain a high energy store of carbohydrate as starch. Farmers select those plants which carry a high proportion of seed to stem. The harvested crop of seeds is grain and is the additional energy part of the horse's ration.

The amount given should be balanced with the roughage to enable the horse to carry out the work required of him.

Concentrates may be:

- Cereals such as oats, barley or maize.

- Legumes such as peas and beans.

- Other foods such as linseed, fish meal, milk pellets, or sugar beet. Lucerne hay is particularly high in soluble sugars and energy rich cellulose.

- Compound mixtures, made up by a manufacturer to include any of the foregoing. Other foodstuffs may be added, such as minerals and vitamins, but manufacturers are not at present obliged by law to declare the content of their mixtures. Mixtures may be produced as cubes, coarse mix, micronised or extruded.

A declaration of the analysis of protein, fibre, oil and ash is mandatory, given as percentages. The rest is starch and sugar, i.e. 'energy', and is not legally required to be stated.

Cereals

These include:

- Oats.
- Barley.
- Wheat (bran middlings and breadmeal).
- Maize.
- Sorghum, rice, millet etc.

The last three are of secondary importance although they may make up part of compound rations. Cereals are the best providers of carbohydrate energy but the protein they contain is deficient in amino acids such as lysine and methionine. Cereals are also low in calcium.

Cereal grains may be bruised to make them easier to eat but should not be ground into powder as this reduces their shelf life and makes them less chewable. Cereal grains can have a 'hotting up' effect. It is thought that this is caused by fermentation by the micro-organisms in the large gut, which causes a rapid rise of glucose and lactic acid. As well as exciting the animals, too much may result in an undesirably high level of muscle glycogen and can be a possible contributory cause of rhabdomyolysis (azoturia) and filled legs. (Pre-cooking is said to reduce these risks.)

Too sudden an intake (e.g. stolen grain) is a common cause of laminitis.

The commercial methods of cooking are:

* Steam pelleting.
* Steam flaking.
* Extruding.
* Micronising.

Extruded grains have been super-heated with steam at 120°C (248°F) for one minute. Micronised grain has been subjected to infra-red radiation. The rapid internal heating causes the grains to swell and gelatinise. The process increases the digestibility of the grain, which is often added to coarse mixes.

All these methods will add to the cost. However, the advantages may include:

* Improved palatability.
* Improved digestibility and so avoidance of high starch concentrations in the large intestine.
* Destruction of natural toxins etc.
* Longer shelf life.

Oats

Oats have been the traditional food for horses over the years and are one of the best sources of carbohydrates with a good ratio of starch to protein to fibre. Top racehorse trainers who still feed only oats and hay go to considerable trouble (and expense) to make sure that the feed is of the finest quality, often buying from abroad.

Oats should be plump, clean, sweet-smelling and heavy. The heavier they are, the more weight of kernel to husk.

They are a safer feed than other cereals in that they are low in density and high in fibre. They can be fed whole to horses over a year old unless the teeth are in need of treatment or the horse is very old.

Oats can be rolled, crimped or clipped. Clipping gives them a lower fibre content. Once treated, they should be used within three weeks as they start to lose their food value.

Oat protein contains slightly more lysine than other cereals. They do, however, have a poor calcium to phosphorus ratio so, if fed alone, must be supplemented by ground limestone or a similar calcium supplement to balance the minerals of the roughage.

Naked oats are grown 'huskless'. They are high in oil and protein and favoured by racing yards where 5–6kg (11–13lbs) per day may be used. Lack of husk reduces volume but also fibre in the diet. This must be considered and balanced elsewhere.

Barley

Barley is more readily available than good oats but, as the grain is too hard for horses to break themselves, it must be cracked, crimped or rolled. It has a higher energy value than oats per kg/lb but is consequently lower in fibre and should therefore not form more than half the concentrate ration. Like oats, it has a poor calcium to phosphorus ratio. It can be boiled – see page 178.

Flaked barley has been heat-treated to improve the digestibility of the grain, but the price is usually higher than for other forms of barley.

Maize

Maize is usually fed cooked and flaked. It can be fed whole, although the grains may be too hard for some horses. It is usually more expensive than oats or barley. Its digestible energy is higher per kg/lb than oats but it is lower in fibre. It is a useful feed in winter in small quantities. It has a high concentration of starch and must not be overfed.

Wheat

Wheat is not usually fed to horses as grain as it can easily cause digestive upsets. Bran, middlings and breadmeal are by-products of wheat and have been widely used for horses. Wheatgerm oil is a rich, natural source of vitamin E.

Bran

Bran was once one of the traditional foods, along with oats, but its use has now become controversial. The following points should be considered:

• It is high in fibre.

• Because it can absorb much more than its weight of water, it can act as a laxative.

• It is not easily digested.

• It is very low in calcium and high in phosphorus and phytate. Therefore the use of large quantities may cause bony abnormalities in young horses and may weaken durability in adult competitive horses, as well as upsetting electrolyte levels.

• It is very expensive.

• The protein it contains is not of good quality.

If bran is fed, it is essential to balance the calcium/phosphorus ratio in the whole ration.

Linseed

The seed of the flax plant, linseed, is another traditional food for horses. Although it is high in protein and oil, the quality of the protein is poor, being low in lysine. It must always be cooked before feeding so that the boiling inactivates the poisonous hydrocyanic acid (or prussic acid) which is present in the seeds. The seeds are extremely hard but are softened by cooking.

If linseed is being fed it should be given daily at a rate of not more than 100g (3oz), weighed before cooking. Linseed can improve coat condition, but as a source of energy there are better types of oil.

Cottonseed and Sunflower Seed

These are rich sources of vegetable protein and of value in the feeding of foals and young stock. They are usually processed into compound foods.

Sugar Beet

Sugar beet is a root vegetable similar in appearance to a turnip or swede. It is processed to extract the sugar (molasses) and the remainder is then made into shredded pulp or cubes. In either form, it is recognised as an extremely valuable food for horses.

It has almost as high a digestible energy as oats but, it does not 'hot' horses up due to the gradual release of glucose into the system. The fibre content is more digestible than most other foods and is of course an additional source of energy.

It is rich in calcium, salt and potassium and, because of this, can be used to offset the cereal imbalance.

Added to short feeds, it will promote digestion by causing the horse to eat more slowly and to masticate the hard food more efficiently. It must be soaked before feeding.

It is particularly useful for mixing into dried foods and bulking up rations. Many people use it as an 'opener' (bulk food) in concentrate feeds. It can also be used as a medium to administer medicines, additives and worm doses.

Up to 2kg (roughly 5lb) per day can safely be fed to horses. It should be weighed before being soaked.

Leguminous Protein Sources

Peas and English Field Beans

These are both high in lysine and are therefore a useful high-protein food. They should be fed cracked or kibbled and may be cooked. The field varieties grown in Britain are safe for horses and are generally used in compound mixes. The energy value is also high.

Soya-Bean Meal

A meal processed from cooked soya beans. The cooking destroys the toxic substances contained in the raw beans. Soya-bean meal is very high in protein. It is normally used in cubes and coarse mixes although small amounts can be added as a high protein addition to the diet if required.

OTHER FOODSTUFFS

MILK PELLETS are useful for feeding to weaned foals and young horses up to one year old. They are high in protein, particularly lysine, but contain little fat. They are sometimes added to cubes made for young horses. Older horses are unable to digest milk properly.

WHITE FISH MEAL is a rich and valuable high-protein food, containing lysine, minerals and trace elements. It also contains vitamin B12. It is often part of high-protein cubes, especially for foals and young horses.

BREWERS' GRAINS are the residue of barley after it has been malted and mashed. It is usually sold dried and can form a useful addition to a horse's ration, but should only be used in small quantities. The grains have a laxative effect when mixed with water.

BREWERS' YEAST is sometimes fed as a supplement as it is a good source of the B-group vitamins. It is said to have a calming effect, but is very expensive.

LAWN-MOWINGS are **NOT** safe for horses as they compact and heat up very easily when placed in heaps. (Ponies gaining access to garden dumps are liable to severe colic and/or laminitis.)

CUT GRASS, provided it is long and has been scythed, is safe provided it is fed immediately it has been cut. Only a small amount should be given to horses who are not already receiving a daily supply of grass.

HYDROPONIC (BARLEY) GRASS is a means of producing a daily supply of fresh green food. Soaked barley seed is placed in trays in a controlled warm, damp, well-lit environment and the 'grass' is ready for feeding in a few days. Not all horses like it but it can be a useful way of providing fresh grass all the year round.

MOLASSES is unrefined black treacle. It is much relished by horses and is a good vehicle for masking the taste of unpleasant substances such as medicines, worm doses, calcium, etc., which have been added to the feed. It is now included in many processed foodstuffs.

TIT-BITS such as Polo mints or sugar cubes are relished by horses and, used in small quantities, will do them no harm, apart possibly from teaching them to nip! Chocolate-based tit-bits should not be given to competition horses as they contain banned substances.

CARROTS, APPLES, CABBAGE LEAVES AND PEA PODS are often appreciated by horses which live in all the time and they will do no harm in small quantities. Carrots, in particular, are rich in carotene and make a valuable addition to the diet. They should be cut lengthways to avoid the risk of choking.

BRANCHES AND PRUNINGS from the garden will be enjoyed by horses (and more particularly ponies and donkeys), being more naturally grazing and browsing animals, especially when they are living out and the grass is rich and lush. The varieties offered should be chosen with care: no evergreens should be included, nor poisonous shrubs like laburnum and box. Beech, lime, sycamore and rose prunings are all safe and much liked.

TURF is a useful source of grass for stabled horses and many people, including racehorse trainers, cut a turf of grass about 30cm (1ft) square for each horse in their care every week. The whole turf is offered – grass, roots and earth – and is usually entirely eaten. The turf should be put on the ground in a corner of the box.

Herbs

Herbs will be selected by horses from whatever grazing is available. Usually these herbs are high in valuable minerals and especially trace elements. When land is re-seeded for grazing horses, it is recommended that a strip be sown with herbs.

Dried herbs can be mixed into the horse's food. If minerals have been added to dried herbs, it is necessary to check the balance of the ration.

GARLIC is usually fed as a powder and is more readily accepted by horses when mixed with molasses. It is useful for horses with respiratory problems and is said to be helpful in the control of sweet itch. Its regular use is effective as a fly repellent. It can be used to disguise the taste of medicines.

COMFREY is known as the 'healing herb' and has been used for many years. It is the only plant known to metabolise vitamin B12 from the soil. It has a high content of easily absorbed calcium.

DANDELIONS, both leaves and roots, are much relished by horses. They are high in minerals and vitamins.

NETTLES, if cut and allowed to dry, are often appreciated. They are a good source of minerals, especially iron.

SEAWEED is said to improve the condition of bad doers. It is high in iodine so the amount used should be carefully monitored. The nutrients are 'organic'.

Compound Foods

These are a mixture of foodstuffs chosen by the manufacturerand presented in different forms.

- CUBES are made from selected foods, which are ground up, steamed and pelleted. They may contain any variety of mixes to make them up to the stated protein, oil, fibre and ash levels.

- COARSE MIXES may be anything from plain cereal mixes to high protein mixes containing soya-bean meal, peas and beans. The contents may be rolled, flaked or micronised. Molasses or corn syrup is usually added to make them very palatable.

Compound foods generally contain added minerals and vitamins. Unfortunately, it is not always possible to find out exactly what raw constituents are in a compound mix. Manufacturers (understandably) often alter the constituents according to availability and cost.

Compound foods include:

- A complete mix, usually in cube form, which is said to need no hay added to the diet.

- A cube mixture of concentrates and supplements to be fed with hay. These usually start with about 10% protein sometimes called 'Horse and Pony Cubes' or 'Country Cubes' – and work up through 14% protein

for competition horses, 15% protein for broodmares to 18% protein for foals.

- A concentrate mixture as a coarse mix, to be fed with hay.

- A combination of two or three foodstuffs such as lucerne and barley, to be fed with hay.

- A fibre mixture as an alternative to bran, chaff or sugar beet, to be mixed with other foods.

- A balancer to be fed with oats or other cereals.

Compound foods for competition horses must be free from prohibited substances. There are some foods, such as 'Racehorse' and 'Event' cubes, which are regularly tested and made under rigidly supervised conditions. Reputable manufacturers realise the importance of these tests and are insured against claims. The onus remains with the horse's owner/trainer, however.

When selecting a compound food, always check the details on the bag. If a horse is receiving only the compound food as well as hay, there should be no need for supplements. However, if part of the concentrate ration is straight food, it may be necessary to work out if any supplement is needed to balance the ration. Some vitamins, such as vitamin E, lose their potency after a time and, if one is included in the compound mix, it is important to check the manufacturer's 'use-by' date.

SUPPLEMENTS

The horse owner is at present deluged by manufacturers advertising all sorts of mixtures to add to horses' and ponies' feeds.

If horses in medium work are being fed their concentrate ration as a compound food plus good quality roughage they should require nothing extra except possibly salt (NaCl).

Horses on home-mixed diets should have their rations checked. Many such diets are low in salt and calcium so competition horses should certainly have salt added to their rations and may need calcium, lysine and methionine. Vitamins A, D, E and folic acid may be necessary in the winter or where horses are permanently stabled.

However, vitamins and minerals cannot be considered in isolation. There are complicated interactions between them and between other nutrients, especially between vitamin D, calcium and phosphorus, and vitamin E and selenium, and the B vitamins and carbohydrates. The specific manipulation of vitamin levels in the diet should only be undertaken with professional guidance. Many of these interactions are not yet clearly understood or identified.

Supplements include:

- Minerals including trace elements.
- Vitamins.
- Herbs.
- Bacterial cultures (probiotics).
- Feed balancers.

Reputable manufacturers of compound food will incorporate the necessary minerals and vitamins in their products and provide details of their analysis. Horses living out on land known to be deficient in some minerals will need supplements unless bought in feeds contain the specific minerals.

Common Salt

Salt may be given by any of the following methods:

- Rock salt in the manger.
- Salt-lick on the wall.
- Added to the feed.
- Salt block in the field.

There are advantages and disadvantages to all methods.

Rock Salt in Manger

ADVANTAGES
- It is cheap to buy.
- It stops horses from bolting their feed.

DISADVANTAGES
- Horses may not gnaw at it every time they have a feed and so will not ingest sufficient salt.

- Metal feed pans become corroded.

Salt Lick on Wall

ADVANTAGE
- Salt is always available.

DISADVANTAGES
- Some horses do not use it.

- It is very messy on walls and tails.

- Empty holders can be physically dangerous.

- There is a danger of infection if horses with coughs and colds are moved from box to box.

Salt Added to Feed

ADVANTAGE
- The amount given can be monitored.

Salt Block in Field

ADVANTAGE
- It is readily available.

DISADVANTAGES
- Some horses may not use it.

- If placed on the ground it will eventually dissolve and kill the grass on which it lies.

- It needs a special non-metal holder, with drainage holes; or it can be hung on a fence, thereby limiting the grass damage.

Probiotics, Antibiotics and Feed Balancers

PROBIOTICS are supplements designed to promote the performance of the existing digestive micro-organisms and also to add to their number. They are now used, under guidance, to assist in the better use of the food fed to some horses.

Their usefulness and perhaps their risks are not yet fully evaluated. It is wiser not to consider them as a supplement but as a therapeutic or prophylactic medicine (e.g. to counteract the stressful effects of worming or travelling).

Similarly with ANTIBIOTICS, including those used against ringworm which are 'in feed' medications. Horses on antibiotics may have to be given extra vitamins. This is because oral antibiotics can kill the good 'bugs' in the gut. The horse will take some time to build up the flora. During this time some degree of indigestion may occur.

FEED BALANCERS (such as Blue Chip or Equilibra) are designed to be added to the chosen ration. They come as small cubes which contain a balance of vitamins and minerals, some probiotics, and perhaps some good quality protein.

CHAPTER

21 DECIDING ON A RATION

Art Versus Science

Although the science of horse feeding has increased with new-found knowledge, it is important not to forget the practical aspects.

Weighing or measuring horses can be very useful, but nothing quite takes the place of the person who, just by looking at his horse, can alter its feeding a little to keep it in top condition. This takes much practice. The changes in amount should still be done by weight.

It is useless saying that the horse must have X amounts of this and Y amounts of that if (a) the horse does not look or do well on it; or (b) he does not like the food. Two horses of the same type and size both eating identical rations may look completely different because one is a good converter of food and the other a bad one. This is where the practical horsemaster comes into his own.

It is almost impossible to give exact amounts of food for different horses doing different jobs. For example, it is possible that one three-day event horse may be fit to run for his life on 5kg (11lbs) of concentrates while another of the same size and type may need 9kg (20lbs). However, **as long as the horse is not losing condition it is better to under-feed than over-feed**.

Deciding what to feed can seem very complicated to the more inexperienced person. The simplest way of doing this is to decide on a maintenance ration by considering the horse's size, age and type. Then it is necessary to work out what energy ration the horse will need to carry out its task, either for work or growth or both. The tables on pages 160 and 162–7 give a guide, and the approximate amount of the concentrates the horse will need can be calculated from them. It is then necessary to decide on the type of concentrates, bearing in mind their energy and protein value.

A more detailed approach to rationing is given in Appendix 1.

Maintenance Rations

A maintenance ration is the food that is necessary to keep the horse that is not working in good condition. This does not cover broodmares and growing young stock as they need extra food, for late pregnancy and lactation in the former, and growth in the latter, all functions which are in fact work.

A maintenance ration may be made up from grass in the spring and summer, or from hay and sugar beet in the winter with some additional food as necessary.

Beware additional feeding of non-working horses at grass in some autumns and early winters. All factors must be taken into consideration.

CALCULATING RATIONS

The following considerations should be taken into account when deciding on a ration:

- Present state of condition and health.
- Size and type.
- Age.

- Weight.
- Amount of work.
- Temperament.
- Ability of rider.
- Weather (is horse out by day, day and night, or only for exercise?).
- Economy (cost-effectiveness).
- Availability of foodstuffs.
- Feed storage and handling facilities.

Condition

The condition of the horse is assessed by using a scoring system. This measures the weight distribution over the neck, back, ribs and quarters.

Fat should not be confused with muscle bulk, or vice versa. Often, apparent muscle bulk is, in fact, excessive fat which has 'hardened', such as on the crest of ponies or overweight stallions. (Note: When, standing relaxed, the horse's muscle, when palpated, feels soft and pliable.) Contrary to popular belief, fat will not turn into muscle with work.

Horses vary in their bulk of muscle, and the type of work they do can also influence their shape. For example, the muscle fibres in the horse trained for stamina and endurance, such as a three-day eventer or long-distance horse, are less 'bulky' than those of a sprinter, dressage horse or show jumper. The fit eventer will look lean and streamlined, whilst the show jumper will have a more rounded appearance.

The muscle groups in a fit horse are clearly delineated.

Excessive weight puts unnecessary strain on the heart, lungs and limbs, and can shorten the useful working life of the horse or pony.

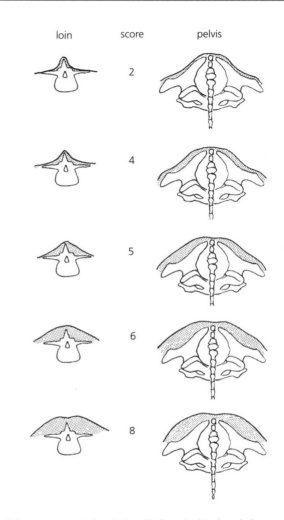

Condition scoring – judged visually from behind and above, and by palpation. Shaded areas are flesh masses, with variable thickness of fat between muscle and skin. In the horse, most 'points' except the buttocks (tuber ischii) are not covered by muscle. Fat is, however, deposited between the skin and the bony prominences (points) but never to any extent.

Condition Scoring

SCORE 1

Starvation level. Croup (sacral) and hip bones (pelvis) prominent and very sharp, especially over points. A marked hollow in front of the withers. A very deep temporal fossa and possibly a sunken eye. Spinous processes of the vertebrae are well defined. Rib cage prominent with all ribs showing from behind foreleg muscles. Skin tight.

SCORE 2

Ribs easily visible. Contour sunken either side of backbone. Spinous processes well defined. Quarters sunken, pelvis and croup points well defined. Deep depression under and either side of the tail. Bones palpable.

SCORE 3

Withers, croup and backbone processes and points still clearly defined. A little more muscle definition but still hollow in front of the wither. Slight cavity under tail.

SCORE 4

Front half of rib cage covered, back half still visible. Neck beginning to fill up in front of the wither. Spinous processes still palpable.

SCORE 5

Approaching normal for degree of fitness or rest. Withers, croup and hip bones palpable with pressure but some muscle definition developing.

SCORE 6

Normal. Firm, muscled neck. Ribs just covered but palpable. Haunch, croup bone and buttocks covered but easily felt. Muscles well defined.

SCORE 7

Beginning to carry too much weight. Slight crest development. Ribs well covered, requiring firm pressure to palpate. Pelvis and croup well covered.

SCORE 8

Fat. Definition of bones, except at points, lost. A hollow gutter from croup to tail. Neck becoming hard and cresty.

SCORE 9

Obese. Ribs, quarters and back buried in fat. Deep palpation necessary to feel croup and hip bones. Loaded shoulder fat; beginnings of a hollow from wither to croup.

SCORE 10

Very obese. Marked crest. Neck very wide and firm. Deep hollow from wither to tail. Back broad and flat. Huge pads of fat on shoulders and quarters. Pelvis and croup buried. Skin distended. Lumbar region 'raised'.

Size and Type

Size and type should be considered together, as the more 'common' type of animal, although heavier, will usually have a better food conversion rate than the more Thoroughbred type of horse.

Size (height and weight)

The height and weight of the horse can provide a guide to feeding. Determining the height is not a problem. To discover the weight the following methods can be used:

- Weighing the horse on a horse weighbridge.

- Driving the horsebox or trailer on to a public weighbridge with and without the horse.

- Using a specially calculated equine weigh tape (available from tack shops and feed merchants).

- Calculation using the following equation:

$$\text{Weight (kg)} = \frac{\text{girth}^2 \text{ (cm) x length (cm)}}{11900}$$

Girth is measured around the barrel immediately behind the elbow, after the horse has breathed out. Length is measured from the point of the shoulder to the point of the buttock.

Once the body weight of the horse has been determined, it is important to decide whether the animal is overweight, under weight or showing a lack or loss of muscle tissue, i.e. wasting.

As a rule, weight loss or gain relates to too much or too little body and subcutaneous fat. If wasting is obvious, or if an animal is plainly obese, weight as such is of little consequence. Over-feeding or under-feeding of the average-conditioned horse has no (immediate) effect on muscle mass.

It has been suggested that for the resting horse, i.e. one on a maintenance ration, 1½% of body weight is required in food; for light work, the horse requires 2% of his body weight in food, and for medium to strenuous work from 2½% to 3%. Young horses and pregnant mares in the last three months of pregnancy need between 2½% and 3% as dry matter weight. However, these rough guide percentages depend on the type of food supplied. Horses do vary in the use they make of their food according to their type and individual metabolism. There is also a minimum requirement of 1% of the body weight as long fibre for correct gut function.

Percentages of body weight as total dry food requirement

Height in hands	Girth in cm	Girth in inches	Body weight in kg	Body weight in lbs	2%/3% in kg	2%/3% in lbs
12hh	145cm	57ins	260kg	573 lbs	5.2–7.8	11–17
13hh	155cm	61ins	320kg	705 lbs	6.4–9.6	14–21
14hh	165cm	64ins	390kg	860 lbs	7.8–11.7	17–26
15hh	180cm	71ins	490kg	1080 lbs	9.8–14.7	22–32
16hh	195cm	76ins	580kg	1279 lbs	11.6–17.4	27–38
17hh	205cm	81ins	640kg	1411 lbs	12.8–19.2	28–42
18hh	215cm	85ins	710kg	1565 lbs	14.2–21.3	31–47

Age

The age of the horse will have a bearing on the rations chosen. Young horses who are working as well as growing will need more food to cover the extra requirements for carbohydrate and protein.

Older horses may also require extra rations, but this can be decided by assessing their condition. Very old horses whose teeth may be poor or who are living out and taking very little exercise, will require an easily masticated and digestible ration of high quality food. Old horses who are unable to graze properly or to keep themselves warm should never be left out in the winter. Keepers of these horses should appreciate that it is far kinder to have them painlessly destroyed than to allow them to linger on when they are not happy and cheerful.

Temperament

Placid, easy-going horses that 'take little out of themselves' will be easy to feed but must not be over-fed and allowed to get fat. They are likely to turn extra 'energy' food into fat rather than energy. However, it is important to differentiate between naturally 'laid-back' horses and those with an illness or suffering from a deficiency. Any alteration of temperament requires investigation.

Horses with a difficult, excitable temperament can be a problem to feed as they lose condition through excessive energy expenditure. Giving them more food to replace lost condition can often make them more excitable. They require plenty of roughage and are usually better on energy foods that do not contain too much cereal, e.g. sugar beet and lucerne.

The Weather

The weather must be considered when deciding on the amount of food that horses require. Food is used to maintain body temperature. In very cold weather horses will use more food to keep themselves warm. This is particularly so for clipped horses. If no extra food is given they will use stored fat and thus start to lose condition.

Under-fed horses, because of loss of insulating fat, will more readily lose body heat by radiation and convection and will, of course, feel cold. This results in shivering and further energy expenditure.

Food as such does not produce heat, but the act of eating does produce muscle activity heat. The cellular work necessary for digestion, absorption and metabolism and other vital activities are all heat-producing. It is the risk of loss of body fat in cold weather which necessitates extra winter feeding.

Economy (Cost-effectiveness)

Unless funds are unlimited, price must be considered when deciding what foodstuffs to buy. This is especially true for commercial establishments. However, it is a false economy to feed poor quality food, however cheap it is. It is cheaper in the long run to buy sensibly and so avoid ill-health costs. However, equivalent value foodstuffs should be costed and compared in price.

The length of time taken to make up complicated feeds should also be considered, especially in a large commercial yard. Moreover, mistakes in mixing and serving are more likely.

Availability of Constant Supplies of the Same Quality

This particularly applies to hay, and, to a lesser extent, cereals such as oats or barley. If storage is available, enough forage should be bought to last for several months. If not, arrangements for storage may be made with the supplier, but this, of course, will make the food more expensive.

If compound foods are bought it should be ensured that further supplies will be available when required.

Supplementing the Ration with Grazing

Some daily grazing will benefit the horse's health but will make the balancing of rations more difficult. The quality and quantity of grass will fluctuate with the seasons and this must be taken into consideration. The richness of spring grass can easily lead to feeding too much protein and carbohydrate, with too little long fibre. Conversely, winter grass has lots of long fibre but little in the way of carbohydrate and protein. The possibility of infestation by the various forms and degrees of parasitism must also be considered.

Ease of Handling

The considerations here are:

• The staff.
• The layout of the feed and hay shed.
• Access.

Other points might include:

• Size of bales.
• Size and weight of bagged concentrates.

• Movement of loose food (such as oats).
• Time taken to bring food to feed shed.
• Time taken to make up feeds.

HOW TO CHOOSE WHAT TO FEED

The main points to consider when choosing what to feed are:

• Energy.
• Protein.
• Fibre.
• Calcium/phosphorus ratio.

Measuring the Value of Foodstuffs

Food should always be fed by weight. The weight of a 'scoop' of oats or cubes should be known. As an example, a scoop of oats will weigh less than a scoop of cubes. Hay or vacuum-packed grass should be weighed before being fed unless it is being fed ad lib, i.e. to appetite.

The nutritional value of the food should also be known: straight foodstuffs such as oats, barley or hay can vary enormously in quality and for competition horses it is worth having these analysed.

The value of the food is expressed as:

• The digestible energy (DE) of the food.

• The crude protein.

• The Ca/P ratio.

• The vitamin and mineral content.

Concentrates made up by food manufacturers have their values listed on the label and these should not alter, although the actual constituents may vary in amount owing to the economics of the market.

Digestible Energy

Digestible energy (DE) is computed in megajoules (MJ). (A joule is a measurement of energy. Energy is expressed as the number of joules in a kilogram of food.

Nutritive value of some common foods

	Crude Protein %	Digestible Energy MJ/kg
Cubes		
Horse and pony	10	10
Racehorse	13	13
Cereals		
Oats	12	14
Barley	11	15
Maize	9	17
Protein foods		
Soya meal	49	16
Dried milk	36	17
Linseed	26	24
Field beans	26	16
Grass meal	12-18	13
Intermediate foods		
Wheat bran	16	12
Sugar beet pulp	9	14
Forage		
Good grass hay	9	11
Average grass hay	8	9
Poor grass hay	4	8
Grass silage	13-17	12
Haylage	16	11-12
Hydroponic barley 'grass'	16	17

NB: This table is a guide – in practice a lot of variation will be found.

Because there are so many joules in a kilogram they are measured in megajoules.)

For example, if oats are to be measured they are likely to have digestible energy levels of around 14 megajoules per kilogram; this is written as DE = 14MJ/kg. (See the table above.)

As a rough guide for maintenance, a horse or pony needs 18MJ of digestible energy plus an extra 1MJ for each 10kg of body weight. For work, add 1–8MJ for each 50kg of body weight, according to how hard the horse is working. Add extra MJ for lactation, pregnancy and growth, as shown later. Thus a 500kg horse doing 1 hour of work per day (including cantering and jumping) requires:

$$18MJ + \frac{500MJ}{10} + \frac{4 \times 500MJ}{50} = 108MJ \text{ DE/day}$$

i.e. maintenance + work   = energy required

Protein

Suggested percentage of protein in the total ration:

• Light work: 7.5–8.5%.

• Medium work: 7.5–8.5%.

• Hard work: 8.5–10.0%.

• Strenuous work: 8.5–10.0%.

Adult horses not above 10% in total ration including roughage.

Suggested protein percentages required for broodmares and young stock:

• Pregnant mares: 8–10%.

• Mares in last three months of pregnancy: 11–13%.

• Lactating mares: 14% gradually decreasing to 12%.

• Foals not receiving enough milk or spring grass: 16–18% in a compound ration. On good grass with sufficient milk from their dams no extra food is necessary. (Thoroughbreds will require special consideration, especially if bred for flat racing.)

• Weaned foals 6 months old: 14.5–16%.

• Yearlings 12–18 months old: 12–14%.

• Two-year-olds: 10–12%.

• Three-year-olds: 8.5–10%.

NB: There is some evidence that over feeding foals and yearlings may cause epiphysitis and/or contracted tendons. The significance of excess protein, excess energy and/or imbalanced minerals is not yet resolved but must be important as distinct from just one nutrient being blamed. See also page 143.

Work

The work the horse is required to do, the rider he has to carry and his own temperament all have to be considered under this heading.

Work per day

Work can be defined, on a scale of 1 to 8, as follows:

Scale points for levels of work

Category	Scale Point	Activity
Light work	1	1 hour walking.
	2	Walking and trotting.
Medium work	3	Some cantering.
	4	Jumping, schooling, dressage, driving.
Hard work	5	Eventing, cross-country, driving, hunting, endurance riding.
	6	Hunting 2 days a week, three-day eventing.
Fast or energetic work	7	Racing.
	8	Racing.

Feed requirements for work

The following table is an approximate guide and must be altered to suit the horse. However, the 'bulk food' ration should never be less than 30% of the total ration. This ratio should only be sustained for short periods around a specific time of competition or strenuous work. If other types of roughage are used in the place of hay, the concentrates may well have to be decreased.

A horse at rest will require only a maintenance ration but if the horse is working, not only is it necessary to feed him so that he has enough energy for the job, but also so that he remains sensible enough to carry it out without harming himself or his rider.

A guide to feed requirement for work

Work category	% Hay	% Concentrates
Maintenance	100	0
Slow light work	85	15
Light work	80	20
Medium slow work	70	30
Medium work	60	40
Hard work	40	60
Race training	30	70

An Example of Practical Application in Deciding a Ration

Size, type and temperament

Addington Dazzler is a 15.3hh, seven-eighths-bred horse with a good temperament. He weighs 500kg (1100lbs).

Age

He is 8 years old, so has stopped growing.

Health

He is in good condition, has been blood-tested, is regularly wormed and has his teeth done.

Work

He is required to take part in local riding-club competitions at weekends (these do not include fast work). He goes for a short hack daily. His rider is of average ability.

Economy

He must be fed well but not extravagantly.

Ease of handling foodstuffs

His one-horse owner, who has a job, finds ready-mixed compound food easy to deal with.

Availability of constant supplies

This has been checked with the feed merchant. Enough hay to last the winter has been bought.

Does part of the ration come from grazing?

Addington Dazzler is occasionally turned out for a short time when he is not ridden. It is autumn and as there is little grass at this time of the year, it is not added to the diet sheet.

Weather

As it is autumn and it is not yet cold, Addington Dazzler is trace-clipped and wearing a rug. The weather at the moment will not affect the ration.

Suggested ratio of hay to concentrates during fittening work

Week 0	The horse is on maintenance alone	100% hay or grass
Week 1	Walking exercise 1 hour	Hay to concentrate 90%:10%
Week 2	Walking exercise about 1½ hours	Hay to concentrate 85%:15% to 80%:20%
Week 4	1½ hours walking and trotting	Hay to concentrate 80%:20% to 70%:30%
Week 6	1½ hours walking, trotting and cantering	Hay to concentrate 75%:25% to 60%:40%
Week 8	2 hours walking and cantering	Hay to concentrate 70%:30% to 50%:50%
Week 10	As before but including some fast work	Hay to concentrate 60%:40% to 40%:60%
Week 12	Starting competitions	Hay to concentrate 50%:50% to 35%:65%

Hay has been used in these tables as the 'roughage' ration. If other forages, such as vacuum-packed grass or lucerne, are used the protein level will need to be checked and the ratio altered to balance the ration.

Having considered all the above points the ration is worked out as follows:

- Because he weighs 500kg (1100lbs) it is estimated that he needs about 13kg (29lbs) of food daily (2½% of his body weight).

- Because of the work he is doing he probably requires a concentrate:roughage ratio of about 25%:75%, or roughly a quarter of his ration in energy food made up with threequarters of hay.

- His suggested ration might be:

  2.6kg (6lbs) Horse and Pony cubes at 10% protein

  0.5kg (1lb) sugar beet weighed before soaking at 9% protein

  9.5kg (22lbs) quality hay at about 8% protein

- As a compound food has been used no extra supplementation should be necessary.

- According to condition, behaviour and the work being done the ratio of concentrates to roughage may be altered.

(For the sake of simplifying calculations in this chapter, figures have been 'rounded up'.)

RATIONS FOR DIFFERENT TYPES OF HORSE

Horses Living Out and Not Working

A horse who lives out and is not working requires maintenance rations, i.e. it must be given enough food to maintain it in good condition.

Spring and Summer Maintenance

From May to October grass is an adequate ration for all horses and ponies if in sufficient quantity and quality. Native ponies may need their grazing restricted if the grass is very good, owing to the danger of laminitis. All stock benefit from additional long fibre when on lush grass.

If the grazing is limited, or of poor quality, supplementary feeding will be necessary.

Winter Maintenance

Horses and Thoroughbred type ponies will require supplementary feeding. The grass will have little feed value and they will need food to keep themselves warm. Oats, sugar beet with a mineral/vitamin supplement and hay ad lib is a suitable ration. A ratio of between 10% to 20% concentrate to 90% or 80%, hay should be adequate, but a careful watch should be kept on the animal's condition. Winter coats can easily disguise any loss. If an 'oat balancer' (a concentrate mix specially made to feed with oats) is fed with the oats no mineral supplementation should be required as this will contain the necessary minerals.

Native ponies grazing on a sufficiently large area may not need any extra food but those on restricted or possibly horse-sick grazing will need hay. In years when hay is difficult to obtain, clean oat straw and sugar beet is an acceptable alternative.

Competition Horses

The horse in hard competition work will require more energy food than the same horse in medium work. However, it should be appreciated that each horse will have a level of required energy food and feeding above that level will not produce extra fitness. Food does not create fitness.

The fittening chart on page 164 should be studied, but note that fitness can go on increasing after the level of food remains the same.

It used to be thought that horses doing strenuous work required considerably more protein than other horses. This has been found not to be so. The level of protein in the whole diet should not add up to more than 10%. (National Hunt and three-day event horses may, as individuals, benefit from a determined amount of extra protein such as from soya.)

The rations for competition horses doing fast or energetic work are all very similar although the type of work is often very different.

Blood-testing at the beginning of the fittening programme and then again when the horse is fit, is a sensible investment. Some successful racehorse trainers test as often as once weekly.

Competition horses on a restricted hay ration become less bored and digest their food better if their bulk food is chaffed.

Prohibited Substances

Racehorses in Britain and horses running under FEI Rules anywhere in the world may be tested for prohibited substances. This also applies to some national competitions. It is therefore very important that these substances do not get into horses' food by mistake. Contamination usually occurs during transit or at the milling plant. Cocoa is the most likely contaminant.

Event Horses

Event horses should be fed so as to maintain trained muscle bulk without becoming fat, yet so that they have enough energy to do their work in a sensible, economical way. Feeding them so they are jumping out of their skins is not only a waste of food but may cause them to injure themselves or their riders.

Keeping them on a very low hay ration can cause them to run up too light and lead to digestive problems.

Event horses are subject to doping rules, so if they are fed cubes or a coarse mix it is important that the supplier is prepared to state that these products are free of prohibited substances.

Event horses need very little extra protein. The normal rules for feeding should apply and the total protein ration should not go above 10%.

Advanced, energetic competition horses should have their hay/roughage analysed and properly balanced rations made up in conjunction with quoted or determined analyses for concentrates and supplements.

Novice event horses are often given too much concentrate food. They are not being asked to do a

Two possible example rations for 16.2 hh event horses weighing 600kg

Example 1	Example 2
Oats 4kg (9lbs)	Event cubes 6.5kg (14½lbs)
Sugar beet 1kg (2¼lbs)	–
–	Hay 6.5kg (14½lbs)
Vacuum-packed grass 6.5kg (14½lbs)	–
Grazing 1hr	Grazing 1hr
Supplements including calcium	–
Salt (amount depending on supplement content)	Salt (1–2 tablespoonful, according to sweating)

(The amounts should be varied according to fitness, condition and behaviour)

great deal and should be fed according to the rules based on the recommendations given earlier. Most novices do not need more than 2.5–4kg (6–9lbs) of event cubes of, say, 11% protein, plus hay. For some horses this may have to be stepped up, or for others decreased.

Advanced three-day event horses probably do not need more than 4.5–6.5kg (10–14½lbs) of concentrates. Several Badminton winners have had less than this. Too much concentrate food can result in toxicity problems and the horse refusing his food. Feeding per se will not produce fitness: this must be built up through work. The energy spent must be replaced by appropriate feeding.

Horses must be watched to see they are not losing condition and the ration adjusted accordingly.

Roughage/concentrate ratio – event horses

	Roughage	Concentrate
Novice eventer	75% to 50%	25% to 50%
Advanced 3-day eventer	50% to 40%	50% to 60%

The roughage part of the ration should not be less than 40%

Endurance Horses

Endurance horses need more energy than any other type of competition horse. Although they are not working at high speeds they have very long distances to cover at a consistent speed.

These horses must not be allowed to become fat. They should be very well muscled, but without the round outline of the dressage horse or show jumper, and should carry no excess weight.

They require easily digestible high-energy food with not too much bulk food. The addition of oil (such as corn oil) to their daily ration may be of benefit. Usually 100 ml (2–3 fl oz) is fed but much more has been given successfully.

Roughage/concentrate ratio – endurance horses

60% to 40% Roughage: 40% to 60% Concentrates

If a grain diet is fed a broad-based mineral supplement should be given. This should not be given with compound foods as they already have added vitamins and minerals. Salt (100 mg) given daily is important. Endurance horses lose water and electrolytes since they sweat for much longer (distance/time). NaCl (salt) are two of the electrolytes lost. More advanced horses will benefit from a more comprehensive mixture of minerals (precursors of electrolytes) but specialist advice should be taken.

The following rations have been recommended by successful long-distance riders.

Endurance horse working 2 hours daily but not yet competition fit

Bulk (hay etc.) ad lib

Oats rolled
Barley micronised or flaked — 2.5–3.5kg (6–8lbs) daily
Sugar beet

Possible additions — 25–50ml (1–2 fl oz) corn oil
25g (1oz) salt
mineral supplement

NB: Lucerne pellets and straw are better than poor hay

Endurance horse competing

Between 3.5–6.5 kg (8–14½lbs) concentrates and roughage to balance, possibly made up as follows:

6.5kg (14½lbs) best quality hay

2kg (4½ lbs) oats

1.5kg (3½ lbs) barley

1kg (2lbs) hunter cubes

Sugar beet, soaked

75ml (3 fl oz) corn oil

Mineral supplement

Dressage Horses

The dressage horse needs to produce slow power and does not need speed (it must be fit for this job but must not carry too much fat). It must be sensible and obedient, yet have enough energy for the job. The more advanced the training the more energy the horse will require. Dressage horses, unlike hunters, have to remain fit throughout the whole season, so a careful check must be kept on their condition and behaviour.

Below is a typical home-mix diet for a 16.1hh, six-year-old, threequarter Thoroughbred/quarter British warmblood of Advanced level, being worked hard for approximately an hour a day by an international rider.

16.1hh Advanced dressage horse

7kg (16lbs) hay

1kg (2¼ lbs) rolled oats

1kg (2¼ lbs) micronised barley

1kg (2¼ lbs) coarse mix 12% protein*

1kg (2¼ lbs) cubes 12% protein*

0.5kg (1lb) sugar beet

Broad-based supplement with additional salt and vitamin E

1 hour turned out in the field

Total: 11.5kg (26lbs)

Roughage/concentrate ratio approximately 60%:40%

* Guaranteed free from prohibited substances

Point-to-Point Horses

Getting the point-to-point horse fit for racing is rather different from getting other horses fit because the point-to-pointer will have been hunted and will be fairly fit but may well be rather lean. After hunting he may benefit from being let down slightly for one to two weeks with slow exercise. During this time he should be on reduced rations but not if it is judged that he requires to put on some fat before serious fittening starts. (See also Chapter 40.)

Because he is now partly fit his feeding can start as at the sixth week with a ratio of 50% roughage:50% concentrate.

The traditional method of feeding point-to-point

horses was good quality oats and hay. This is still practised with much success. However, it is suggested that if this diet is chosen an oat balancer and some sugar beet should be added. Salt and a general-purpose supplement will help to make up for any deficiencies in the diet.

If compound foods are preferred it is important that they are free of banned substances.

By the time he starts racing the horse should be on a ratio of 40% roughage:60% concentrates, or even 35%:65% briefly around the race date.

Show Jumpers

Show jumpers are required to make a considerable muscular effort for a short time. They need to stay fit for a good length of time as the season is now very long. When horses are travelling for so much time it is sensible to feed compound foods. This will mean that it is necessary to carry only one kind of concentrate food and hay.

It is important that feeding times are carefully adjusted to fit in with jumping times.

Polo Ponies

Polo ponies have to work very hard and fast for short intervals. They need to be muscularly very fit and need high energy food to enable them to achieve this.

Because of the usual time of play, the afternoon, the feeds on playing days need adjusting. Polo ponies should be fed four times daily if possible – morning, early lunch, tea and late night.

On a non-playing day the larger feeds should be lunchtime and late night. On a playing day only a very small feed should be given at lunchtime. Hay should be given first thing in the morning and taken away at least three hours before playing.

Any pony who stops eating should have his mouth carefully examined. Owing to the severe bitting used on some ponies there may be damage to the mouth which will make it reluctant to eat.

Usually cubes or coarse mix are more suitable than a straight cereal ration as although polo ponies need to be very fit they must be sensible.

Hunters

Hunters should be fed the same sort of rations as event horses but with more bulk food. They require between 3.5–5.5kg (8–12lbs) of concentrates and good hay. Thoroughbred horses need hay or bulk food ad lib but the usual rules still apply.

They must be fed so that they are well behaved, can gallop for short distances, and remain out of their stables all day, probably for three days in every fortnight from November until March. Once fit it is not easy to keep them looking and feeling well throughout the season. Their rations will need careful adjustment so that they do not lose condition. A late night feed, if it can be given regularly, can be of great benefit.

On rest days after hunting, their concentrate ration should be reduced and their roughage ration increased. They should be walked out and, if possible, grazed in hand.

Traditionally a bran mash was given after hunting, but nutritionists now consider this to be contra-indicated. (See bran mashes, page 178.)

Riding School Horses

In this case economy has to be a prime consideration. It is essential to shop around and compare the prices and contracts that are available from feed merchants and farmers. Large discounts are often offered for yearly contracts. However, low cost must not be the only criterion. The horses and ponies must look well, their manners must be impeccable and they must stay healthy and content.

Good, clean, affordable hay is the basis for feeding riding school horses; good hay will lower the necessity for large quantities of expensive concentrates. Meadow hay is usually cheaper than seed hay and if well saved is just as nutritious. In years when good hay is scarce and very expensive, alfalfa pellets fed with clean oat straw make an acceptable alternative.

Most commercial riding schools have found that horse and pony cubes provide the best type of energy food. These cubes are low in protein and high in fibre and are ideal for animals doing 2–3 hours of slow work daily. They will already contain the necessary supplements. Soaked sugar beet, although sometimes difficult to manage in large quantities and rather messy, is a good non-heating addition.

It is impossible to give amounts of food without knowing the weight, size, type and the work the horses are going to do. If students are being trained it will be necessary for them to experience the use of some other kinds of foodstuffs – possibly these could be used for special liveries or any competition horses at the school.

The 'One-horse Owner', Non-competitive Horse

For the one-horse owner whose animal is not going to do anything very strenuous it is sensible to choose a reputable manufacturer and feed their compound food, either as cubes or mixture. In this way all the necessary nutrients will be automatically included in the diet. Usually manufacturers are apt to be too generous in the quantities of food that they suggest for the various sizes and types of animals, so, as always, it is necessary to consider the condition of the horse and its behaviour.

Between 70% and 80% roughage should be fed, as little energetic work is being done.

Children's Ponies

It is of prime importance that children's ponies are fed so that they are quiet and sensible. They should look well, but their manners are paramount.

They can be divided into groups and then divided again as to the work they are doing:

GROUP 1 – The small woolly native pony, such as the Welsh or Dartmoor types.

GROUP 2 – The larger native pony, such as Highland, Dales, Fells and Welsh Cobs.

GROUP 3 – The more finely bred pony, such as Connemara, Arab or Welsh riding ponies.

GROUP 4 – The Thoroughbred or nearly Thoroughbred pony.

All these groups blend into one another but it should be remembered that Groups 1 and 2, the native ponies, are much better converters of food than their aristocratic cousins. If being fed with the pony's weight as a guide, Groups 1 and 2 will probably only require 2% of body weight. The ever-present danger of laminitis in over-fat ponies in Group 1 must be kept constantly in mind.

Groups 3 and 4 will need about 2½%, of body weight. The roughage:concentrate ratio will depend on the other factors already discussed on page 163 but particularly on the work they are doing. This can be divided into:

- **Beginners' ponies** – Whatever their size, as long as they have sufficient roughage, they usually require little or no concentrate. If it is necessary to give them a tit-bit to catch them or some food in the stable so that they are anxious to come in, this should be a small handful of the lowest possible energy food. If they start to lose condition, the roughage should be increased or some sugar beet and dried grass or alfalfa added. The last should not make them excitable and will help to put condition on again.

- **Family ponies** – The energy food required will depend on the amount and kind of work being done. Their concentrates will probably be between 5% and 20% of the total ration. A compound mix of about 10% protein is most useful.

- **Riding-school ponies** – Those working 2–3 hours daily may need between 10% and 20% of the total ration fed as concentrates. Their concentrates are likely to be whatever is fed to the other school horses.

- **Hunting ponies** – Those living in at night and hunting every Saturday may need between 15% and 30% of their total ration as concentrates. A compound mix is better for them than a ration based on oats.

- **Competition ponies** – These may vary from top-class JA jumping ponies to show ponies in lead-rein classes and their rations must depend on the type of competition for which they are used.

It must be remembered that cereals such as oats have a hotting-up effect. A horse and pony type of mixture is a safe food. It will contain a reasonable amount of fibre and have all the necessary vitamins and minerals added. Soaked sugar beet pulp is a good source of digestible energy and does not have the same hotting-up effect as the cereals.

Mares, Foals and Young Stock

Mares
Although pregnant and nursing mares should be well fed with good quality, easily digestible food they should not be over-fed. The size of the foal is determined by the genes from the sire and dam and not by the amount of food the mare is given. A higher protein ration should be given in the last three months before foaling (especially if foaling early) and while the foal is on the mare. If the grass is very good this extra protein may not be necessary. During the last three months feed a specialist mix or compound developed specifically for 'in foal' mares. There are many good products now available.

Foals
The feeding of the foal should depend on the response of the foal to the dam's milk. Some mares give a large quantity and others very little. In the latter case the foal should start eating concentrates as soon as possible. Early weaning may cause a loss of condition so foals who are to be weaned early should be eating well and will have to have their energy and protein ration increased when they are taken away from the mare. Foals who are weaned at five to six months should need little extra if they are eating concentrates and grazing.

Mares who are not in foal again may be allowed to run on with their foals. As the mares dry off they will wean the foal themselves when they may be separated with little trauma.

A balanced diet must be fed to growing animals if their full growth potential is to be reached whilst achieving healthy bones and teeth. Incorrect or over-feeding of foals may cause epiphysitis or contracted tendons. A good appropriate pre-mixed feed, specially prepared for young stock from a major horse feed manufacturer and fed at recommended levels, is really the most efficient way to feed foals and young stock. Foals cannot digest whole oats and need quality foods suitable for their limited digestive ability. Due to the immature nature of the foal's hind gut, proteins produced by the gut flora in the adult are not available to the youngster.

Young Stock
Young stock will require supplementary feeding from August right through to the end of April, and possibly longer if there is little grass. They need extra protein for growth and carbohydrates for maintenance. They should be fed so that they always look well covered but not too fat. Too much protein may cause lameness problems. Their diet must be well balanced with sufficient minerals, especially calcium, to ensure steady healthy growth.

Young stock, particularly weaned foals and yearlings who are living out in the winter, will require extra feeding to make up for the loss of condition through adverse weather conditions. Given sheltered conditions and good grazing they can be left out until the end of December and then brought in at night until the weather starts to improve in the spring. However, even Thoroughbreds, provided they have enough proper food and available shelter, will do perfectly well kept out all the time. The disadvantage is that they do not have daily handling and may object strongly when they are eventually brought in. Whether they are in at night or out all the time, good hay, sugar beet, stud cubes of about 14% protein, plus a good supplement will provide a good ration.

Growing horses require a higher level of protein than mature horses and the younger they are the more they will need.

Stallions

Stallions may be divided into two groups:

- Those used only as stud animals.
- Those that are ridden or competition animals as well as used at stud.

Stallions Doing Stud Work Only

Stallions should be on a rising plane of nutrition as the stud season (spring) approaches. During the season and depending on the number of mares they are to cover, they should have high-energy food but must not be allowed to become fat. When the season is over they should be let down gradually and returned to a maintenance ration.

Stallions Being Ridden and Used at Stud

Sometimes the competition stallion is retired to stud for the season and afterwards continues with his ridden work. In this case, depending on the work that he has been doing there is usually no necessity to change the rations. However, if the ratio of roughage to concentrates is altered the foods should remain the same. For the stallion who continues to do his ridden work as well as his stud work, his concentrate ration will have to be increased. In all cases, the condition and behaviour of the stallion will determine the changes that should be made.

22 FEEDING PROBLEMS AND NON-ROUTINE FEEDING

Sudden Refusal of Food

If a horse who usually eats normally refuses to eat, it may be a sign that something is seriously wrong with the horse. Should refusal persist for more than two feeds, especially if hay is also refused, the veterinary surgeon must be called. If other signs of illness appear, this professional help should not be delayed.

However, there are other reasons such as:

- Stress – horse usually shows little desire for food and won't eat.

- Over-feeding – horse usually shows little desire for food and won't eat.

- Lack of clean, fresh water over several days – horse usually shows little desire for food and won't eat.

- Contaminated food – horse will usually show a desire to eat but appear fussy.

- Dirty feeding utensils – horse will usually show a desire to eat but appear fussy.

- Acute tooth problem, sore mouth, tongue etc. – horse will usually show a desire to eat but appear fussy.

In the absence of clinical illness check on the above.

'Bad Doer' (Thin Horse)

A bad doer is the horse that eats all he is given but does not look well. The more common reasons are:

- Worms.
- Sharp teeth.
- Lack of adequate food.
- Poor quality food, especially hay.
- Unbalanced diet.
- Poor metabolism, often caused by being incorrectly fed or starved in the past.
- Over-stress.

The bad doer should be checked by the veterinary surgeon.

There are serious reasons for horses being thin or losing weight. They include:

- Heart abnormalities.
- Chronic pain.
- Anaemia.
- Ragwort poisoning and other liver pathologies.
- Cancers.
- Kidney disease.
- Worms (internal parasites). See also Chapter 15 and consult your veterinary surgeon.

The Sick Horse

It is essential that veterinary advice is taken when there is doubt about diagnosis and especially treatment.

Illness can be acute or chronic. The acute form could be colic, sudden viral infection or something resulting in a high temperature. The chronic form could be liver damage, anaemia, or cancer of some part of the digestive system. Some of these can be diagnosed by blood-testing.

The veterinary surgeon will advise on a ration in cases of acute illness.

Recent work suggests that feeding certain types of fats may stimulate healing mechanisms.

For ill and convalescent horses, feeding little and often is of prime importance. Any food not eaten within 10 minutes must be removed. Containers must be washed with plain water between each feed. Many sick horses prefer to eat from the ground.

They require nourishing, easily digestible food and should not have large quantities of cereal-type food. Anything that they are prepared to eat is generally the answer, and molasses is often of assistance to tempt them. Some freshly cut grass, if it is available, is usually acceptable and many horses enjoy dandelions, which are full of minerals. Chopped dried lucerne is often appreciated and is a valuable food. A mash may be made of soaked, dried grass cubes and sugar beet.

Enforced Rest

For fit horses who are going to be off work for perhaps only a few days, reduce the concentrates the first day by 75–80%. Give more roughage and cut out any oats after three days, and adjust down to maintenance rations.

Horses who have to be kept stabled with no exercise for some time, perhaps from lameness, should be given a maintenance ration with no additional energy food. They must have all the necessary minerals and vitamins to aid recovery, and the diet must be sufficiently laxative. Hay or its equivalent plus sugar beet and some additives should be sufficient. A slight loss of condition is of less importance at this time than the maintenance of a calm attitude and prevention of conditions such as azoturia.

Stress

Excessive stress may result in loss of condition. The causes of stress can be:

- Strange stable.
- Disturbance in yard.
- Poor handling.
- Travelling.
- Pain.

Discover and minimise the cause as far as possible. Give most of the food at night. If this is only a short-term problem, do not worry; although the horse may run up a bit light he will not be harmed by not eating very much for two or three days.

The horse suffering from long-term stress may show this by weaving, box walking or other stable vices. He needs to be kept in a quiet box and have a strict work and feed routine. A slightly larger, late-night feed, that he can digest quietly, may be of help. The more he can be turned out, even if the grass is poor, the better.

Travelling

Water and food should always be carried and enough time should be allowed for stopping to water and feed the horses en route. They should be given time to finish their feeds before proceeding. Some horses will not eat when on the move and there is a risk of choking should they do so.

When travelling very long distances, such as across Europe, care must be taken during the journey that the food is easily digestible and is not likely to cause metabolic upsets. The quantity of concentrates should be halved and the roughage increased. Extra electrolytes may be necessary. It is essential when travelling abroad to find out what feed is and is not allowed into the countries that the horse is travelling through. (See also Dehydration, Chapter 17.)

PROBLEMS AND AILMENTS CONNECTED WITH DIET

There are several ailments that may be directly caused by incorrect feeding and others on which incorrect feeding may have a bearing. They include:

- Under-feeding.
- Over-feeding.
- Colic.
- Diarrhoea.
- Azoturia.
- COPD.
- Electrolyte losses.
- Anaemia.
- Choking.
- Lameness, including laminitis.

Under-feeding can cause:
- Starvation – resulting in death.
- Chronic malnutrition.
- Weakness and inability to work.
- Deformity of bones in young horses. Stunted growth in young horses.

Over-feeding can cause:
- Bone disorders in young horses – mainly epiphysitis. (See feeding young horses, page 170.) Contracted tendons in young horses.
- Reduction of fertility; problems in foaling mares.
- Problems in strenuous exercise owing to carrying excess weight.
- Lameness from added strain on the legs and feet.
- Inability to carry out proper work.

The over-feeding of concentrates is particularly common with event horses, who may suddenly go off their food. The answer is to cut the concentrates by half for one day and by another half the next day and then gradually alter the diet so that the quantity and balance are correct.

Colic

The word 'colic' simply means abdominal pain of some sort. It may be severe, sudden or gradual in onset, or of a low level of pain. There are many different kinds of colic and many causes, including those listed below:

- Incorrect feeding, e.g. wrong times, variations in type, wrong amounts.

- Wind sucking, which may cause indigestion.
- Mouldy food.
- Poor teeth.
- Too large feeds.

It is now accepted that feeding mistakes are less commonly the cause of colic but the risk should not be ignored. A sudden change to cereals will readily produce large gut acidosis producing electrolyte imbalance, bacterial proliferation and colic as well as laminitis.

A sudden change to long fibre is more likely to induce colonic impaction.

Diarrhoea

Can be caused by:

- Worms.
- Stress.
- Oral antibiotics.
- 'Upset' micro-organisms in the gut, due to sudden change to cereal or lush grass feeding.

In all cases of diarrhoea there will be dehydration, particularly in acute cases. Fluids containing electrolytes may be required. Veterinary advice should be sought; this is particularly important in the case of foals, and in adults if it persists for more than 24 hours, or less if general signs of illness are present.

Azoturia, Monday Morning Disease, Set Fast, Tying Up

These are all names for more or less severe symptoms of the same process. Equine rhabdomyolysis is the name used to encompass the whole syndrome. This may range from muscular cramp to actual degenerative changes in the muscles.

It is usually associated with exercise after a short period of rest on full rations, thus the name Monday morning disease. Symptoms range from stiffness and sore muscles to a refusal to move, severe sweating, high pulse and respiration rates and obvious pain.

The horse should not be moved and veterinary assistance should be sought immediately. Permanent damage to muscles can occur if the horse is moved.

Mares are said to be more prone to azoturia than are stallions or geldings.

Horses should always have their concentrate ration reduced on their rest days and should be warmed up slowly before work and allowed to unwind gradually after fast work.

The following have all been suggested as possible predisposing factors:

- Faulty metabolism.
- Vitamin and/or mineral deficiencies.
- Hormone abnormalities.

Some success has been achieved in the treatment of constantly recurring rhabdomyolysis by feeding extra calcium and/or salt.

Feeding a Horse with Wind Problems (COPD)

This horse will almost certainly be allergic to the fungal spores in hay and straw and also to dust. As far as possible he should live in a dust-free environment, be bedded on shavings, sawdust or paper. His food should be chosen or treated so that he does not inhale any mould spores or dust. The bulk food should be either hay that has been immersed in water for 10–20 minutes, or vacuum-packed grass. He should be fed on cubes and sugar beet and if other food is fed it should be damped. Good ventilation is of prime importance (see also pages 335 and 342). He should be taken outside to be groomed and for mucking out. The more he can be outside, the better he will be. The addition of garlic to the ration has proved beneficial in some cases.

When travelling, straw should not be used on the floor of the vehicle and the haynets of any travelling companions should also have been soaked.

Lameness

Lameness which can be caused by incorrect feeding includes:

- Laminitis.
- Epiphysitis.
- Contracted tendons.
- Poor hoof formation.
- Lymphangitis.

23 PREPARING FOOD

The means by which horses are fed are important. Food is expensive and it should be carefully stored and prepared so that none is wasted.

HAY

Hay can be fed:
- Loose on the ground.
- In haynets.
- In hay mangers.
- In hay racks.

Loose

ADVANTAGES
- Natural for the horse to eat off ground.
- Labour-saving.
- Safe.

DISADVANTAGES
- Difficult to soak but can be tipped out of soaking net.
- Difficult to weigh but slices of a known weight bale can be measured off reasonably accurately.
- Wasteful.
- Untidy in yard.

Haynets

ADVANTAGES
- Economical.
- Can be prepared in advance.
- Easy to weigh.
- Tidy to handle.
- Easy to soak.

DISADVANTAGES
- Time-consuming in labour, and if soaked, difficult to lift unless a pulley is fitted.
- Can be dangerous, especially in young-stock boxes.
- Expensive to buy – poor 'life'.
- Unnatural eating position.

Hay Mangers
Have the same advantages and disadvantages as loose hay on the floor. They may be less wasteful.

Hay Racks
Due to the possible danger of hay seeds entering the horse's eye there are reservations about the use of hay racks, particularly those which are high.

Soaking Hay

Mould Spores
Certain horses are allergic to the mould spores found in hay, straw and chaff. These spores are a danger to these horses, not because they are eaten and swallowed, but because they are inhaled. All hay and straw contains some spores, but visibly mouldy products contain considerably more. Horses fed and/or bedded on mouldy material run a greater risk of inhaling more spores than the natural mechanism of their respiratory system can

cope with. Sooner or later some spores reach the bottom of the airways (alveoli). Although the healthy horse may be able to cope with this, in the allergy-prone horse a disease state is stimulated.

It is advised that such horses should be fed soaked hay. Soaking hay keeps the spores' surface tension high and so prevents them from escaping from the eaten fibre into the airway. They are masticated and swallowed and do not enter the lungs. Another school of thought suggests that soaked spores swell to become too big to gain access to the sensitive alveolar area. When exhaled they are wiped off on to the throat and swallowed. Of course, the welfare of a known COPD horse should be discussed with a veterinary surgeon.

Soaked Hay

The prevalence of allergic coughing has encouraged the feeding of soaked hay. This is hay soaked in water for about 10–20 minutes before being fed to the horse. The easiest method is to put the hay into a net, which is then immersed in a tank of water. The haynet is then hung up and allowed to drip before it is given to the horse. It should not be allowed to dry right out. Another advantage of soaking is that it washes out much of the dust, which in many horses causes 'hay cough'. It must not be soaked for longer than 30 minutes maximum or valuable soluble nutrients will be lost to the water in which the hay is immersed.

In a large establishment where all the hay is soaked (a sensible precaution) a long tank not more than 1–1.25m (3–4ft) high is essential. The nets are lifted from the water and allowed to drain before being hung up. They are very heavy and a pulley system saves the staff considerable strain. In a one-owner yard a dustbin makes a suitable receptacle for soaking.

When travelling, boiling water poured on to hay in a plastic sack will be a suitable alternative to soaking. This is not practical for large numbers of horses.

All hay should be weighed before being soaked.

Feeding Hay Loose on the Ground or in a Manger

Hay to be fed loose should be shaken up and placed on a sack that has handles at each corner. The sack can then be picked up by the handles, weighed on a weight hook and carried directly to the horse's box. Hay should not be carried loose in the yard; it is both wasteful and untidy.

Feeding in Haynets

The hay should be shaken up and then put into the nets; these are then weighed.

Some yards colour-code the nets so that if they are prepared in advance they do not need to be re-weighed.

Several nets can be carried to the boxes at one time.

Nets, either full or empty, should never to left on the ground. Bad accidents have been caused by horses getting caught up in nets. In the box they should be hung as high up as possible and tied with a quick-release knot.

Haynets are unsafe for foals and young horses, who should be fed loose hay on the ground.

CONCENTRATE FEEDS

The correct mixing and preparing of concentrate feeds is important. All equipment must be kept clean and washed before fresh food is mixed.

SOAKING SUGAR BEET

Cubes must be soaked in cold water for 24 hours. Flakes and pulp must be soaked in cold water for at least 12 hours. They should be put in a bucket and just covered with water. Sufficient water should be used to make them soft without making them too watery.

Hot water should not be used as it will cause the beet to ferment.

In cold weather the soaked beet must be protected or it will freeze.

In hot weather it should be covered and kept in a cool place. Fresh supplies should be mixed daily, especially in warm weather, as fermentation will occur if beet is left soaked. Any surplus liquid should also be used up.

'OPENERS'

It is good practice to mix all compound or grain food with some form of bulk, usually sugar beet or some form of chaff (bran can still be favoured for this). These have the effect of making the horse chew his food more thoroughly and eat more slowly.

Sugar Beet
The sugar beet is soaked as above and mixed with the remainder of the concentrate feed. Once mixed it should be fed immediately.

Chaff
In the past all stables were equipped with a chaff cutter through which hay or good quality straw was passed. It was cut into short lengths and a double handful was mixed into each food.

In the absence of a chaff cutter, it is possible to buy chopped hay or straw mixed with molasses. This is an expensive method of adding chaff but for the one-horse owner or in special circumstances it may be worth it.

Short-cut Molassed Grass or Lucerne
This comes already chopped. It is usually a high protein product (12%-16%) and although it will improve the digestibility of the food, it may also alter the protein and energy levels adversely. It requires careful assessment and balancing with the other foodstuffs.

Grass or Lucerne Cubes
These will probably have a similar value to the packed short-cut material but will not be as useful in making the horse chew and eat more slowly. Some makes of these cubes need soaking so it is important to read the instructions on the bag.

MASHES

In the horse world it has been traditional to give horses a bran mash the night before a rest day or after strenuous work. Modern nutritionists now consider this practice to be contra-indicated for the following reasons:

- There is a sudden change of diet with which the gut bacteria cannot deal. Until these bacteria have re-established themselves the new rations cannot be fully utilised.

- A mash which is based on bran is not a balanced food.

If, however, some bran is fed every day, a bran mash will not have such a detrimental effect as the gut bacteria are accustomed to it. Mashes can be made from bran, barley or linseed, dried grass cubes or sugar beet.

Making a Mash
A bran mash is made by putting the required amount of bran in a bucket and pouring sufficient boiling water on it to form a crumbly consistency without being wet. It is then covered and left until cool enough to feed, which usually takes about 20 minutes. Salt and a handful of rolled oats or flaked barley may be added to make it more palatable.

A linseed mash is made in the same way as a bran mash but the liquid from cooked linseed is substituted for the boiling water. (See cooking linseed, below.)

A barley mash is made by pouring boiling water onto flaked barley in the same way as for a bran mash. Alternatively, whole barley is cooked in water until it is soft and fed when cool.

If the horse is accustomed to having some grass cubes and sugar beet, an acceptable mash can be made from them. It is prepared by soaking the grass cubes and sugar beet in the usual way and then mixing them together. A tablespoonful of common salt may be added.

COOKING

To Cook Oats or Barley
These may be cooked in a boiler designed for the job or boiled in water in a large pan. It is important that sufficient water is added to prevent burning, but not so much that all the goodness is lost in the water. The

grain should feel soft when pressed between finger and thumb and should have absorbed most of the water. Common salt may be added.

Whole barley will require 3–4 hours' steady simmering; oats have a less hard husk and will cook in less than half the time.

To Cook Linseed

The seeds should be added to boiling water to make sure that the enzyme linase is destroyed before it has time to release the poisonous hydrocyanic acid, which is in turn destroyed by boiling. It is important that plenty of water is used; when cold this will form a jelly.

Linseed burns and boils over very easily, so needs careful attention while cooking. Once the seeds have come to the boil, the heat may be reduced and the seeds simmered for 4–6 hours. The cooked seeds and jelly are added to the feed when cold. Alternatively, the hot liquid may be added to bran to form a linseed mash.

Linseed and barley can successfully be boiled together.

Oatmeal Gruel

A double handful of oatmeal is put into a bucket and boiling water is poured on while stirring vigorously. It is then left to cool.

Sometimes a handful of sugar is added to make it more palatable.

24 THE FEED SHED AND STORING FOOD

FEED SHED LAYOUT

In the layout of the feed shed the following points should be considered:
- Saving time.
- Cleanliness.
- Storage.

Time
Feeding, especially in a large establishment can take up a lot of time. It therefore makes sense to design a layout that is as labour-saving as possible. Everything should be easily to hand.

Cleanliness
The feed shed should be easy to keep clean, in the interests of hygiene and appearances. It should be possible to sweep the floor easily without having to move any of the contents. The walls should be smooth and easily dusted. Shelves should be of painted wood or melamine.

Storage of Concentrates
All types of concentrates can deteriorate if kept in poor conditions. Ideal conditions are:

- Low temperature.
- Little or no variation in temperature.
- Low humidity.
- Good ventilation.

- No direct sunlight.
- Protection from infestation by rats, mice, birds, insects and mites.

Suitable containers are:

- Metal bins, which should be raised off the floor on small wooden blocks.
- Plastic dustbins – these hold about one 25kg bag of feed.

Storage of Vacuum-packed Forage
Because these bags are airtight, the storage considerations are rather different from those of other foodstuffs. Under no circumstances must the bags be punctured. It is important to protect the bags from:

- Rats and mice, who chew the plastic.
- Sharp edges, which may puncture the bags.
- Sunlight.

Storage of Salt
Salt absorbs moisture from the atmosphere; it is therefore important that it is kept airtight.

It is usually bought in plastic bags and, even when opened, it is sensible to keep it in its bag inside a container such as a dustbin.

A week's supply should be taken out at a time and kept in a large plastic jar. Do not leave it in contact with metal as it is corrosive.

A plastic spoon should be used for measuring.

Storage of Hay etc.

Even well-saved hay deteriorates with age but good storage conditions will keep it palatable for a longer time. The important points to consider are:

- Protection from weather.
- Protection from damp, especially from the ground.
- Good ventilation.
- Protection from vermin such as rats and mice.

Further information on hay storage can be found in Chapter 51.

Equipment in the Feed Shed

Feed Containers

These may be **buckets,** which are then emptied into mangers, or feed pans, which are given directly to the horse.

Buckets should be fairly heavy-duty 10 litre (2–gallon) size with names or numbers on them. These can be painted on or put on with Dymo tape. The contents should be emptied into a more suitable manger in each loose box.

ADVANTAGES
- Easy to keep clean.
- Easy and light to handle.
- Easy to store by stacking.

DISADVANTAGES
- Some food is liable to be left and wasted in the bucket.
- The feed is difficult to mix.
- Buckets may be piled inside one another without checking that the bottoms are clean.

Feed pans should be difficult to knock over yet not too heavy.

ADVANTAGES
- All the mixed food goes straight to the horse.

- Owing to their large size the food is easy to mix.
- They can be fitted into a rubber tyre which prevents bruising and tipping over.

DISADVANTAGES
- They can be difficult to keep clean.
- They are bulky to store.
- They must be removed from the boxes after the feed is finished. This can be a problem with late night feeds. Rubber bowls as distinct from metal ones are not so dangerous but even these can be stood in and kicked up, bruising cannon bones.
- They can be difficult to label.

Water

It can be useful to have a tap and a drain either inside or just outside the feed shed. This can be used for washing utensils, for providing water for soaking sugar beet or for damping feeds. Large establishments will also benefit from a sink.

Power point

This should be well away from the water supply and should be used with a circuit breaker. A power point can be useful for boiling a kettle or for a linseed or barley boiler if used.

Kettle

An electric kettle can provide hot water for washing. If used it must comply with up-to-date requirements for safety at work.

Cupboard

This is essential for storing small items supplements and oral medicines that are currently in use. It should be fitted with a lock and be out of children's reach.

Scales

It is essential that scales are large enough to hold the contents of a scoop. Fold-up, hanging wall scales are tidy and unobtrusive. Spring-loaded balances are required to weigh hay, these may be more appropriately sited in the hay barn or where rations are made up.

A Shelf
A separate shelf for supplements, additives, oil, etc. is advisable.

Other Useful Additions
- A damp-proof container for salt.
- Scissors or snips.
- Knife.
- A big mixing spoon for molasses etc.

- A ml measure for medicine.
- A dustpan and brush for cleaning out bins.
- A short, soft broom for sweeping the floor.

A Feed Trolley
In a large yard this can be a valuable piece of equipment. The buckets or feed pans can be placed on the trolley and the food then put directly into them. The trolley can be wheeled along the yard to the boxes. If a trolley is used, there should be a ramp into the feed shed instead of a step.

SADDLERY AND TACK

CHAPTER
25 THE SADDLE

LEATHER

Most leather used for saddlery comes from the skin or hide of cows. To meet the demands of the saddlery and leather trade, much of it is imported. Pigskin, which is tough, elastic, and of light substance or thickness, is used in the making of the saddle seat. Doeskin is also occasionally used for saddle seats.

Tanning

The hide requires considerable preparation before it is ready for use. The process is known as tanning and currying. It takes many weeks, as much of it is done by hand and cannot be hurried, which is one of the reasons why good-quality leather can never be cheap.

The skin is first cleaned of hair, and then the texture is improved, strengthened and made more waterproof by the tanning process, which involves the use of lime, chemicals and greases. The skin can now be described as leather, and is ready for dressing by the currier. Oils and greases are incorporated into it to improve its tensile strength, flexibility, and water-resistant properties. The hide should have plenty of substance – i.e. thickness – thus ensuring that it will last and will also stay flexible, because of the added fat content of the leather.

Oak bark tanning is a vegetable process using tannin or tannic acid produced from tree bark. It is considered one of the best methods of producing first-class leather.

Texture

Leather has a grain or outer side, and a flesh or inner side. The grain side is shiny, and is sealed to make it more waterproof; it is therefore less absorbent to any soap or dressing. The flesh side is unsealed, and has a rough, dull surface. It is less smooth, and is the side of the leather (except for the underneath lining of the saddle) which is usually next to the horse's skin. In use it loses fat and oil, and this is the side which, being more absorbent, should receive greater care during cleaning.

Colour

The colour of leather can be varied during the tanning process, but is confirmed by the use of dyes during curing. The most widely used colours are **London tan,** which is yellow, **Havana,** which is the colour of a good cigar, and **Warwick,** which is much darker.

Types of Leather

RAWHIDE is cowhide which has undergone a special vegetable tanning process. This makes the leather very strong and leaves a light, untanned central strip. Rawhide is used for stirrup leathers, girth straps, etc.

BUFFALO stirrup leathers are red in colour, greasy and soft to the touch. They are exceptionally strong and last many years, but they usually stretch.

CHROME-TANNED LEATHER is pale blue-grey, and is produced by using salts of chromium. It is very strong, withstands wet, and remains supple. It is often used for New Zealand rug straps.

HELVETIA LEATHER is yellow in colour, and very greasy and tough. It is often used for reinforcing martingales, nosebands, etc.

BLACK LEATHER is also produced for the saddlery trade. It is generally used for harness, but is also fashionable with dressage riders.

Quality

When buying saddlery, it is important to be able to judge the quality of leather. Good leather should have plenty of substance, unless a special request has been made for a lighter weight – e.g. show bridles and racing saddles. This light type of leather wears much more quickly than a heavier sample.

Good leather should feel slightly greasy but firm. When bent in the hand, neither side should form bubbles on the skin. The grain or sealed side should not crack.

When examining a bridle, note that the flesh or unsealed side of the leather should be smooth in texture, with no rough fibres visible. Poor leather, seen in a cheap bridle, has rough edges and may feel spongy. Such bridles often have poor-quality fittings.

STRUCTURE OF THE SADDLE

The Saddle Tree

The tree is the framework around which the saddle is built. Trees were traditionally made from beech wood, but this has now been superseded by laminated wood, which is bonded under pressure, and then moulded to the required shape. The resulting tree is both lighter and stronger. Other materials, such as fibreglass and plastic, are also used, though these are not as easy to work with as laminated wood. Fibreglass is

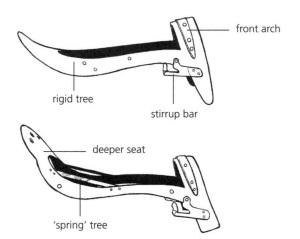

front arch

rigid tree

stirrup bar

deeper seat

'spring' tree

Saddle trees.

exceptionally strong and light, and is used in some racing saddles. Plastic appears to lack strength, and another disadvantage is that nails or tacks cannot be driven into it.

Trees are made in different shapes and sizes according to the intended use, the shape of the horse, and the preference of the rider. The shape of the tree dictates the main outline of the saddle.

Trees can be either rigid or sprung. The **rigid tree** has a larger, more solid framework than the spring tree.

The **spring tree** has two flat panels of steel, which run from the underneath of the pommel to the cantle, and are set to lie about 5cm (2ins) on the inside of the broadest part of the seat. These steel panels are thin enough to give greater resilience to the seat of the saddle, and a more direct communication between the seat of the rider and the horse's back. On occasions, there can be a risk of damage to the back, and it is advisable always to use a numnah. The pommels of both types of saddle are reinforced on top and underneath with steel plates. For very heavy work, they can be reinforced further.

The Stirrup Bars

The bars, which should be of hand-forged steel, are riveted to the points of the tree.

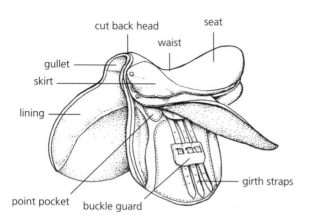
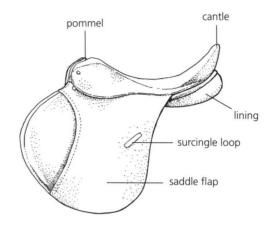

General-purpose saddle.

'Cast' stirrup bars are found on cheap saddles, and their strength is suspect. The words 'forged' or 'cast' can be found stamped on the bar The bar itself should be of the regulation type and straight.

There has been discussion as to whether the safety-catch hinged on the bar is of any value. It is suggested that the length of the bar should be slightly extended, and with a slight upward curve. As yet, the Saddlers' Association has not come to a decision.

Recessed bars are used on most spring tree saddles, and on some rigid trees. They are placed under, rather than on top of, the tree, so there is less bulk to the stirrup leather under the rider's leg.

The bars on jumping saddles are placed well forward, and on dressage saddles they are positioned further back.

The Seat

Pre-strained webbing is fixed from the pommel to the cantle, and over this is placed stretched canvas or linen. Bellies – short pieces of felt or leather – are placed round the broadest part of the seat to support the edge. Serge is stretched down to form the seat shape. Wool, plastic foam or foam rubber are added to give resilience and extra cover to the tree, so that the rider cannot feel it through the seat. Of these three materials wool is most popular, and lasts indefinitely. In time, foam

rubber tends to disintegrate.

Pigskin or hide is then stretched on as the final seat cover with attached skirts. The **saddle flap**s are attached, followed by the **girth straps**. Two of the girth straps should be fixed to the web strap. The third strap is fixed independently. Girth straps should be of best quality leather, as its strength and attachment are vital to the safety of the rider.

The Underneath Panels

These panels vary in design. The **full panel** gives a greater weight-bearing surface, and is probably the most comfortable for the horse as it makes the waist wider; however, this can be uncomfortable for the rider. The **side panel** must be kept thin to allow the rider a close contact with the horse's sides. **Continental-type panels** are narrow at the waist and allow the rider a much closer contact. They go by various names and are used on all modern deep-seated saddles, together with knee and thigh rolls.

The **short**, or **half**, panel is now mainly used in the cheaper pony and cob saddles. It has no knee or thigh rolls and does not help the rider as much as the other types to establish a balanced position. As there is less bulk under the leg, allowing a closer contact than a full panel, it is also used for show and polo saddles.

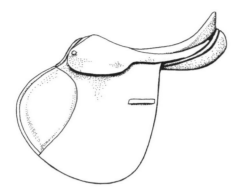

Forward-cut jumping saddle.

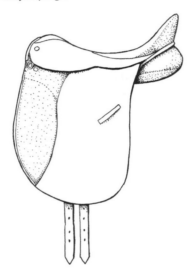

Dressage saddle.

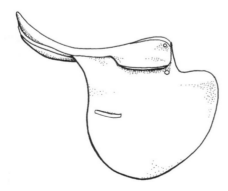

Racing saddle.

Girth Straps

Many modern saddles, in particular those for dressage, have long girth straps to avoid bulk under the knee and thigh. These saddles require a shorter girth. The design is not as suitable for cross-country riding and hunting, as the straps are difficult to adjust from the saddle and are more likely to cause sores from mud around the buckle and guard area. They must be adjusted so that the buckles are clear of the elbow.

Linings

LEATHER is easy to clean and wears well, as long as it is well looked after. Most good modern saddles are now lined with leather. Note: when first put on, a leather lining can be cold to the back of a sensitive horse.

SERGE AND LINEN LININGS are rarely seen in modern saddle manufacture; they may still be seen on side-saddles. Serge is warm, absorbs sweat well but does not wear well. Linen is easier to clean and longer-wearing than serge.

MAIN TYPES OF SADDLE

Traditional Jumping

This saddle is shaped to help the rider stay in balance when riding in a forward position over a jump. It has forward-cut flaps to accommodate the knees of the rider with his relatively short leg position, and it has appropriately placed knee and thigh rolls to assist in maintaining the leg position. The depth of seat varies: the choice depending on the individual's style of riding – some riders favouring a much deeper seat position.

Close-Contact

Many show jumping and event riders now favour the close-contact saddle. These have a flat seat and are designed with the minimum of padding between rider and horse. The aim is to allow the rider maximum contact with the horse and versatility in his position in any situation.

Dressage

This saddle helps the rider to sit with a deep seat and a long leg. The saddle flap is cut much straighter, with only a slight curve in front. The stirrup bar is placed further back than on a jumping saddle. The saddle flap may need to be correspondingly longer, so that it does not catch the top of a tall rider's boot. The seat is deep, and the knee and thigh rolls are placed so as to assist the leg position.

General Purpose

This saddle is a combination of the jumping and dressage types, with a less forward-cut flap. It is the saddle used by the majority of riders who do not wish to specialise.

Endurance

Special saddles are now available for endurance riders, and are designed to spread the rider's weight over as wide an area as possible so that pressure points are minimised. The various designs incorporate features of cavalry and Western saddles, and they are placed over very thick, specially designed numnahs.

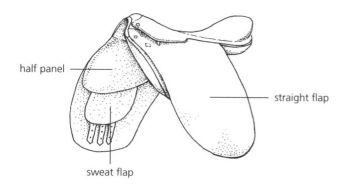

half panel

straight flap

sweat flap

Show saddle with half panel.

Showing

This saddle has a very straight-cut flap to show off the horse's front and the seat is relatively flat. It is always used without a numnah, so the fitting is of even more than usual importance.

Racing

This is a lightweight saddle with a sloping head and very forward-cut flaps. Exercise racing saddles are built more substantially, although the design is similar.

Polo

This saddle may have a reinforced pommel. It is built with a short panel, extra long sweat flaps, and no knee or thigh rolls.

Pony Club

This saddle is Pony Club approved and manufactured by Cliff Barnsby. The saddle was originally produced in a general-purpose design. The Pony-Club approved saddle was initially designed with the young person in mind, to ensure that children were able to develop a riding seat that was assisted not hindered by the saddle on which they rode. This concept has been perpetuated and the general-purpose saddle is now accompanied by a dressage model.

Competition Saddles

Many top competition riders now design saddles to their own personal specifications. These are then marketed by saddlers under the rider's name.

ACCESSORIES

Girth

The following types can be used:

LEATHER GIRTHS if well cared for, last much longer than

any other type. The main types are: three-fold, Balding and Atherstone. The latter two are shaped so that there is less risk of causing girth galls.

SHORT GIRTHS OF LEATHER OR MAN-MADE FABRIC suit dressage and jumping saddles with long girth straps.

NYLON is a good general-purpose material, which is easy to wash. Care should be taken that the girth does not get twisted.

STRING is similar to nylon but of softer texture.

LAMPWICK girths are soft and comfortable but not commonly seen these days. They are not suitable for demanding work.

WEBBING is generally preferred for racing, and by some riders for eventing. It is less restricting than leather. Two girths should always be used, as one can break without warning under strain.

WHITE TUBULAR SHOW GIRTHS can be used on hacks and ponies.

LEATHER OR WEBBING GIRTHS WITH ELASTIC INSERTS at the buckle ends are more often seen on racehorses and eventers. They are less restrictive to a horse when he is galloping than the conventional girth. The elastic needs regular checking for stretch and wear.

ELASTIC INSERTS, which are sometimes used for racing, require careful adjustment. Sweat rots them in time.

SYNTHETIC GIRTHS ('COTTAGE CRAFT' TYPE) are soft and comfortable yet hard-wearing and durable. They are versatile and widely used.

Stirrup Leathers

These should be made of the best-quality leather. They should be chosen to suit the size and weight of the rider. It is dangerous to put a heavy leather on a lighter stirrup iron because if the rider falls, the iron will not easily be freed from the foot. Lightweight leathers are often more comfortable for the rider and easier to handle. For racing, hunting, eventing and jumping,

buffalo or rawhide leathers are the most suitable as they are less likely to break.

Stirrup Irons

Materials used are:

- Stainless steel, which is the best.

- Composite, which is satisfactory.

- Nickel, which is dangerous, as it bends and breaks. It can be recognised by its yellow colour. It requires regular polishing to stay bright.

The normal basic pattern stirrup iron is suitable for all purposes. The important factor is that the stirrup should be of suitable weight and size.

With the foot in the stirrup, there should be 12mm (½in.) clearance on either side. In the event of a fall, a too light stirrup and/or small stirrup may jam on the foot. Conversely, too large a stirrup can permit the rider's foot to slip right through. The correct size and weight of iron are therefore important, and should always be checked by the rider and/or attendant.

Alternative Designs of Stirrup

THE BENT TOP IRON This is designed for the rider who keeps his foot well forward in the stirrup. Fitted so that it slopes forward, it removes pressure from the front of the leg. Fitting sloping backwards the reverse action occurs. The stirrup leather is placed through the iron with the curve away from the leather.

KOURNAKOFF The eye of the iron is offset to the inside, and the sides slope forward with the tread sloped upwards. This helps to keep the heel lower than the toe, and the knee to be tight, with the lower leg off the horse. As it does not encourage an effective position of the rider's lower leg, it cannot be recommended. If it is used, however, it is essential to put the irons on correctly, and it is advisable to mark the left and the right.

SAFETY IRONS These have a thick rubber band replacing the metal on one side of the iron. The band should always be to the outside. A leather loop fitted to the bottom of the iron helps to hold the rubber band and to prevent its being lost if it comes off the top hook. As the weight is uneven, these stirrups do not hang straight, but they are much safer than standard irons, and as such are advisable for young children.

RACING IRONS These are usually made of lightweight stainless steel or aluminium. Because of their lightness they are not easily dislodged during a fall and are not suitable for general use.

CAGED STIRRUPS These stirrups, used by endurance riders, are fully enclosed to prevent the foot from sliding through the stirrup. This allows the rider to wear soft footwear without a heel, e.g. trainers, which are more comfortable for long hours in the saddle and give the rider the option of being able to run more easily beside the horse if required.

Treads

Rubber or plastic treads are used to help the rider maintain his foot position. They are also warmer to the foot than metal.

BUYING AND FITTING SADDLES

Buying

Always buy the best you can afford, and one made by a reputable saddler. Good saddles, well looked after, last for years and retain their secondhand value. A cheap saddle may look suitable, but it has a limited life and trouble can be expected from the tree. The leather of the girth straps, saddle flaps, and seat may rapidly show signs of wear. A good secondhand saddle is usually a better buy than a cheaper new one, as long as the tree is undamaged.

Nowadays there is a much more holistic approach to the choice and fitting of saddles. Far more is known about the effect and distribution of weight over the horse's back, and this can be taken into account when choosing a saddle for a specific discipline.

If you are buying for a riding school, remember that

rigid-tree saddles are stronger and stand up to hard wear and heavy weights better than spring-tree saddles. They are also likely to be less expensive. A wide-seated saddle, cut generously to spread the rider's weight, is kinder to the horse's back. The general-purpose type is preferable.

When buying for private use, many people prefer a spring-tree saddle as it is more comfortable for the rider and gives a better feel of the horse's movement; also the seat aids are more directly felt by the horse.

Before deciding on a saddle, it is essential to ensure that it fits both the horse and the rider, and suits the latter. For the average rider, a general-purpose saddle is the most practical.

Fitting

A good saddler, who is a member of the Master Saddlers Association, should always be happy to come out and fit a new saddle. He can then advise as to whether the tree is the correct shape and width, and can often adjust the panel-stuffing on the spot.

When you are having a saddle fitted, remember that the outline of a horse's back can change according to his management and work, and that the padding of a new saddle flattens during use. It may therefore have to be set up (padded) by the saddler.

A good fitting and a comfortable saddle is essential if a horse is to perform well and avoid such problems as pressure and girth galls. A badly fitting saddle can easily be a cause of poor performance.

Procedure for Examining a Saddle

- Test the tree for signs of breakage or distortion (see below).

- Up-end the saddle and examine the underneath for any unevenness of padding or outline.

- Place the saddle on a saddle horse and check that the padding is level and even, both from behind and in front.

- Make sure that the gullet is wide enough – 63–75mm

(2½ to 3ins) – so that there can be no pressure close to or on the spinal vertebrae. This must be checked again when the saddle is on the horse's back.

- To avoid marking the leather, a stable rubber should be placed on the horse's back under the saddle. To avoid marking the saddle flap, buckle guards should be fitted on the girth straps. The saddle should sit level and even, with no tilt either towards the cantle or the pommel. Either would affect the rider's position and make it difficult for him to stay in balance with his horse.

- The saddle should fit the horse with the weight evenly spread over the lumbar muscles, not the loins.

- It must not hamper the movement of the horse's shoulder. This can occur with a forward-cut jumping saddle.

- A rider, preferably the future owner or one of similar height and weight, should be legged up on to the horse. In both upright and jumping positions it should be possible to place four fingers under the pommel without feeling any pressure.

- There should be ample clearance under the cantle and along the gullet.

- When looked at from behind, it should be possible to see a clear channel of daylight.

To Test for a Damaged or Broken Tree

The Front Arch

The first indications of trouble are usually when the arch of the saddle widens and comes down on the withers. There may or may not be a squeaking or grating noise when the saddle is used; such symptoms are more likely to be heard with a heavy rider.

Place your hands on either side of the pommel and try to widen and move the arch. Or hold the cantle and grip the pommel firmly between your knees.

Any movement or clicking sound usually indicates damage.

The Waist

Rigid Tree Place the pommel against the stomach and, holding the cantle with both hands try to press it upwards and towards the pommel. A movement in the waist indicates a break or crack. If the tree is seriously damaged, the rough edge can usually be felt when the gullet is examined. Damage can occur either on both sides of the tree or on one side only.

Spring Tree There should be some give in the seat of a spring-tree saddle, but when it is tested, as for a rigid tree, it should spring firmly back into place. A weak or broken tree feels slack and without spring.

It can be difficult for an inexperienced person to detect damage in a saddle with a spring tree. If you are in doubt, ask a saddler for advice.

Damage to a saddle caused by careless mounting by the rider can be observed by putting the saddle on a saddle horse and checking if there is a twist and/or if the saddle does not sit evenly.

Cantle This should feel rigid. Any movement indicates damage.

Points As these are the continuations of the front arch, they can be tested in a similar manner to that of the arch. Some saddles have flexible points, which should not be confused with damaged points. If the damage is below the stirrup bar, it is possible to see it.

To Measure a Horse's Back for a New Saddle

- Obtain a piece of soft lead or flexible wire, e.g. a coat hanger about 46cm (18ins) long.

- Place this just behind, and at right angles to, the horse's withers – in the position where the saddle would rest.

- Shape it into the required outline, then trace round it on a large sheet of paper.

- The next measurement should be taken about 23cm (9ins) further back and similarly traced.

- A third outline should be taken from the top of the withers along the spine.

- When ordering a saddle, it is helpful to give the rider's height, weight and length from crotch to knee.

Horses with Problem Conformation

The following are problems which need special attention in the fitting of saddles:

HIGH WITHERS A narrow tree is necessary, well padded at the withers, and preferably 'cut back'. The saddle may have to be set up at the rear to give a level seat.

STRAIGHT SHOULDER The saddle is likely to work forward, with a possible risk of girth galls. This is more likely if the horse is at grass and has a big belly. The rider feels insecure. The saddle needs careful padding to encourage it to stay back. A point strap can be fixed (see below).

FLAT SIDES These encourage the saddle to slip back, which can be a problem when horses become fit. Careful padding of the saddle and the use of a breastplate should resolve the problem.

FAT PONIES WITH LOW WITHERS The only satisfactory solution is to use a crupper. A point strap can also be fixed to the points of the tree, and the girth attached to this and to the first of the girth straps. This may help prevent the saddle sliding forward over the withers and on to the neck.

Typical Examples of Ill-fitting Saddles

- **Too much padding.** This causes the saddle to rock, with a risk of saddle sores, particularly on either side of the backbone to the rear of the saddle.

- **Too wide in the pommel.** This causes sores on the top of the withers and possible damage to the spine.

- **Points of the saddle too long.** This causes undue pressure on either side of the withers.

- **Uneven padding or a twisted tree.** This results in a crooked rider; uneven weight distribution, and the

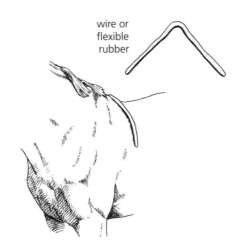

wire or flexible rubber

Measuring the shape and size of the back for saddle fitting.

development of pressure and friction sores. It may also damage the muscles of the back.

- **A crooked rider** can also cause uneven muscle development of the horse's back, thus aggravating the crookedness.

SIDE-SADDLES

TREE AND CONSTRUCTION Most side-saddles used today are of pre-1939 construction. The tree is wooden, reinforced with metal, and with the top pommel built into it. The tree makes it heavy, and because of the shape and angle of the pommel or horn, it is easily damaged. Well-known makes of saddle are Owen, Champion & Wilts, Mayhew and Whippy, but unfortunately these makers no longer exist, having been closed down or bought up by larger concerns.

THE SEAT should be compact and flat, with the off side slightly set up to prevent the rider's seat from slipping over. A doeskin-covered seat gives more grip than a smooth leather seat.

The POMMELS or HORNS are spaced wide apart. The angle of the lower pommel or 'leaping head' can be adjusted by tightening the horn pin. On some saddles, an alternative screw-in socket is provided. The adjustment is necessary according to the build of the rider, and whether she is riding on the flat or jumping. On some saddles, a small leather strap secures the pommel at the required angle.

The UNDERNEATH LINING should be of linen or serge, which is more comfortable for the horse, and less likely to cause a saddle rub. New saddles may have a leather lining.

If small round holes can be seen in the leatherwork, it is a sign that the tree may have been attacked by WOODWORM or small spiders. When storing a saddle, cover it with fresh newspaper, as the printers' ink will deter small insects.

Buying a Side-Saddle

- Before buying a secondhand saddle it is advisable to have it checked over by a saddler experienced in this type of work. Repairs can be expensive, and in some cases the saddle may not be worth buying.

- Modern or post-World War 2 saddles are usually built on a tree of laminated wood, and are of simpler design. The tree is both stronger and much lighter: the saddle weighing around 5.5kg (12lbs). Older saddles weigh between 10kg (22lbs) and 12kg (26lbs).

- Before buying a saddle, make sure that it fits the horse. A good saddler can adjust the underneath padding as necessary. He may also be able to alter the padding and the set of the pommels to help the rider. If the saddle is made for a tall person it can never be altered sufficiently to enable a short rider to sit correctly. The reverse also applies.

Stirrup Attachment

All manufacturers of side-saddles had their own patented stirrup attachment, with the leather fitting made to match. The essential feature was a quick-release mechanism. Some of these can be difficult to manage if the user is inexperienced with a particular make. They should all be kept well oiled.

Stirrup Leather and Iron

These have the necessary fitting at one end, with a metal hook attached at the other. The hook is put through the enlarged eye of the side-saddle iron, and then hooked back on itself. A leather sleeve slips down to cover the adjustable fastening.

The stirrup iron must be big enough for the rider's foot: the heavier the better.

A specially designed collapsible safety iron can sometimes be bought secondhand – although with a quick-release attachment it should not be necessary. Such irons can also be unreliable.

Girth and Balance Strap

The girth is usually of the three-fold leather or Atherstone type. Lampwick is suitable for showing.

- The girth is first buckled on the near side.

- The leather balance strap is fastened on the near side of the saddle, either with a buckle or strap, according to the make of the saddle. It is attached to the back of the saddle on the off side, and helps to keep the saddle steady. Nowadays, short balance straps are often used, sewn to the girth on the off side and then fastened to the saddle. Opinion is divided as to which is the most efficient.

- On some saddles with girths which buckle under the off-side flap, there is a strap to hold down the flap. The flap strap is sewn to the bottom of the near-side saddle flap, and is fastened with buckles under the horse's belly. A hook on the near-side flap then fastens it to the strap.

- The girths of many saddles are now fastened to girth straps on the outside of the off-side flap, which make the extra strap redundant.

26 BRIDLES

Leatherwork must be of a suitable size and weight. Show and dressage bridles can be of lightweight leather. Bridles for hunting, eventing, show jumping and riding-school work should be of a heavier type.

Bridles can be made of nylon, which is cheaper, hard-wearing and does not require the care needed by leather. Good quality leather, well cared for, remains the first preference. All buckles should be of good quality metal.

Parts of a Snaffle Bridle
- Headpiece and throatlash.
- Browband.
- Cheek pieces.
- Noseband – plain cavesson or specialist type depending on requirements of the horse.
- Reins.
- Snaffle bit.

Parts of a Double Bridle (see illustration, page 274)
- Headpiece andthroatlash.
- Browband.
- Cheek pieces.
- Plain cavesson noseband.
- Bridoon headpiece (or 'sliphead' – has no throatlash).
- Bridoon cheek piece.
- Bridoon rein, often thinner than a snaffle rein. It may be laced, plaited or covered with rubber.
- Curb rein (thinner than the bridoon rein).
- Bridoon.
- Bit (curb).
- Curb chain (metal or leather).
- Lip strap (optional). Some bits do not have the necessary Ds.

Parts of a Pelham
- Headpiece.
- Browband.
- Cheek pieces.
- Plain cavesson noseband.
- Bridoon rein and curb rein as for double bridle above. Or Pelham rounding and one rein.
- Pelham bit.
- Curb chain.
- Lip strap.

REINS

Reins should be chosen according to intended use. A full size rein should be 1.5m (5ft) in length: i.e. the longest strip which can be cut from a hide. Reins for show jumping or flat racing – when a shorter hold is taken – can be 1.4m (4½ft). Pony reins can be 1.3m (4ft 3ins) or shorter.

Types of Rein

LEATHER REINS can be plain, plaited, laced or rubber-covered. The latter when used for steeplechasing or

Snaffle bridle with loose-ring bit and plain cavesson noseband.

eventing should have enough plain leather at the buckle end to enable them to be knotted. If the rider should slip the reins to the buckle end, the strain can be taken on the knot, not the buckle.

OTHER MATERIALS are webbing, linen, plaited cotton and nylon. These are all popular for jumping. Some have leather slots stitched every 13–16cm (5–6ins) to provide a more secure hold.

FITTING A SNAFFLE BRIDLE

When the bridle is on the horse, check the following points:

- The fit of the **browband.** This must be comfortable. If it is too tight it will pinch the ears; if too loose, the headpiece may slip back causing the browband to look untidy.

- The **headpiece** and **cheek pieces** should be an even height on both sides, and preferably of such a length that the buckles lie just above eye level; there should also be room for adjustment either way.

- The **snaffle bit** should be adjusted so as to slightly wrinkle the corners of the horse's mouth and not to

protrude more than 5mm (¼in.) on each side. The horse's mouth should be opened to make sure that the bit is high enough in his mouth to clear the tushes. The bit should be pulled straight in the mouth to measure its width.

- The **throatlash** should be buckled so that it is possible to place a hand between the leather and the horse's jaw when the head is flexed. Some thick-jowled horses may need it looser than this.

- The **cavesson noseband** should be fitted two finger breadths or 2.5cm (1in.) below the cheek bone. When it is fastened at the back, allow two fingers' width between the jaw and the leather. During training it may, if necessary, be tightened.

- The **reins** should have about 43–51cm (15–20ins) spare when the rider is holding them. They must never be so long that the slack part can be looped over the rider's foot. They should be buckled together at the ends.

FITTING A DOUBLE BRIDLE

- The fit of the **browband, headpiece, throatlash** and **noseband** should be similar to that of a snaffle bridle.

- The **bridoon bit** should be fitted on a separate headpiece, which should buckle on the off side a little below the buckle of the main headpiece, at a height which will slightly wrinkle the horse's lips.

- The **bit** should lie below the bridoon so that both can act independently.

- The **curb chain** should be hooked on the off side of the bit, so that when twisted clockwise to the right the lip strap ring will hang down. The flat ring of the curb chain should be put on the near-side hook with the index finger underneath and the thumb on top. The selected link should be taken up and also attached with thumb nail on top and maintaining the twist to the right. If it is shortened more than three links, take up an equal number on both sides.

- The end links of a **leather-covered chain** should be put on in a similar manner.

- The **curb chain** should lie flat in the chin groove, and remain flat when the cheeks of the bit are drawn back. There must be no risk of rubbing the horse on this very sensitive part of the jaw. Double link or leather curb chains are preferable. The curb chain must never be fitted through the rings of the bridoon. Its action must be quite independent of the bridoon. The curb chain should begin to act, exerting pressure in the chin groove, when the cheek of the bit forms a **45° angle** with the horse's mouth.

- The **lip strap** should be put through the loose ring of the curb chain and buckled on the near side. It is decorative, rather than functional. If used, there is less danger of losing a curb chain when the bridle is being carried.

- A **bridoon rein** may be slightly shorter than that of a snaffle bridle. If so, the bit rein should be 20–30cm (8–12ins) longer than the bridoon rein.

FITTING A PELHAM BRIDLE

This is fitted so that the bit lies close up against the corners of the lips without wrinkling them. The curb chain, when fitted, should lie in the curb groove. The curb chain should be hooked on as for a double bridle and should have a lip strap. The chain can be placed through the top ring of the Pelham, which prevents any friction from the curb hook on the sides of the horse's face. Alternatively, it can go directly from hook to hook. If the upper cheek of the Pelham is too long, the action of the curb chain will be too high up.

FITTING A KIMBLEWICK

This is fitted in a similar manner to a Pelham. The correct fitting of the curb chain depends on the length of the horse's jaw. The action of the Kimblewick can be

Simple double bridle.

quite severe and vice-like because the 'square eye' attachments for the cheek pieces exert stronger poll action.

NB: Curb chain hooks must open outwards, away from the horse's face.

NOSEBANDS

Drop Noseband

This is used to prevent a horse opening his mouth so wide that he can evade the action of the bit.

Correct fitting is essential. If it is fitted too low, it interferes with the horse's breathing and can cause distress. The front of the noseband should lie on the bony part of the nose. The strap at the back should drop down to be fastened below the snaffle bit. The noseband should allow the horse to flex and move his jaw, but should prevent him from opening his mouth wide.

Rings which connect the front nosepiece to the back strap have small hooks which, when the leather is firmly stitched, hold up the front nosepiece and prevent it dropping down. Alternatively, short leather straps can be attached from the nosepiece to the headpiece.

Some nosebands have adjustable fronts to ensure a better fit over the nose.

Common Faults in Manufacture

- The front strap is too long, which makes fitting difficult, and the horse's face and lips can get pinched.

- The buckle end of the back strap is too long, making adjustment difficult. The buckle end should be short and the strap end long.

- The front strap is not on fixed rings, and therefore drops down.

- The headpiece is too long, resulting in an untidy and unnecessary length of strap.

Flash Noseband

This is a cavesson noseband on to which a light strap is sewn, or is passed through a small loop on the front of the cavesson. The strap fastens below the bit, and acts in a similar way to a drop noseband. It is easier to fit than a drop, and is generally more comfortable for the horse.

Grakle, Figure-of-Eight or Cross-over Noseband

This is made of lighter-weight leather than a drop noseband. Its action is similar, but it is also effective in preventing a horse from crossing his jaw, as it acts over a wider area of the head. It is also less likely to affect the horse's breathing. It acts from a headpiece, which ends immediately above the horse's cheekbone or just below. Two straps cross over the bridge of the nose,

Flash noseband.　　　　Grakle noseband.

where they are stitched together and are usually padded underneath. They then pass round the horse's nose, and are buckled above and below the bit. They can be adjusted as required, but must allow some freedom to the mouth, although not so loosely that the lower strap drops down under the horse's chin. A small adjustable strap can also prevent this.

Kineton Noseband

This may help in the control of a strong horse. It acts by transferring some of the rein pressure from the mouth to the bridge of the nose. Note that it can also restrict breathing. It consists of two metal loops, one end of which is attached to a short, adjustable, centre strap, the other to a long headpiece. It is worn with a snaffle bit. The metal hoops fit round the mouthpiece between the bit ring and the horse's face. The centre strap should rest on the bony part of the horse's nose.

CHAPTER 27

OTHER SADDLERY

MARTINGALES

In theory, the educated rider on a well-schooled horse should not need auxiliary aids such as martingales. There are occasions, however, when they can be used with benefit, to prevent the horse or pony putting his head above the angle of control, and so to help the rider to manage his horse as well as to improve the safety factor.

Running Martingale

The running martingale consists of a thin neck strap buckled on the near side and with a loop at the bottom. A wide strap is put through the loop. At one end it has an adjustable loop through which the girth is threaded. At the other end it is divided into two, each end having a ring through which the reins are threaded.

This acts via the reins and should be adjusted so that it only comes into action when the horse puts his head up. It also can be used by a novice or unbalanced rider who carries his hands too high, as this ensures that the action of the rein is kept lower, which is less likely to upset the horse.

A rough guide to fitting is to hold both the ring ends of the martingale together and to draw them towards the withers when the loop end is through the girth. If they reach the withers, the fit is loose. When adjusted too tightly they can cause insensitive hands to damage a horse's jaw – as they have a very severe action with

strong downward pressure on the bars of the mouth. As a running martingale affects the rein contact with the horse's mouth, two reins may be used, with the martingale attached to the bottom rein, thus leaving the other rein free to maintain direct contact (although this is a practice rarely seen). When attaching the reins: to ensure that they can slide smoothly through the rings, stand on the near side, hold the rings up with the right hand, and thread each rein through separately. Then buckle the reins together.

Reins not sewn on to the bit should have rubber or leather stops to prevent the martingale rings catching on the fastenings. The stops will also be necessary if a martingale is used on the curb rein of a double bridle or Pelham on which the bit rings are smaller than the martingale rings. This practice is not recommended. A running martingale on a bridoon rein is more effective and less severe.

A strong rubber ring or stop should also be used to secure the neck strap to the martingale.

The girth strap loop should have two keepers which should be adjusted to hold the loop flat.

The running martingale detracts from the correct use of the open rein.

Standing Martingale

A standing martingale prevents the horse from putting, or throwing, his head up above the angle of control: a bad habit which can result in the rider receiving a bang on his nose.

The martingale is secured to the neck strap and girth in a similar manner to the running martingale, but it is then attached directly to a cavesson noseband. It should never be attached to a drop noseband. It is adjustable either:

- At the girth.
- By a buckle under the neck.
- At the noseband.

When the horse's head is in a normal position, it should be possible to push the martingale up into the horse's gullet. It should not be fitted so tight that it exerts continual pressure, as this will encourage the horse to lean up against it. It can also restrict a horse's action, particularly over a jump.

Combined Martingale

This is a combination of a standing and running martingale and has the action of both. It is more often used for show jumping.

Bib Martingale

This is similar to a running martingale, but with a centre piece of leather between the two rings. It can

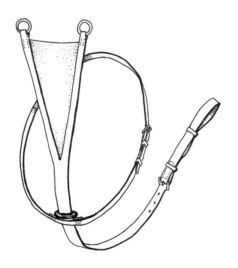

Bib martingale.

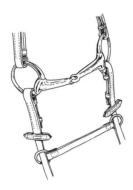

Irish martingale with rein stops.

be used on an excitable or overbent horse who is likely to catch his teeth in the strap or ring of an ordinary running martingale. It is more restrictive, as an opening rein cannot be used.

Irish Martingale

This consists of a short leather strap about 10–12cm (4–5ins) long, with a ring at either end. The reins are threaded through the rings, and if the reins are not sewn on, stops must be used. It is positioned under the horse's neck. In the event of a fall, it helps to prevent the reins going over the horse's head. It is more generally used on racehorses. It can also help a horse who shakes his head and gets the reins over his head when being ridden. It can adjust the angle of the rein contact down if the hands are too high.

BREASTPLATE

This is used to prevent the saddle slipping back – usually on horses who are very fit or on those whose conformation fails to prevent the saddle from slipping back (i.e. flat-sided or herring-gutted horses). It is also used when hunting or in very hilly country.

There are two types of breastplate:

1. A leather neck strap is attached first to the girth by

a leather strap and then to the 'Ds' of the saddle by two narrow leather straps on either side of the withers. The neck strap has buckles on both sides so that it can be evenly adjusted. It should fit comfortably and not be tight around the neck.

If a martingale is used, a short attachment can be added, either in the design of a running or standing martingale. This can be buckled to the centre ring of the neck strap and then put on to the reins or the noseband as required. This centre ring should have a leather 'safe' or protecting pad to avoid any rubbing of the horse.

2. 'Aintree breastgirth' – more often used on racehorses or eventers. It consists of a band of webbing or wide elastic which is passed round the horse's chest and attached to the girth on either side by adjustable leather loops. Another strap is passed over the horse's withers, which prevents the breast girth from dropping down. The breast girth must be carefully fitted so that neck movement and the windpipe are not affected.

SPECIAL GIRTHS

Surcingle or Overgirth

This is essential equipment for racing and eventing, but is also used in show jumping. It is a high-quality webbing strap, long enough to fit over the top of a saddle and weight cloth. It helps to hold everything in place, and acts as an extra girth in an emergency. Some may have a small but very strong elastic inset, which makes for easier fitting.

Foregirth

This has a metal arch positioned in front of the saddle to stop it slipping forward. It is attached to a surcingle, which must be buckled up tightly before the saddle is put on. It is used particularly in dressage to ensure that the saddle does not come too far forward, thus putting the rider's weight on to the shoulders.

WEIGHT CLOTH

This is necessary when a rider has to carry weight to conform to competition rules, e.g. in racing. It consists of a strong linen or man-made fibre saddle cloth secured to the saddle by leather straps, and with stitched pockets for holding lead. The pockets can be placed in front of the saddle on the shoulders, or behind the saddle flaps and against the horse's ribs. When fitting a leaded weight cloth, care must be taken that it does not press down on the withers, or interfere with the rider's position .

NUMNAHS

These are protective pads worn under a saddle to add to the horse's comfort and to protect his back from friction and pressure. They are especially valuable:

- When first riding the horse after a long rest.

- When riding for a long period of time with the rider's weight continually on the horse's back.

- With learners, who find it difficult to sit quietly and in balance, particularly if they are of heavy build.

- With experienced or novice riders who do not sit straight and square in the saddle. The uneven distribution of weight can encourage the saddle to rock, and creates additional friction.

- With a modern saddle which has a spring tree, narrow waist and deep seat, and which concentrates pressure over a smaller area.

- When the shape of a horse's back changes. A numnah will then make the saddle fit more comfortably. This should only be a temporary measure, as the stuffing of the saddle has to be altered.

- When jumping. A numnah under the saddle may add to the horse's comfort and encourage him to jump with a rounded back.

NB: The saddle should fit the horse's back before the numnah is put on. Except in emergencies, it should not be put on to make a badly fitting saddle usable.

Types of Numnah

SHEARED SHEEPSKIN is a natural fibre which absorbs sweat. It can be washed, but afterwards the leather side requires treating to keep it soft. This type of numnah is expensive.

NYLON 'SHEEPSKIN' is cheaper and not so absorbent, but it is very easy to wash and quick to dry.

LINEN or good-quality COTTON with a PLASTIC FOAM LINING can be satisfactory and is sweat-absorbent.

PLAIN FOAM RUBBER is cheap and easily cut to shape, but unless covered with cotton or linen, it does not wear well and tends to draw the horse's back, causing sweating and ruffling of the hair. Many new synthetic materials are now on the market.

A WOOLLEN BLANKET folded in four makes an effective back protector if a saddle has insufficient padding.

NATURAL FIBRES – wool, linen, cotton – are the best choice as some horses are allergic to man-made fibres.

Numnahs are best washed with soap, as again some horses are allergic to detergents. Numnahs must be kept very clean – even if it means daily washing.

Numnahs are secured in place by:
- Light leather straps and buckles.
- Nylon loops, which fit round the girth straps.
- Velcro straps.
- Strong elastic.
- Leather loops, which are the most satisfactory type for hard or competition work.

Saddle Pads

Saddle pads come in a wide variety of synthetic materials and range from gel pads, through Polypads to riser pads. They are used under the saddle and are purchased according to individual preferences or needs. Most are designed to increase the horse's comfort and provide protection; however, riser pads can affect the balance of the saddle by raising the area under the cantle.

WITHER PADS

These can be of woollen cloth or foam rubber. They are used under the front arch of a saddle, which would otherwise press on the horse's withers. They should be considered an emergency remedy, as their use often upsets the balance of the saddle and encourages the rider's seat to slip back. The saddle should be restuffed to make the use of such pads unnecessary.

They can also be used successfully under stable rollers and breaking rollers.

CRUPPER

This is usually worn by small, fat ponies or donkeys to prevent the saddle or roller from slipping forward.

The crupper consists of an adjustable leather strap put through a metal ring or 'D' on the cantle of the saddle. The other end of the strap is a padded and rounded leather loop, through which the tail is placed. This may or may not have a separate fastening.

When fitting the crupper, place it over the tail first, well up under the dock. Then connect it to the saddle and adjust it to a comfortable length. Care should be taken when first fitting, as ponies not used to a crupper may buck or kick. If they are worn frequently, the skin under the tail may become sore, and this area should be regularly examined. The crupper loop must be kept soft and supple.

GRASS REINS

These are adjustable narrow leather, nylon or cord reins which are attached round the girth straps or clipped to the saddle 'Ds'. They are then crossed over the withers and attached to the bit rings, or they may be fitted through the browband loops and down to the bit. They are used to prevent a pony from putting his head down to eat, thus pulling off a small rider. They also help to control a pony who bucks. Note that a crupper may be necessary to prevent the saddle from being pulled forward.

NECK STRAP

This is a strap which is fastened round the horse's neck and buckled on the near side. It can be a stirrup leather. It is used (a) to give unbalanced, novice riders something to hold on to rather than the reins, and (b) to help more experienced riders on young or fresh horses who might buck. If required, it can be attached to the front 'Ds' of the saddle and/or to the girth between the front legs.

HEADCOLLAR

A good-quality headcollar is an expensive item because of the amount of leather and stitching involved. Various designs are available: brass-mounted with buckles on both sides being the smartest (as well as the most expensive). A cheaper headcollar with galvanised or tin fittings and less stitching is perfectly adequate; in many cases the stitching is now replaced by rivets. On some designs the front may unbuckle, which makes it easier when brushing the face and when putting on a bridle.

Nylon headcollars are a cheap and suitable substitute for leather but they can be dangerous as the nylon is virtually unbreakable.

A headcollar should be fitted so that the noseband lies 5cm (2ins) below the cheek bones and allows a hand to be held between the leather and the jaw bones. Tight headcollars cause discomfort and result in sore patches.

Foal slips should be made of leather, with adjustable

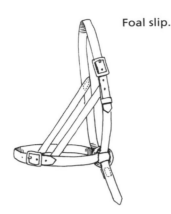
Foal slip.

buckles on the headpiece and noseband to allow for growing. Nylon foal slips are unsafe. A 16cm (6ins) strip of leather, which a hand can hold, should be attached to the back of the noseband.

ROPES

These are made of:

HEAVY JUTE, which is strong and stands most wear.

COLOURED COTTON, which is attractive but can snap.

NYLON, which is strong but which should be discarded when frayed; a frayed rope, when knotted, can be impossible to untie.

PLAITED BALER TWINE. This, if carefully plaited, with no loose ends or knots, is effective and is also easily and cheaply replaced.

Ropes are made with an eye, or are fitted with a spring hook which when attached to the headcollar should face away from the horse's chin and towards the neck. On occasions, spring hooks have been known to catch in a horse's face.

HALTERS

These are made of webbing or rope, and usually have the rope as an integral part of the halter. The

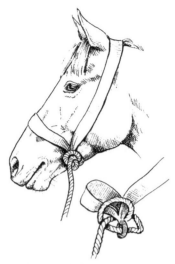

Halter with knot to ensure security.

'Yorkshire' type of webbing halter is very strong and has a string throatlash which prevents the halter from being pulled over the head. Rope halters are adjustable and can be altered to fit any size of head

It is important to ensure that the nose section of the halter cannot pull tight and panic the horse. After fitting, a knot should be tied by putting the lead rope round the noseband and pulling it back through the loop. This prevents the halter from tightening and also from working loose.

LUNGEING EQUIPMENT

Cavesson

Modern lungeing cavessons are lightweight and made of either leather or man-made fibre. They must be well padded in the nose area, so that pressure will not cause a sore. There are two patterns:

1. The nose piece is made like a drop noseband with the back straps attached to rings, so that they can be fitted either above or below the bit.

2. The nose piece is made like a cavesson noseband, and fits round the nose above the bit.

Both patterns have a jointed metal plate on the front of the noseband with a central swivel-mounted ring. They may also have small rings on either side. They should be firmly fastened or they will cause rubbing on the front of the nose.

A securing strap is set about halfway down the headpiece. This passes over the cheek bones and fastens on the near side. It holds the cavesson in place and must be firmly fastened. If it is set too high, its firm hold causes discomfort and affects the horse's ability to flex at the poll.

Lungeing Roller or Pad

This is made of leather, man-made fibre or strong reinforced webbing approximately 10cm (4ins) wide, and should be well padded over the back. It can be shaped in the elbow area to avoid rubbing the horse's sides. It should have two buckles on either side to ensure that when the roller is fitted the rings on either side are at the same height. It is also safer to have two buckles when dealing with a young horse who may play up as the buckles are tightened.

Rings should be fixed on to the roller:

- On the front and back of the top.

- Approximately 5cm (2ins) below the top on either side.

- On the girth section near the buckles.

- Underneath of the girth.

- 'Ds' should be fixed as for a rug roller, to hold the breastplate.

If required, a pad may be put under the roller. For a young horse, who may buck, the pad should be attached to the roller.

Breastplate

This is a leather or webbing strap placed round the chest and fastened to the lower rings or 'Ds' of the roller to prevent it from slipping back. For lungeing young horses it should always be used, even with side reins.

Side Reins

These are adjustable leather straps about 1.8m (6ft) long, which run from the roller or saddle to the bit. They are used:

- To teach the young horse to accept the bit.

- To help to keep the horse straight, and if lungeing in a paddock to prevent him from snatching at a mouthful of grass.

- To increase control over a fresh or strong horse.

- To improve the trained horse's way of going.

- To teach a horse piaffe from the ground.

- To lunge a rider and so help to keep the horse in better balance and give the rider a better feel of the paces.

There are two types:

1. At one end the leather strap fits through the rings on the roller (or round the girth of a saddle), and returns to buckle on itself. To allow for adjustment, there should be at least twelve numbered holes. At the other end, strong spring clips facing outwards and away from the horse are clipped to the bit rings, or on occasion to the side rings of the cavesson.

2. Similar to (1), but with a loop to go round the roller or girth and an adjustable fastening in the middle.

Some side reins are made with rubber rings or strong elastic inserts. If the elastic becomes stretched, it must be replaced.

Some trainers prefer not to lunge in side reins, because they feel that the horse should be free to adjust his head and neck position according to his natural balance. This applies particularly in the early stages.

When leading the horse in and out of the stable or when the horse is not being worked, side reins should be unfastened and put over the withers or clipped to the side 'Ds'. Side reins should not be attached until the rider has mounted the horse, and they should be fitted with care. Whether training the horse, or when lungeing the rider, the side reins must never be fitted so tight that free forward movement is hampered. It is advisable to lunge an unfamiliar horse without side reins for a few minutes before fitting the side reins.

Lunge Reins

A lunge rein, unless for ponies, should not be less than 10m (33ft) in length. It should have a loop at one end and an attachment to the cavesson at the other.

MATERIALS FOR LUNGE REINS

- Tubular or flat linen webbing.
- Nylon.
- Rope.

Webbing is strong but may be a little heavy. Nylon is strong but very light – this can make it difficult to handle, and even with gloves on it can cut into or burn the hand. Rope is strong, and lighter than webbing, but can be more difficult to handle and may burn the hand.

CAVESSON ATTACHMENTS FOR LUNGE REINS

- Swivel joint with buckle or strong clip.
- Strong clip.
- Leather strap and buckle.

If the cavesson has a swivel it is not necessary to have a swivel on the buckle or clip, as this makes the attachments heavy for a young horse. When a horse is jumping on the lunge it is cumbersome and can bang his nose.

Lunge Whip

This should be carefully chosen. Most whips are now made of fibreglass, with a thin plaited thong about 2.4m (8ft) long, with a lash at the end. The whip must be well balanced, and not so heavy that it will be tiring and difficult to manipulate. It must be long enough for the lunger to influence the horse and, when necessary, to touch him with the lash. To give extra length, a short lash can be replaced by a long leather bootlace.

Crupper

The use of a crupper when lungeing is optional. It can be a helpful extra discipline for an obstreperous young horse. It should be of leather, with a well-padded dock piece and a separate buckle for fastening round the dock.

Long Reining

Equipment for this is similar to that for lungeing, but it includes two reins which are often lighter and shorter than a full lunge rein.

NB: When lungeing or long reining it is essential for the lunger to wear gloves and stout, sensible footwear. A hat is advisable.

GADGETS

Gadgets are auxiliary aids which, by positive action, govern the position of the horse's head. In experienced hands they have their uses in the restraining of spoilt or 'nappy' horses. They can also be used, on veterinary advice, in the building up of back and hindquarter muscles after injury. In the hands of inexperienced riders they may produce more problems than they cure and can result in severe muscle strain. They have no place in the classical and systematic schooling of the horse.

Running or Draw Rein

This is the simplest and most commonly used auxiliary aid. It consists of a double-length rein made of leather, webbing or nylon, with a loop at either end. It may or may not have a buckle in the middle. It is fitted by passing the ends of the reins through the bit rings (from the outside to the inside) and back to the girth. The loops may go between the horse's front legs and through the girth, or be put through the girth on either side of the horse at the required height. The lower the rein the more effective and also more severe the action.

Market Harborough

This is a type of running martingale, and is the least severe of the auxiliary aids. The divided pieces of the martingale go through the rings of the bit. They either clip back on to 'Ds' stitched on the reins, or are attached by an adjustable buckle. When the horse's head is in a normal position it has no restrictive influence, but it comes into use if the horse raises or throws up his head. It can therefore be effective on headstrong horses. It is seldom seen today.

Chambon

This is used for lungeing. It consists of a divided rein which goes from the girth through rings on a special fitting attached to the headpiece behind the horse's ears, and from there to the bit where it is attached by clips.

A horse wearing a chambon is worked on the lunge in a circle of 20m in a slow trot, and with the equipment loosely fitting until the horse is familiar with it. It is gradually tightened, so that the horse moves with a lowered head. He will then accept the restriction without worrying.

Schooling time should start with two to three minutes on each rein, and should never exceed a total of 20 minutes.

The horse should not be expected to move in a working trot. If this is demanded, it can only have the affect of forcing the horse on to his forehand.

The horse should never be ridden in a chambon.

In the hands of an expert, lungeing with the chambon can be beneficial, but unskilled use can easily cause damage.

There are various other types of gadget on the market,

most of which, working on some form of lever principle, persuade the horse to lower his head and go in a rounder outline. Some gadgets can be attached to the tail: but it is worth remembering that the tail is an extension of the horse's spine. If the horse is not permitted to carry his tail in a natural manner, the use of the back muscles can be adversely affected and the muscles of the tail may suffer permanent damage.

One gadget, **The Balancing Rein**, introduced by Peter Abbott-Davies, can be placed in three positions: (1) mouth to girth;

(2) mouth to tail, with a rope passed through a soft sheepskin sleeve; or

(3) mouth to behind the ears by means of a soft rubber connection.

To summarise: all gadgets need understanding and experience for correct use. In most cases the value of a schooling aid is in the retraining of a spoilt horse. Muscles must always be allowed time to work hard and develop. Force will never achieve results and a stressed, overtired horse will resist and work will be wasted.

28 BITS

The purpose of a bit is to enable a rider to control and guide his horse. The success of this control also depends on the effectiveness of the rider's seat and leg aids. The bit functions by bringing pressure to bear on the horse's mouth and head, either by direct action of the bit or by the indirect action of the bridle. This action will be affected by the horse's mouth, the angle at which he carries his head, the height of the rider's hands, and the design and fit of the bit.

STRUCTURE OF THE MOUTH

The horse has an upper and lower jaw, the comparative length of which varies in different breeds and can affect both the action of the bit and that of the curb chain.

The parts of the lower jaw bone lying between the incisor teeth in the front and the molars at the back of the mouth are called '**the bars**'. They are covered with a layer of skin and flesh containing nerves. In a young horse, this area is very sensitive, but can become deadened by rough use of the bit.

The top of the mouth is formed by the **palate** – a very sensitive area.

In stallions and geldings, **tushes** grow on either side of the jaw between the incisors and the molars. If present in mares they are usually rudimentary and barely visible. **Wolf teeth** may grow at the front and to the side of the first molars. They can cause bitting problems and should be removed.

The areas on which the bit and/or bridle act are:

- The corners of the mouth and lips.
- The bars of the mouth.
- The tongue.
- The roof of the mouth.
- The side of the face.
- The chin groove.
- The nose.
- The poll.

PROBLEMS IN THE MOUTH

CAUSES
- Sharp molars.

- Wolf teeth.

- Sensitivity of the skin covering the bars of the mouth.

- The length of the jaw, or the position and size of the tushes.

- The shape of the jaw relative to the size of the tongue.

- A badly fitting or badly fitted bit.

If after inspection the problem does not appear to lie in the mouth, other possible causes are:

- Physical discomfort and/or pain caused by ill-fitting tack other than the bit.

• Pain from unsoundness in the back or legs.

• The hands of an insensitive or unbalanced rider.

If the horse has discomfort or pain in his mouth he may try to ease this by altering his head position and opening his mouth, which brings pressure to bear on a less sensitive surface, i.e. the corners of the mouth rather than the bars. If the latter become deadened because of insensitive treatment, then all too often the rider will seek a stronger bit and a more severe method of control. If mouth problems are encountered, the cause should first be investigated. In many cases, a change of bit to a milder type often solves the problem.

Types of Problem

• A too narrow bit pinches and rubs the corners of the mouth. A too wide bit runs through the mouth when one rein is pulled, and a joint may even go across the mouth and rest on the bars. If jointed, it may press on the roof of the mouth, when the rider has a less direct feel of the mouth.

• A horse with a shallow mouth space between the upper and lower jaws, or a thick tongue, may find a thick bit uncomfortable because of tongue pressure. The joint of the bit may press up into the roof of the mouth, and a bit which is too wide aggravates the problem. With such horses, make sure that the bit fits the mouth, i.e. that it only extends just to the outside of the lips. Use a tapered mouthpiece, i.e. thick at the ends and thinner towards the centre. Try a double-jointed or French snaffle, as the smooth centre plate should lie comfortably on top of the tongue and there is no nutcracker action.

• A horse with a normal mouth space and sensitive bars may find a thin mouthpiece sharp to the tongue, and uncomfortable on the bars. Use a thick-jointed or double-jointed mouthpiece. This can be of the German hollow mouth type – which is light – or it can be covered in rubber or leather.

• A horse can double back his tongue and try to put it over the bit. This may be due to discomfort, especially if the bit is relatively low in the mouth, when there will be excessive tongue pressure. A mullen mouth snaffle can be more comfortable, and if fixed high in the mouth may also discourage or check the habit of tongue over bit. Alternatively try a single- or double-jointed snaffle fixed high in the mouth.

• A dry mouth is the sign of an insensitive mouth, and the horse is often very strong. He must be encouraged to salivate, and bits which stimulate this are leather and copper types, and key-ring snaffles.

• A horse often tries to avoid the action or message from the bit by opening his mouth and/or crossing his jaw. Various nosebands can help this problem (see pages 197–8).

CONSTRUCTION OF THE BIT

The mouthpiece of a bit can be made of metal, vulcanite, nylon, rubber or leather. The metals most commonly used are stainless steel, composition alloy, chromium-plated steel and nickel.

Types of Material

STAINLESS STEEL is the safest and best-looking metal. The second choice is one of the named composition alloys.

CHROMIUM PLATED STEEL is safe, but the chrome has a limited life and flakes off, causing a roughened surface and exposing rusty steel.

PURE NICKEL BITS are still in common use, but they can be dangerous, as they snap without warning. Nickel is a soft metal which wears quickly. It is used in snaffles of the flat-ring type, where the ring goes through the bit. It often wears, becomes sharp, and can rub a horse's mouth. The centre joint of such snaffles can wear through, at first causing discomfort, and then coming apart. The colour, depending on care, ranges from dull to bright yellow.

ALUMINIUM can be used if a light but strong bit is required, as in racing.

COPPER is a soft metal. It is suitable for covering steel mouthpieces, but where there is any strain, it should not be used on its own. Some horses are known to go well in copper-covered bits, as the taste of the metal can encourage salivation.

VULCANITE is rubber hardened by heat. It makes a thick, heavy mouthpiece, which is kind in action.

NYLON is very light but strong.

RUBBER makes a very soft but kind bit. It should always have an internal mouthpiece of linked steel.

RAWHIDE LEATHER or LEATHER STRIPS can be stitched over an internal mouthpiece of steel or copper wire.

'HAPPY MOUTH' or 'NATHE' bits are mild, and made of a synthetic but hard-wearing material.

'SWEET IRON' bits are also popular and much used. They encourage the horse to salivate.

Bit Fittings

At each end of the mouthpiece are the fittings to which the headpiece and reins are attached. They can be of various designs. Loose- and flat-ring snaffles allow a certain amount of movement of the bit within the rings. Eggbutt, D-shape and cheeked snaffles limit this movement. Examples of the different types are found in the following text.

To Measure a Bit

Lay the bit flat and measure the distance between the inner sides of the bit rings or cheek.

Bits are made in sizes ranging from 9cm (3½ins) to 15cm (6ins), and go up in 0.5cm (¼in.) to 1cm (½in.) stages.

To Measure a Horse's Mouth

Hold up a piece of thick string in the mouth. Mark with your fingers where the sides of the corners of the lip occur. Remove the string and measure between the marks. It can often be difficult to buy the exact width of

bit required, particularly in the smaller sizes, and it may be necessary to approach several suppliers. Remember that the height of the horse will not reflect the size of the required bit. This will be governed by type and breeding.

TYPES OF BIT

There are five main types of bit or bridle, all of which have many variations. These are:

- the snaffle;
- the double;
- the pelham;
- the gag snaffle;
- the hackamore/bitless bridle.

NOTE: There are restrictions on bits used in competitions governed by the FEI and the relevant competitive discipline's rules. Illustrations of permitted bits can be found in the relevant rule books.

THE ACTION OF THE SNAFFLE BIT

There are many variations of the snaffle, but the main ones are:

- Solid mouthpiece with no joint.

- One joint with two arms.

- Two joints with two arms and a middle link, which encourages more play in the mouth and has no nutcracker action.

- A straight bar or curved (mullen mouth) mouthpiece with no joint works on the lips and directly on to the bars of the mouth. If the tongue is large there is also considerable tongue pressure. If the tongue is small the bars will take more pressure.

- With a single joint mouthpiece there is a nutcracker (squeezing) action when the joint closes. It also acts on the bars of the mouth and the lips, but there is less tongue pressure than with a straight bar.

- A snaffle with two joints acts on the bars of the mouth and the lips. It has no nutcracker action. It may drop down on the tongue.

- A thick mouthpiece is milder than a thin one.

- Eggbutt rings are less likely to pinch the lips or corners of the mouth than are loose, flat-ring or cheek snaffles.

- A loose-ring snaffle allows more movement in the mouthpiece and encourages a horse to champ the bit.

- The severity of the bit is increased by any uneven or squared-off bearing surface, i.e. twisted, plated, roller.

- A straight bar snaffle puts pressure on the tongue and then the bars.

- A cheek or Fulmer snaffle correctly fitted can give more control over a strong or uneducated horse unwilling to accept brakes or direction. The cheeks do not pull through the mouth, and when a direct rein is used, pressure is brought to bear on the other side of the horse's face and on the bars of the mouth. All cheeked snaffles help to keep a horse straight.

- A cheek snaffle used without keepers puts more pressure on the corners of the mouth, lips and tongue and less on the bars.

TYPES OF SNAFFLE

Single Joint

FLAT RING There is some risk of pinching the lips.

EGGBUTT Prevents rubbing at the corners of the lips.

GERMAN LOOSE-RING HOLLOW MOUTH The mouthpiece has a broad, thick design. The hollow mouthpiece makes it light in weight. The cylindrical loose ring reduces pinching.

D-RING RACE SNAFFLE This is less likely to be pulled through the mouth

LARGE RING RACING Will not pull through the mouth.

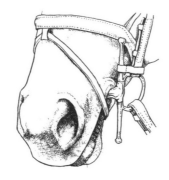

Cheeked snaffle, sometimes known as a Fulmer. Note how the cheek is attached to the bridle with a keeper.

FULMER OR AUSTRALIAN Keepers are attached from the top of the cheek to the bridle, which keep the bit upright and still in the mouth. They also prevent the joint from dropping down on the tongue. This bit is made of solid metal and can be heavy. The action is more 'fixed'.

CHEEK This should have keepers so that the joint is kept off the tongue. It can be a useful bit on a young horse. All cheek snaffles help to keep a horse straight.

SPOON This can have full cheeks or half cheeks, which can be either above or below the mouthpiece. The latter type is used for driving.

HANGING CHEEK This suspends the bit in the mouth and gives more room for the tongue. It can be useful for horses who put their tongue over the bit

TWISTED This is severe because of the ridged surface.

Double Joint

FRENCH BRIDOON This has a rounded plate which joins the two parts of the snaffle. The plate lies flat on the tongue. There is no nutcracker action from this bit, and many horses – particularly if a little set in the jaw – relax and go better. If this bit has square arms, or a sharp-edged plate, it is not allowed in dressage tests.

DICK CHRISTIAN This has a small ring joining the two arms which removes the nutcracker action and gives room for the tongue. Some horses resent the pressure of the small ring.

DR BRISTOL This has a plate with squared edges fixed between the two arms. The bit is severe, but when the flat plate is on the tongue it is less sharp than when the bit is turned the other way, and the plate goes across the tongue. It should never be used reversed.

Straight Bar

The following four bits are more comfortable for the horse, and have a milder action if they are of half-moon or mullen design.

> **VULCANITE LOOSE RING**
>
> **NYLON** This has a half-moon or mullen mouthpiece.
>
> **RUBBER** This is a very mild bit, but not durable.
>
> **METAL LOOSE RING**

LEATHER There are several types of this bit on the market. One is made from strips of leather stitched over a lining of steel or copper wire. It has a soft, broad bearing surface, and is a kind bit, useful for a horse with a very sensitive mouth or with mouth problems, e.g. bruised tongue, sore bars.

Severe Snaffles

The bits below have a severe action, and none is permitted in dressage tests. Before resorting to such bits, expert advice should be sought. These bits should only be used when other methods of control have failed.

WILSON This has four rings, the two attached to the mouthpiece take the reins, the other two take the headpiece. A severe squeezing on the face results when the reins are pulled.

MAGENIS This has set-in squared-off arms with a small set of rollers set horizontally with the arms. It prevents the horse from taking hold of the bit and running away.

CHERRY ROLLER This prevents the horse from catching hold of the bit and running away. The uneven surface is sharp to the tongue and bars.

'Y' or 'W' The two mouthpieces can give a pinching action to the lips and tongue.

SCORRIER This has a twisted mouthpiece, and two rings which act as 'Wilson Rings' (see above).

CORNISH or **WATERFORD** This is composed of several thick jointed plates which are similar in shape to those of a French bridoon.

CONTINENTAL SNAFFLE/DUTCH GAG or **BUBBLE BIT** This may have a single- or double-jointed mouthpiece. The rein can be attached to any of the three rings – the lowest gives greatest leverage with the first ring giving only a direct snaffle action.

ROCKWELL, NORTON, NEWMARKET These are forms of snaffle bridle which by various means produce pressure on the nose as well as on the mouth. They can be of some use on strong horses with uneducated mouths who have not been taught to obey, or who fight against more legitimate methods of control.

SPRING MOUTH This attachment is clipped to the rings of a snaffle giving a double bit action.

CHAIN SNAFFLE The mouthpiece is made of chain links.

TWISTED WIRE This has two mouthpieces of twisted wire.

THE GAG SNAFFLE A gag snaffle has two holes in the bit rings or cheeks, through which run roundings attached to the headpiece and ending in the reins. The mouthpiece is jointed and can be smooth, twisted or with rollers. It is generally used on its own, but can be operated as the bridoon part of a double bridle if greater control is required. Its action brings considerable pressure on the corners of the mouth and the poll. It can be effective on horses who put their heads down and pull, as its leverage action makes it possible for the rider to raise the horse's head. It should always have a second rein attached to the rings of the bit, which allows it to be used as a normal snaffle when the gag action is not required.

Used in the wrong hands this bridle can prove dangerous, making some horses rear. It should never be used by inexperienced children or novice adults.

The use of such bits may indicate failure in the correct training of the horse. Frequently the bits create more confusion in the horse's mind, thus compounding the problem.

Snaffles for Use on Young Horses

KEY-RING MOUTHING BIT This is still used by some trainers. The keys encourage the horse to play with the bit, but may also encourage him to put his tongue over the bit. Modern training methods require that the horse should learn to hold the bit quietly in his mouth with a relaxed jaw. For this purpose, a jointed or mullen mouth snaffle is now more acceptable.

TATTERSALL YEARLING BIT This is a circular bit sometimes used on yearlings to give greater control when being shown. The bit is attached to a show headcollar by short straps, and encircles the lower jaw with the key section in the mouth. The lead rein is attached to the back ring and when used the bit rises up in the mouth.

Special Bits

CHIFNEY ANTI-REARING This acts in a similar way to the normal ring bit, but if the horse should attempt to rear, the U shape puts greater pressure on his tongue: therefore discouraging the habit. It is made of thin metal and is severe .

HORSESHOE CHEEK STALLION BIT This is a showing bit and is buckled to a show headcollar. It has horseshoe-shaped cheeks, and a straight-bar or mullen-mouth bit which acts on the corners of the mouth, the tongue, and then the bars.

Fitting a Snaffle Bit

The bit should fit the mouth. It should be neither too small, when it will pinch, or too large, when it will pull through the mouth. It should not protrude more than 0.5cm (¼in.) on each side. A straight-bar snaffle should not wrinkle the corners of the lips; a jointed snaffle should do so slightly.

THE DOUBLE BRIDLE

This has two metal bits. The bridoon is a snaffle bit with small rings. The curb has no joint, and comes in a variety of designs, but it must have a curb chain. Some types do not have 'Ds' for the lip strap.

The double bridle is used:

- On horses which, having been trained in a snaffle, are ready to understand and accept a lighter and more refined aid from the rider's hand.

- In the showing ring, where its purpose is not only to add to the horse's turnout but also to increase control.

- In the hunting field and competitive jumping. It helps keep the horse in better balance, and acts as a more effective brake than the snaffle bridle.

Action of double bridle bits on horses trained in a snaffle:

- The bridoon acts in a similar way to the snaffle. It asks the horse to work on the bit, going forward with an even contact on both hands. It can raise the head.

- The curb bit, more correctly called 'the bit', gives a lighter and more refined aid. It helps to position the head as collection is developed through the horse's progressive training. It also helps balance and control. It should lie immediately below the bridoon, but not so low that it touches the tushes. Both bits must be able to act independently.

A curb bit with double-link curb chain and lip strap.

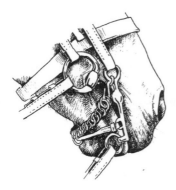

Correctly fitted double bridle.

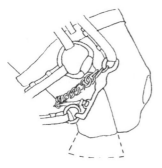

If curb bit and chain are fitted correctly, when a contact is taken on the rein there is sufficient play for the cheek of the bit to form an angle of 45° with the mouth.

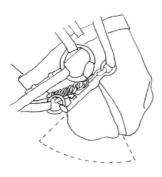

Too much play – curb action will cause resistance and mouth opening.

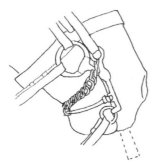

Not enough play – curb chain too short.

The mouthpiece of the bit acts on the tongue and the bars of the mouth, and as a fulcrum for the upper and lower cheeks of the bit. When the curb rein is used, the curb chain tightens and applies pressure in the chin groove. As this occurs, the upper cheek piece moves forward (its position governed by the tightness of the curb chain), and brings slight pressure to bear on the poll. The longer the lower cheek, the greater the pressure and severity of the action, both on the chin groove and the poll. A shallow tongue groove or port gives added room for the tongue; a high port presses on the roof of the mouth, and in the wrong hands can cause discomfort and pain.

The curb chain should be adjusted to ensure that the chain comes into action when the cheeks of the bit are at an angle of 45° to the mouth. If adjusted more loosely, both the bit and curb chain lose their true action and effectiveness. An adjustment tighter than 45° gives a more effective action, and requires a more precise use of the curb rein.

Types of Bridoon

These can be loose ring or eggbutt. They can have hanging cheeks and either one or two joints. The usual variations are:

- Ordinary bridoon with loose rings.
- Double-jointed bridoon with eggbutt rings.
- Hanging cheeked bridoon with one joint.

Types of Bit

These may have a tongue groove or port of variable height. The cheeks may also vary in length.

FIXED CHEEK with a low port and medium-length cheeks. This has a more immediate and precise action than the sliding mouthpiece.

WEYMOUTH SLIDING MOUTHPIECE with a low port and longer length cheeks. The mouthpiece moves up and down. The cheeks go through the mouthpiece, so that increased contact on one will bring the other into play.

THICK MOUTH GERMAN This has a thick, fixed mouthpiece which tapers in the centre to allow room for the tongue. The mouthpiece can be hollow. The cheeks can be straight or curve forward. Many thoroughbred horses are uncomfortable in this bit owing to its thickness.

HALF-MOON OR MULLEN MOUTH This has short, fixed cheeks and a half-moon mouthpiece which allows some room for the tongue; though when the curb rein is used, pressure falls more on the tongue than on the bars.

BANBURY The mouthpiece is a round bar which is slotted through the cheeks. It has no port. It can revolve to give an independent action to each cheek, and to the corresponding curb rein. It may also be made to move up and down. It is now rarely used.

Curb Chains

These can be of metal, with either single or double links. They can also be of leather or elastic. Rubber or leather curb guards can be fixed to metal chains. For fitting see pages 196–7.

THE PELHAM

This bit is a combination of a snaffle and a double bridle. It has one bit, the mouthpiece of which can be jointed, straight or half-moon (mullen), either with or without a port. It has a lower cheek, which takes the curb chain. It usually has two reins: one attached to the rings on the mouthpiece and the other to the rings on the lower cheek. It may be fitted with roundings on the bit and one rein. Some pelhams do not have a lip strap. The longer the cheek, the greater the leverage.

The pelham can be said to have a rather indefinite action, but many horses ,and particularly ponies, appear to work happily in it. It cannot be considered as a bit likely to improve a horse's mouth, but its use by an uneducated or heavy-handed rider probably does less harm to the horse's mouth. With a long-jawed horse it can be difficult to get the curb chain pressure in the chin groove.

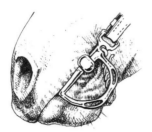

Square attachment for cheek piece increases severity of poll pressure.

A Kimblewick with curb chain and two rein positions for varying action.

Types of Pelham

MULLEN MOUTH This is widely used. The mouthpiece can be made of metal, vulcanite, nylon or rubber.

JOINTED This can be smooth or twisted, and has a more severe action than the mullen or vulcanite.

VULCANITE A mild type of pelham, which seems to suit a variety of horses whose owners find them too strong in a snaffle.

KIMBLEWICK A single-rein bit with a curb chain – neither a pelham nor a snaffle. The mouthpiece can be of metal or vulcanite. It is a strong bit, which some sensitive horses resent. The strength and severity of the Kimblewick is sometimes underestimated, and as such it is used on ponies who are too strong for their child riders in a snaffle. As with all strong bits, its continual use soon deadens a horse's mouth. The child rider may be better using a pelham, which is less severe.

SHOW HACK or GLOBE CHEEK These are curb bits used without the bridoon, and only have one rein attached. They are seen on show ponies and occasionally show hacks. They are severe, and need to be used with a light hand.

RUBBER This is a soft version of the vulcanite. It can encourage horses to chew and then tends to get worn through.

ARMY REVERSIBLE OR UNIVERSAL This was designed for army use to suit as many different types of horses as possible. It is reversible, allowing either the smooth or the twisted side of the bit to be in action. The top rein is attached to 'D'-shaped rings; the lower rein can be used in either of two slots according to need. It does not have a lip strap.

THREE-IN-ONE or **SWALES** This is a very severe bit. It acts without poll pressure, the headpiece being supported by rings on the mouthpiece. When the curb rein is used, very severe pressure is put on the chin groove. If the top rein only is used, it is much less severe and many horses go well in it.

THE BITLESS BRIDLE OR HACKAMORE

This bridle is designed to control the horse by applying pressure on the nose without the use of a bit. It can be effective on horses with damaged mouths, or on those who through bad handling resent and fight a normal bridle. It gives good control, but steering can be difficult.

It consists of a bridle head, cheeks, and padded noseband held in place by a strap or chain. Reins are attached to rings on the noseband, and these apply pressure via the strap or chain. Considerable leverage is obtained on the nose, and to a lesser degree on the poll, the chin groove and the lower jaw. The precise action depends on the design of the bridle. Some hackamores can be very severe and should only be used by riders aware of their action and capable of using them with tact and discretion. They should not be used on ponies or by uneducated riders.

Bitless bridles are not permitted in dressage competitions.

A simpler form of control – but which is only suitable for exercising in a confined space – consists of two rings attached to a well-padded and firmly fixed dropped noseband.

BIT ATTACHMENTS

RUBBER BIT GUARDS fit on each side of the bit, and protect the face and corners of the lips of a sensitive horse.

TONGUE GUARDS are made of metal, and are held in place by a separate head strap. They are fitted into the horse's mouth above the normal bit to prevent a horse putting his tongue over the bit.

RUBBER TONGUE GRIDS fit on to a narrow straight-bar snaffle and may help to prevent a horse putting his tongue over the bit. It is important to check that they cannot be easily dislodged, as they may either be spat out by the horse or swallowed.

AUSTRALIAN CHEEKER This is a rubber device running from the headpiece down the front of the horse' s face. Attached to each side of the bit, it helps to keep the bit high in the mouth, and prevents a horse getting his tongue over the bit. It is most often used on racehorses.

29 BOOTS AND BANDAGES

BOOTS

Boots are used to protect a horse's legs and joints from injury, either self-inflicted or from an outside source. Self-inflicted injuries are the result of:

- Faulty conformation.
- Faulty action.
- Poor shoeing.
- Lack of balance in a young horse.
- Poor condition and lack of muscle.
- Fatigue.
- Careless riding.
- Galloping in deep going.
- Pecking over a fence.

Types of Injury

BRUSHING This occurs on the inside of the leg below the knee or hock. It is inflicted by the opposite foot, usually in the area of the fetlock joint.

SPEEDICUT This occurs when a horse is galloping and cuts into the leg just below the knee or hock with the opposite front or hind foot.

LOW OVER-REACH This is a bruise or cut on the heel of a front foot. The front foot stays on the ground too long, and the inside edge of the hind shoe strikes down into the heel. It usually occurs over a jump, when an extra effort has been made and the jump was unbalanced, or the landing in deep going. However,

big-moving horses can lose their balance on the flat and over-reach.

HIGH OVER-REACH This is an injury to the back tendon caused by the horse when over-jumping or losing balance when galloping. The front of the hind foot or front outside toe edges strikes into the back of the front leg below the knee. It is always serious, even if there is no cut. When caused by another horse it is referred to as 'struck into'. All over-reaches must be treated seriously as they entail internal bruising and a possible seat of infection.

TREAD This can be self-inflicted, when a horse treads on the inner coronet of one front foot with the other front foot. It can occur during travelling or when a horse is weaving. Injury from another horse treading on the outside of the coronet can occur:

- When travelling with another horse if there is no partition.
- During a game of polo.
- Among a restless group of horses, e.g. out hunting.
- When riding one horse and leading another.
- Riding too close behind another horse.

Treads can be serious if they cause internal bruising.

BRUISED SHINS These are caused by rapping either front or hind shins on a jump.

POLO KNOCKS These can occur anywhere on the legs;

but they happen more usually on the front legs below the knee. They are caused by a blow from a polo stick or from the ball.

USE OF BOOTS

When schooling valuable horses of any age it is a sensible precaution to fit protective boots according to the work to be done. This often saves an unnecessary injury, which could lay the horse off work.

Boots are not advisable when out hunting or on long-distance rides, as mud works up under the boot. Pressures and friction over several hours will cause a sore place, which can quickly become infected and cause acute lameness.

Boots are designed to give protection to a specified area of the horse's leg. In some cases, it may be necessary to have them individually made to alleviate a particular problem.

MATERIALS

- Boots are made of a wide range of synthetic materials, leather (usually lined with sheepskin or foam rubber), rubber or heavy cloth such as Kersey or box cloth (a lighter material).

- Regardless of material, it is important that after use, boots are either brushed clean, or if wet, washed.

- Leather boots and fastenings must be kept supple.

- Leather or cloth boots, if not well cleaned, become hard and cause pressure sores on a horse's leg. These can quickly become infected, resulting in a lame horse.

- Synthetic material is less likely to cause this problem, but cleaning after use is still essential. Synthetics are lighter, more comfortable for the horse, and easier to clean or wash. They are generally preferable to boots made of traditional materials and are extensively and predominantly used today.

FASTENINGS

Boots should always fasten on the outside of the leg, with the end of the strap pointing towards the rear. Fastenings can be buckles, clips or Velcro. Front leg boots should have three or four fastenings. Hind leg boots, which are longer, should have four or five. A short type of boot may have only one fastening.

Leather boots always have sewn-on leather straps and buckles. These are secure, but are time-consuming to fasten, and the stitching needs regular inspection. Kersey, box cloth and synthetic material have buckles, clips or self-holding Velcro fastenings.

Fastenings are often set on strong elastic, which makes the boots more comfortable for the horse. The elastic must be regularly inspected. If it becomes stretched and loses its strength it must be renewed.

If Velcro is used, make sure that the straps are wide enough and long enough, so that the boot is held securely. The Velcro must always be kept clean. If it wears smooth it must be renewed. As Velcro fastenings are easily dislodged in rough conditions, they are not suitable for hacking through mud or for cross-country work. They can be made more secure for these purposes by wrapping adhesive tape several times around the Velcro fastening. Care must be taken, however, not to pull the tape tighter around the fastening and so exert uneven pressure.

TYPES OF PROTECTIVE BOOT

BRUSHING BOOT This is full size and protects the whole of the inside of the leg below the knee, including the fetlock joint. The boot should be placed on the leg rather higher than in its final position. It should be firmly fastened, starting from the top, and then eased a little down the leg so that it protects the fetlock joint. The boots should then be checked to see that they are firm, cannot move about, but are not too tight.

FETLOCK BOOT A short boot, usually of leather, with one strap. It gives protection to the fetlock joint only.

YORKSHIRE BOOT This is usually put on a hind joint. It is a piece of heavy cloth about 30cm (12ins) long and 20cm (8ins) broad, with wide tape or Velcro stitched about two-thirds down its length. It is then wound round the leg, fastened on the outside above the fetlock joint and eased down over the joint. The cloth above the tape is then folded over to form a double thickness. Note that in deep going the boot can be sucked off. Yorkshire boots are rarely seen today, having been superseded by synthetic fetlock boots.

RUBBER RING This is a hollow ring with a leather or metal fastening running through its centre. It can be used round the horse's pastern or above the fetlock joint, and protects either area.

SAUSAGE BOOT This is a large, leather-covered, padded ring. It is fastened round the pastern, and prevents a horse bruising an elbow with the heel of the same front foot when lying down.

SPEEDICUTTING BOOT This is similar to a full-length brushing boot, but is cut to come higher up the leg and to give greater protection to the inner, lower part of the knee or hock.

OVER-REACH BOOT There are a number of different types of over-reach boot including the well-known bell-shaped boot made of rubber. This type is pulled on and off over the foot, which can be difficult. Care should be taken not to twist the foot when applying or removing these boots otherwise damage to the pastern joint can result. Over-reach boots made of individual rubber 'petals' are fastened round the pastern, making them easier to put on. Movement through the individual 'petals' prevents the boots from turning inside out and also allows replacement of damaged or missing 'petals'.

TENDON BOOT This is shaped like a brushing boot, but has extra padding down the back of the leg to protect the tendon from being struck. Open-fronted tendon boots leave the cannon bone area exposed and unprotected and are often used for show jumping.

CORONET BOOT This is a small, semi-circular covering, usually of leather. It fastens round the pastern, and

protects the coronet. It is used for polo or travelling. Horses can easily tread on themselves or one another if travelling without a partition. The bruising may be invisible, but can be quite severe. Bandages and Gamgee give some protection if they are put on to reach down over the coronet, but coronet boots can be a wise precaution.

POLO BOOT This is similar to a brushing boot, but is made of heavy felt, and is longer – coming well down the fetlock. Some types which cover the coronet also strap round the pastern. The boots are put on in the usual way, or they may be secured with firm bandages, the latter being fastened with cotton or self-adhesive tape and the ends sewn for extra security.

HEEL BOOT Racehorses and eventers, when galloping, sometimes injure the point of the fetlock joint around the ergot. A brushing boot, cut lower at the back and shaped to the joint, will prevent this.

EQUIBOOT This is a patented boot made in varying sizes to fit over a horse's foot. It is used:

- To protect the foot if for any reason the horse cannot be shod and is required for work

- In the stable, to keep a foot poultice in place, especially for a horse which chews at bandages and sacking.

Its exact fit is very important. A boot which is too tight can bruise the heels and set up an infection. If it is too loose, it will come off. The fit can be adjusted by tightening or loosening the heel pads at the back of the boot.

POULTICE BOOT This is made of synthetic material or leather. Its purpose is to hold foot dressings in place. It fits over the foot and is fastened up the leg.

TREATMENT BOOT This is similar to a wellington boot, and reaches from the foot to well above the knee. It is used for cold-water treatment to an injured leg.

TRAVELLING BOOT Many people use a well-applied stable bandage with a generous lining of Gamgee or

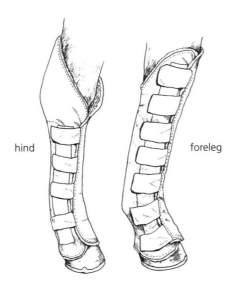

Travelling boots.

similar material. Bandaging, however, takes time, and labour-saving alternatives are leg guards or travelling boots. These are made from synthetic materials, and may have a thick, fleecy-type lining. They are fastened with several wide Velcro straps. They should be long enough to reach from above the knee and hock down over the coronet. It is then not necessary to use knee caps and hock boots. Young or nervous horses need to be familiarised with the feeling of having the whole leg enclosed in a fairly inflexible casing; they can also be alarmed by the noise of the Velcro being pulled apart, and accidents have been known to be caused by their use.

KNEE CAPS These are used for travelling or exercising on roads, and are made of leather, heavy cloth or synthetic material. To be smart, the material should match the rest of the travelling equipment.

The front of the knee cap is covered with a reinforced pad of leather. The straps are leather. The top strap should be set in strong elastic to allow it to be firmly fixed without causing the horse discomfort. It should be fastened well above the knee on the outside. The spare end of strap will then be forwards. The knee cap can then be eased down and tested to make sure it cannot slip over the knee.

The bottom strap must be fastened very loosely, usually in the first hole, so that it cannot interfere with the action of the joint.

For long journeys pad the back of the top strap to prevent a pressure injury.

SKELETON KNEE CAPS These are lightweight knee boots consisting of a top strap and a reinforced leather pad to cover the knee. When hacking out on roads, skeleton knee caps can be a wise protection. For general use they should have two straps. For jumping, to prevent further bruising or to protect an old injury, they should have top straps only.

HOCK BOOTS OR CAPS Hock boots are designed to protect the point of the hock when travelling. They are of similar construction to knee caps. The top holding strap is usually set in strong elastic so that it can be firmly buckled without cramping the joint. It is fastened on the outside, quite close to the front of the hock. The bottom strap should be fastened more firmly than that of the knee cap but not too tightly. The spare end of strap should be towards the rear. The horse should be familiar with the feeling of the hock boots before they are first used for travelling.

BANDAGES

Bandaging the Legs

Whichever type of bandage is used, it is essential for them to be correctly and skilfully applied. Badly or carelessly applied bandages can cause untold damage to a horse's legs.

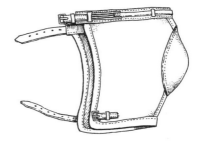

Hock boot.

Various methods of bandaging are acceptable, but the essential requirements are:

- The bandage must be correctly and firmly rolled. First, roll the tapes or fastenings in the position where they are stitched. Then roll the bandage round them.

- The bandage should be applied only as firmly as is needed to keep it in place and to prevent it slipping.

- It should be applied without wrinkles. The pressure should be minimal but even. The tapes should not be tied any tighter than the bandage.

- Tapes should be tied on the outside and the loose ends should be tucked away.

CAUSES OF DAMAGE

- Tapes and/or bandages which are too tight, or a twisted bandage, can cause marks on the leg and possible tendon damage.

- Tapes tied at the back or front of the leg may cause pressure and injury to the tendons. Tapes tied at the front are likely to bruise the cannon bone. Those tied on the inside may be interfered with by the other leg, be less secure and less accessible.

- A poorly applied stable bandage can slip down and become dislodged when the horse stands up after resting. It may then be stepped on and torn. Poorly applied work bandages could cause injury to horse and/or rider if they loosen during exercise.

Types of Bandage

STABLE BANDAGES are used:

- For warmth and to keep the circulation active.

- To assist in the drying-off of wet legs.

- For keeping poultices in position. Crepe bandage can also be used for this.

- To protect the legs whilst travelling.

- As a support for a sound leg when the opposite one is being treated for injury.

- To prevent the legs from 'filling'.

EXERCISE BANDAGES are used as a protection against injury during work.

SURGICAL BANDAGES are used to cover and protect wounds, for poulticing of the leg when there is a likelihood of swelling, and for wounds on joints. They are usually made of a synthetic stretch material.

TAIL BANDAGES are used:

- In the stable to help keep a tail tidy.

- When travelling, as a protection, both for the tail hair and the dock.

Under-bandage Protection

When worn under bandages the following give added protection and warmth:

GAMGEE is cotton wool in a muslin cover. It gives the best protection, but is expensive and easily soiled. If the edges of the Gamgee are machine stitched, it is easier to wash and will last longer.

LEG WRAPS are thick felt pads cut to fit the leg. They are more expensive, but are quick to put on, easy to wash, and durable. Long term they are very economical.

FYBAGEE is a man-made cloth, similar to thick felt. It is easily washed and durable.

FOAM RUBBER is cheap and easy to wash, but can draw the leg, and some horses are allergic to it.

HAY OR STRAW can be used for drying off wet legs. Pummel it well, then wrap it round the leg before bandaging. This is known as 'thatching'.

Bandage Fastenings

TAPE This should be wide and flat. When the bandage is washed, the tape should be smoothed out and, if necessary, ironed flat.

VELCRO This is a rough synthetic material, and is self-

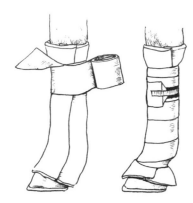

Putting on a stable bandage.

holding. It is efficient and easy to use as long as it is kept clean. In a shavings bed it can become impregnated with dirt, and then will not hold securely. It must also be long enough to secure the bandage.

CLIPS These are satisfactory for stable bandages, but their shape can cause unnecessary pressure.

SELF-ADHESIVE TAPES These are separate from the bandage; they add security to existing tapes. For extra security during competition, racing, or strenuous work, the end of the tapes can be sewn to the bandage. Surgical and crepe bandages are usually fastened with adhesive tape but the tape must not be tighter than the bandage itself.

1. Stable Bandages

These can be made of wool, stockinette or man-made fibre. Thermal bandages are also available: one type keeps the leg warm, and the other, cold; the latter can be useful after an injury. Full-size bandages should be 10–12cm (4–5ins) wide by 2.1–2.4m (7–8ft) long. Unless of double thickness – e.g. Newmarket bandages – they should be always used with padding underneath.

Additional Requirements

- POULTICING THE LEG A double thickness of Gamgee is often used for extra protection should there be a tendency for the leg to swell.

- TRAVELLING The Gamgee or other lining should extend well up over the knee and well down over the coronet. The bandage is applied as described below.

- SUPPORT BANDAGES A horse who injures a front leg is likely to put more weight on the other leg, which can result in a strain. A supporting bandage firmly applied over a thick layer of Gamgee will help to avert this.

To Put on Stable Bandages

- Place the lining around the leg. It should show 1.2cm (½in.) above and below the finished bandage.

- Start bandaging in the same direction as the overlap of the lining, just below the knee or hock.

- If the bandage is too short, start just above the joint, apply down to the coronet, and then back up the leg to the knee or hock.

- Many people prefer the latter method of bandaging as it is more likely to discourage filled legs: the bandage being put on upwards towards the heart.

- Place 10cm (4ins) of the bandage at an angle above the first round, which should be pulled firm, and should hold the flap of the bandage in place. This flap can now be folded down, so that the rest of the rounds of bandage help to secure it.

- Roll the bandage down the leg. About half the width of the bandage should be left exposed on each circuit. It should not be necessary to alter the angle or to pull the bandage.

- When nearing the coronet, unroll the bandage in a similar upwards direction, finishing just below the knee or hock.

- The tapes or fastenings should finish on the outside of the leg. The tapes should be tied in a neat bow with the ends tucked in.

The legs of a stabled horse can fill due to:

- Strenuous exercise.
- Standing in for a day.

- An old injury.
- A stomach upset.
- Old age.
- When first brought up from grass and making the change to living in the stable.

Such 'legs' usually walk down, but well-applied stable bandages at night help to keep the legs warm and the circulation active, which alleviates the problem.

2. Exercise Bandages

These should be made of stretch material: stockinette, elastic, crepe or elastic material. They should be 8cm (3ins) wide and 1.8–2.1m (6–7ft) long.

To Put on Exercise Bandages

- It is usual to put Gamgee, Fybagee or Porter boots under the bandage to give added protection, and to ensure that the bandage and tapes do not mark or injure the leg. Porter boots are preformed synthetic shells designed to protect the leg and prevent the bandage from being applied too tightly.

- The requirements and likely faults are similar to those of stable bandages, except that, being more elastic, and being used for work, these bandages need to be put on much more firmly.

- The flap over at the top should be longer. When folded down under the rest of the bandage, this helps to keep it firm.

- The bandage should extend from below the knee to the fetlock joint. Stockinette or crepe bandages should not extend over the joint, as this may cause interference. Vetrap bandages, expertly applied, can be fitted over the joint for extra support and protection. Exercise bandages should be sewn or taped for extra security when going across country. Elastic bandages when worn for any length of time will contract if they get wet and then dry. This can cause serious discomfort and possible injury to the horse, by restricting circulation. It is therefore essential to use Gamgee or similar lining.

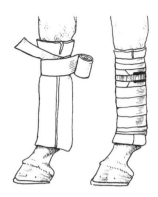

Putting on an exercise bandage.

The Purpose of Exercise Bandages

Exercise bandages are often used to ease the strain on tendons and ligaments in the front leg, but there is little evidence to indicate that they do. Their main purpose is to protect the leg from bruising when hitting a jump, and to avoid cutting and bruising the back of the front tendon by an over-reach or blow on the inner surface of the opposite front leg. If the bandages are used in muddy conditions for several hours, the mud is likely to work up underneath the bandage, leading to friction and possibly infection.

Various new types of reinforced lining (e.g. Porter boots) are now produced which can be cut to fit the leg. Used under an exercise bandage they give good protection from tendon injury for competition horses.

Removing Leg Bandages

- Undo the fastening and, using both hands, unwind the bandage and remove the lining.

- Shake the linings and bandages out.

- If they are dirty, wash them. If they are damp, hang them up to dry.

- When the bandages are dry, roll up the tapes. Position them where they are sewn. Roll the bandages as firmly as possible over them, and tuck in the ends.

Bandages should be put away in twos or fours. Unroll

enough of one bandage to wrap it around one or three other bandages, and secure by tucking in the flap.

Never use dirty bandages. Soiled material is stiff and hard, and is not suitable for either legs or tail. Elastic bandages should be frequently washed to retain their elasticity. Their correct application is then made easier.

When the bandages have been removed and before they are rolled, the horse's legs should be given a brisk rub in an upwards direction to stimulate circulation.

3. Tail Bandages

These are made of stockinette, crepe or elastic. They should be 6–8cm (2½–3ins) wide, and about 1.5–1.8m (5–6ft) long.

Correctly applied they will improve the appearance of the tail. They should only be left on for a few hours in the day after grooming, and should be removed at night. Left on too long they can cause discomfort and restrict circulation, leading to a sore under the tail and in extreme cases to death of the bone and loss of the tail.

When applying the bandage it is important to ensure that it is firm but not pulled too tight, and that the tapes are not tied tighter than the bandage itself.

To Put on a Tail Bandage

- Hold the bandage in one hand and lift up the tail. Put it over the shoulder or hold it up with the other hand.

- Place the spare end of the bandage under the tail and out at an angle at the top of the tail.

- Hold this spare end, and put one or two rounds of bandage as high up as possible round the tail.

- Turn down the flap and roll the bandage firmly down the tail to just below the dock. Tie it firmly, but not too tight.

- To give extra security, when bandaging an unpulled tail, take a small wisp of hair and put it over the first twist of the bandage and under the second.

To Remove a Tail Bandage

Undo the tapes, and with both hands pull the bandage down from the top. It should come away complete. Shake it out and roll it up.

If a tail has been plaited, the bandage must be unwound, not pulled off.

Dirty bandages should always be washed before they are put away.

CHAPTER

30 CLOTHING

RUGS

Clothing includes the various rugs, blankets and sheets worn by the stabled horse. He requires rugs in winter to keep him warm, and in summer to protect him from flies and to keep his coat clean. All stable and day rugs are secured by a front fastening. A leather roller or webbing surcingle can then be fixed round the body. The roller may or may not have a breast girth. Modern rugs have cross-over surcingles and fastenings under the belly, or just leg straps.

Types of Rug

Stable or Night Rug

This can be made of synthetic material, jute, canvas or hemp.

- HEMP RUGS are lightweight and are usually only half-lined with blanket. They are suitable for ponies and small horses, but will not stand up to as much hard usage as jute or canvas rugs.

- The fully lined, traditional, heavy-duty JUTE or CANVAS RUG – although more expensive – lasts longer, stands regular washing and is recommended for the larger type of horse.

- SYNTHETIC RUGS, made of polyester and other man-made fibres, are more expensive but are light, warm, comfortable, easily washed, and more easily kept in

place. They usually have a polyester filling. They offer improved temperature control and better air flow to provide insulation. They come in a variety of weights for different times of the year. Synthetic rugs have taken over extensively from those made of traditional, natural fibres.

Blankets

These are worn under the stable rug for extra warmth. They are usually made from thick woollen cloth. Blankets are sold in different sizes, by weight. They are expensive but last many years and repay the outlay. **Lightweight blankets** are a cheap alternative, but are not as warm, and they tear more easily. Some **man-made fibre blankets** are shaped like a rug with fastenings across the chest.

Sweat Rug

There are many different types available. The large cotton-mesh type is used to help cool off a horse. It works on the same principle as a string vest – by forming air pockets – and effectively insulates the horse against chill. For it to work efficiently, it is essential to put a top rug or sheet over it. Worn on its own, it helps to keep flies away, but it has little other value. Sweat rugs made of cotton towelling do not need a top rug to work efficiently.

Many modern rugs work by 'wicking' moisture away from the skin and allowing the horse to cool down and dry without risk of catching a chill.

Day Rug

This rug is worn for a special occasion, such as for competition work. It is made of wool, and is available in a variety of colours. It is buckled in front. The hems are braided, and it usually has the initials of the stable or owner on the bottom corner.

Summer Sheet

This comes in a variety of colours or checks, and is made of cotton, linen or synthetic material. In hot weather it protects the horse against flies and keeps the coat tidy and clean. In warm weather it is used instead of a day rug for horses travelling to competitions and for special occasions. It can be put over a sweat sheet in summer. As it is light in weight, it is kept in place by a fillet string, which lies round the quarters and under the tail. The string is attached to both edges of the sheet about halfway down.

A summer sheet can be used to provide an extra insulating layer under a winter rug, and as it is easily washable it can be regularly laundered, thus helping to keep the weightier winter rugs clean.

Exercise Sheet or Paddock Sheet

This is usually made of lightweight wool, and matches the day rug in colour. It can be used under the saddle for exercise in dry cold winter weather. For racing, it is worn in the paddock over the saddle. It reaches from the horse's withers and shoulder to the top of the dock, and is kept in place by a lightweight, matching roller or surcingle. Fillet strings are essential to prevent the rug from blowing up and frightening the horse. When used under the saddle, the sheet should have tape loops through which the girth runs, to prevent it from slipping back.

Waterproof Rug

This is made from waterproof fabric, and can be lined or unlined. A lining helps to absorb any sweat and allows the skin to breathe. Waterproof rugs are used at competitions, or sometimes when hacking in very cold, wet weather. A fillet string is essential.

Hoods, Caps, Neck Covers and 'Vests'

Racehorses are often exercised in cold weather wearing a hood, or a cap, which is a shortened version of the hood. They are made of lightweight, coloured cloth, and are of value to well-bred horses on walking exercise in very cold weather. They can be of waterproof fabric, and used in conjunction with a waterproof rug.

Hoods and neck covers are widely available in a variety of materials for warmth and protection from the weather and mud. They are also extensively used in the stable, for exercise and turn-out.

Close-fitting 'vests', usually of Lycra-type or silky materials, can be worn as undergarments. These cover the forehand of the horse and offer protection from rugs which tend to rub with continued wear.

New Zealand Rug

This is worn:

- By horses and ponies living out in the winter, to protect them from wind and rain. It also helps to keep the horses clean.

- To protect stabled horses and ponies who are turned out during the day. The rug keeps them warm and comparatively clean.

The rugs are made of lined waterproof canvas or man-made fibre. Investigation into materials continues, and improved types are constantly coming onto the market. The canvas type is heavier and stands up to harder use, but the lighter nylon rug is less likely to rub or rip, and is easier to keep in place. Canvas rugs in the original design have no roller and are self-righting when the horse rolls. Other makes have a stitched surcingle, which is eased over the backbone to avoid pressure. Both ends pass through the side of the rug. Nylon rugs are either buckled under the girth, or have cross-over surcingles or leg straps. There is now a lightweight material on the market which is waterproof and allows the skin to breathe. The extra expense is justified by the well-being of the animal.

It is essential for a New Zealand rug to fit correctly. It should lie snugly round the neck, so that it cannot work back and rub the top of the withers. It should reach to the top of the tail. Sheepskin pads can be sewn to prevent chafing, though – especially with a fixed surcingle – the pads restrict the flow of air under the rug, resulting in condensation. Leg straps can either be passed round the upper thigh and clipped on to the rug on the same side, or crossed to the opposite side. They should fit neither tightly nor too loosely. They must allow room for the horse to move.

When animals are living out all the time in New Zealand rugs two rugs should be available so that a wet one can be dried.

Always put a New Zealand rug on in the stable. Although it is not intended for wear in the stable, allow the horse to get used to it before turning him out. If he has never worn a New Zealand rug before, he may need to be walked and trotted with the rug on before letting him loose. The following day, the horse should have settled and can be turned out wearing the rug.

Rollers

Rollers are used to keep rugs in place. They are made of leather or webbing, and are placed round the horse just behind his withers. Leather rollers are very expensive, but they last many years. Good quality webbing is less expensive, but it becomes worn comparatively quickly. Cheap jute webbing is only suitable for ponies, and will not stand hard wear.

It is important to buy a width of webbing roller suitable to the size of the horse, as rollers of narrow width quickly deteriorate if worn by a full-size horse. Be sure to choose the correct length. Remember that if a fat horse loses weight or when extra blankets and wither pads are removed, the roller will become loose.

A PLAIN ROLLER has padding on either side of the spine. Even so, a thick foam-rubber pad should always be put under the roller to minimise pressure on the spine and withers.

AN ANTI-CAST OR ARCH ROLLER is used for horses who habitually roll over in their stable and get cast. It can have a fixed or adjustable iron arch, and comes in a

normal size, a small arch, or a giant-size arch. The giant size is more likely to prevent the horse getting cast. The small arch can be useless if the horse has a thick straw bed.

The main drawback to an arched roller is that although it removes all pressure from the spine, it concentrates it on either side of the withers. Even if a thick pad is used, it can have a severe pinching effect. An arch roller should have buckles on both sides, so that if a horse gets cast, it is possible to release the roller and then right the horse.

Rollers used over a day rug, summer or paddock sheet are usually made of matching woollen or cotton web.

Breast Girth or Breast Strap

This is a strap made of leather or webbing. It has a buckle at each end, which is fastened to a leather strap threaded through the 'Ds' placed on either side of the roller. The purpose of the breast girth is to prevent the roller from sliding back, which can easily happen with a fit or thin horse The breast girth should rest close to, but not tightly, around the chest and just above the point of the shoulder.

Surcingles

Webbing surcingles are stitched on to a night rug, and take the place of rollers. They should not be stitched over the top of the back, as the surcingle should be loose in this area. If two surcingles are used, one should be stitched about 46–60cm (18ins–2ft) behind the other. This provides a very secure arrangement for holding a rug in place. The second surcingle need not be fastened tightly.

A most effective alternative is to use two surcingles stitched at an angle on the sides of the rug and crossed over under the belly. They are usually referred to as 'cross-over surcingles' and are fastened with a flat metal clasp. This type of surcingle removes all pressure from the spine and holds rugs in place efficiently. Cross-over surcingles are seen extensively in modern rugs.

Belly Straps

These are wide pieces of matching material used to hold in place a polyester or nylon quilted rug. They may be crossed straps, as previously described. They are attached to the bottom edge of the rug, about 16cm (6ins) behind the elbow, and are fastened under the belly.

Leg Straps

Leg straps, as used for New Zealand rugs, can hold a lightweight man-made fibre rug in place. A roller should not be necessary. For fitting, see page 227.

MEASURING AND PUTTING ON RUGS

Measuring for a Rug

To find a correct size of rug, measure the horse from the centre of the breastbone along the side to the point of the quarters. Alternatively, measure from a point just in front of the withers to just above the top of the tail.

Putting on a Rug

A correctly fitted rug is comfortable for the horse, stays in place better, and lasts longer – as the material will be less subject to wear and tear. It should fit snugly, not loosely, round the neck. If it is too large in the neck, it may slip back, tighten over the withers, and eventually cause a sore. Also, when the horse gets up he can catch a foot in the neck, and in struggling he may hurt himself and tear the rug. If it is too tight, it will cause discomfort through pressure on the windpipe. It should cover the body down to the elbows and extend as far back as the dock. Sheepskin pads on the top inside part of the neck aid comfort, particularly for horses with high withers.

To Put on a Rug and Blanket

Use either a surcingle or roller with attached breast girth:

- Tie up the horse.

- Collect up the blanket – the left side in the left hand and the right side in the right hand.

- Speak to the horse and then gently throw the blanket on to his neck and withers. Alternatively, fold the blanked in half and place it over the withers with the fold facing forwards. Unfold half the blanket up the neck and slide the whole blanket back (with the lie of the hair) into place.

- Repeat this procedure with the rug, being careful not to displace the blanket by needing to pull the rug too far back into place.

- The front part of the blanket should lie well up the neck. If it does not, remove it and start again.

- A rug or blanket must never be pulled forward, as this ruffles the hairs of the coat and causes discomfort.

- The blanket can be turned back on itself. Approximately 16cm (6ins) should show in front of the stable rug. Alternatively, it can be turned back over the front of the stable rug. This helps to stop it slipping back at the withers.

- If using a roller and pad, fit these after doing up the front buckle of the rug. Alternatively, fasten the cross surcingles to secure the rug, making sure that they are not twisted on the off side.

To Take Off a Rug and Blanket

- Tie up the horse.

- Unfasten the front buckle of the rug and the roller or surcingles.

- Remove the roller separately.

- Next, fold both rugs back and then draw them off backwards and over the nearside hip and buttock area.

- Fold up the rugs and put away, along with the roller.

When saddling up in cold weather, the rugs may be turned back sufficiently to allow the saddle to be put on, and then turned back over the saddle. If they have slipped back, do not pull them forward. Take them off and replace them separately. If the horse is left with his saddle on, he must be tied up.

The blanket or rug worn next to the skin should be taken out and brushed regularly, but well away from the horse. As mentioned earlier, a summer sheet next to the skin keeps rugs clean and is easy to wash.

CHAPTER 31 CARE AND CLEANING

CLEANING AND STORAGE OF CLOTHING

- Before clothing is put away in store, it must be washed or cleaned and, if necessary, sent for repair. If it is put away dirty, the ingrained manure rots the material, so the rope and blankets tear more easily and the rollers and surcingles break.

- Woollen day rugs and blankets should be well brushed, and then sent to be cleaned.

- Badly stained blankets may need laundering.

- Stable or night rugs can either be sent to a laundry, washed in a launderette or at home.

- Webbing rollers and surcingles should be washed at home, so that the leather fittings can be kept oiled. If they are soaked in hot water, they will become brittle and will break.

- Leather rollers should be washed clean and then treated with Kocholine, neatsfoot oil or dubbing.

Washing Rugs, Rollers and Surcingles at Home

- Before starting the washing, all leatherwork should be oiled. Alternatively, fittings can be removed and sewn on later.

- The rugs should be soaked in cold water in a dustbin or yard water trough. The water should be changed several times. The rugs can then be laid out on clean concrete and scrubbed with hot detergent or soap. A hose is a useful accessory.

- Rinse them in cold water in a yard water trough which is not used for horse's drinking water.

- Put more oil on the leatherwork and buckles.

- Hang them up to dry.

- With rollers and surcingles, it is necessary only to soak the dirty section and then wash it. Hang up to dry, and oil leatherwork and buckles

Storage

- Rugs should be stored in rug boxes or on shelves.

- Moth preventatives should be put with them.

- The storage area must be vermin-proof.

- Make sure that the room is dry and that the rugs are well aired and clean before putting them away.

CARE AND CLEANING OF LEATHER

For safety, appearance and durability, it is essential that all leather is kept clean and supple. Water, heat, sweat and neglect are its greatest enemies. During use, leather loses a certain amount of oil and fat, which are drawn out by the heat from the horse. These must be replaced. Water – e.g. from rain or saturation during cleaning – also causes leather to be hard and brittle. An overheated atmosphere or drying by direct heat have a similar effect.

Leatherwork should be rubbed clean, using a damp cloth or sponge which must be regularly rinsed out. Only in the case of heavy mud – e.g. after a day's hunting – should it be necessary to use a wet cloth and more water. A chamois leather or dry cloth should then be used to dry off the leather. Similarly, if leather has been exposed to heavy rain, it should first be cleaned, wiped with a cloth or chamois, and then allowed to dry out naturally. Never dry it by placing it near direct heat, or in a drying cupboard, as this destroys the natural oils.

Using a slightly damp sponge the saddle soap should be well rubbed in, particularly on the fleshy or rough side of the leather where it is best absorbed. If leather has been very wet, a suitable leather dressing or neatsfoot oil should be applied to the flesh side when the leather is still damp and the pores are open. It will then be more thoroughly absorbed, will replace the oil and fat which has been lost, and will reduce the risk of any stiffening or cracking of the skin. When dry, it can be soaped.

The outer surface of a saddle should not be greased or oiled, as it will stain the rider's breeches when it is next used. Similarly, unless a numnah is used, too much dressing or oil on the lining of a saddle can stain a horse's back, and a sensitive skin can become blistered.

Leather which is stored often sweats, and becomes covered with mould spores. To prevent this, before putting it away apply Vaseline or leather dressing and store in newspaper and/or cover with cotton or natural fibre. This will also prevent the drying out which can occur when leather is not in use.

Care should be taken not to use excessive amounts of leather dressing or neatsfoot oil, as this can result in the leather becoming spongy and lacking in substance.

Equipment for Cleaning Saddlery

- Saddle horse on which to rest the saddle.
- Bridle hook, preferably adjustable and with four hooks.
- Hooks for leathers and girths.
- Bucket, half full of tepid or warm water.
- Foam rubber sponge or rough cloth for washing.
- Foam rubber sponge for soaping.
- Sponge or small brush for applying neatsfoot oil or leather dressing. The latter can also be applied with the fingers.
- Bar of glycerine or tin of saddle soap.
- Tin of neatsfoot oil or other leather dressing.
- Metal polish and cloth, or Duraglit.
- Duster for polishing metalwork.
- Chamois leather. This is only necessary if the leather has been very wet. To use, wet the chamois then wring it out. Do not try to use it dry.
- Clean stable rubber or saddle cover to place over the saddle when finished.
- Horse hair from mane or tail.
- Matchstick.

Cleaning Saddles and Their Fittings

The following notes apply to saddles with leather linings, Balding or Atherstone girths, leathers, irons and treads.

- Place the saddle on the saddle horse.

- Strip the saddle. Place the girth, leathers and stirrup irons including treads, on hooks. If the irons and treads are muddy, separate them and place them in the bucket of water.

- Up-end the saddle and, using a damp washing cloth, rub the underneath of the saddle clean. Put a clean stable rubber on the saddle horse and place the saddle on it.

- After rinsing the cloth, clean the rest of the leatherwork, particularly any part which has been in contact with the horse's skin. Clean leather – usually the seat of the saddle – will not need washing.

- Wash the leathers and girth- especially the inside of any interleaving parts – the stirrup irons, and the treads.

- With brisk circular movements, using a very slightly

dampened sponge, rub the soap liberally into all parts of the saddle flaps, the girth, and the leathers. Take care to move the buckle guards, so that the girth straps receive full attention. The top surface of the seat and saddle flap should then be wiped over with a dry cloth to avoid staining light breeches.

- Stainless-steel irons can be rubbed over with a dry cloth and the treads replaced. They need regular but not daily cleaning with metal polish. Nickel irons, if they are to look bright, require polishing each day.

LINEN-LINED SADDLES As above, but the lining should be washed after use in a similar way to leather. If the saddle is used for showing, a whitener is often put on round the edge of the lining.

SERGE-LINED SADDLES As above, but if only slightly dirty, the lining when dry can be brushed clean with a dandy brush. A very dirty lining requires soaking and scrubbing. It may take a day or more to dry.

SIDE-SADDLES Clean as above, but oil the stirrup device regularly.

Linings of saddles which have been on clean horses with numnahs should be washed only infrequently; this way they will last longer. Girths and leathers, however, should be carefully cleaned after use.

Top surfaces of doeskin, suede or reversed cowhide, should never be treated with soap or leather dressing. They can be cleaned – i.e. the mud removed – by brushing with a soft clothes-brush or by wiping with a damp cloth. Should the surface wear smooth, it can be scuffed up with a wire shoe-brush or sandpaper; this, in time, will reduce durability.

Girths

THREE-FOLD LEATHER GIRTHS should be thoroughly cleaned, and then soaped or oiled. An inner felt lining must be regularly treated with neatsfoot oil, and because this lining is kept inside the fold, this oil will be absorbed into the flesh side of the leather, keeping it soft and supple.

NYLON, STRING, WEBBING, 'COTTAGE CRAFT' OR LAMPWICK GIRTHS If they are dry, brush them clean. If muddy, and/or stained, soak them in warm detergent (making sure that any leather parts are not submerged), scrub them clean and thoroughly rinse them to remove detergent; if a horse is allergic to detergent, pure soap flakes should be used. Hang them up to dry by both buckle ends to prevent any rusting of the buckle tongues. Leather parts should be soaped or oiled. These girths can all be washed in a 'cool' cycle in a washing machine. It is advisable to put them in an old pillowcase to prevent buckles causing excess noise and possible damage to the drum of the machine. Girth buckles, and the metal studs and stirrup bars on saddles should be regularly cleaned with metal polish.

Pad Saddle and Felt Numnah

- Brush the felt with a dandy brush.
- When necessary, wash the underneath part with soap and a scrubbing-brush. Rinse thoroughly, and dry in a warm atmosphere.
- Keep free from hard knots of felt.
- Beware of moth damage and worn stitching.
- Clean leather and metal as already described.

Numnahs and Wither Pads

There are many different types. They must be kept clean, soft (no lumps), well aired, and in good repair. Clean them according to maker's instructions, or as detailed for washing numnahs.

Washing Numnahs

SHEEPSKIN Wash by hand in soap flakes. Thoroughly rinse and immediately apply warm glycerine or neatsfoot oil to the skin side. This prevents it becoming out of shape and hard.

SYNTHETIC SHEEPSKIN, QUILTED COTTON AND LINEN-COVERED FOAM Wash either by hand or by machine. Thorough rinsing is essential.

PLAIN FOAM RUBBER must be washed by hand.

Leather fittings are likely to become dry and brittle. They should be well oiled and regularly tested. Numnahs with leather straps should not be machine washed.

The skin of some horses will react to detergent, in which case, always use pure soap flakes and wash by hand. Rinse **very** thoroughly.

Putting up a Saddle

For saddles in regular use:

- Put the irons on the leathers and put the stirrups back on the saddle. Run the irons up the leathers, and put the leathers through and under the irons.

- Place the girth flat on top of the saddle. With a three-ply girth the rounded edge should face towards the pommel.

- Place a cloth or cover on top of the saddle, and put the saddle up on its rack. For saddles not in regular use, the girth leathers and irons should be hung separately on adjacent hooks, and the saddle should be put up without its fittings.

Cleaning Bridles

1. If Taking the Bridle to Pieces

- Hang the bridle on the cleaning hook.

- Remove the bits and curb chain. Put them in a bucket after removing the lip strap.

- Hang the reins over a hook.

- Take the bridle apart, memorising the buckle holes and the order in which the various headpieces are put together.

- Hang the leatherwork on a hook, or put it on a convenient work surface where it will not be likely to slip down to the floor.

- Thoroughly clean all leatherwork with a damp cloth.

- Wash and dry the bits and curb chain, and put them on a hook or on the work surface.

- Polish the bits, the curb chain, and then the bridle buckles and billet studs. The mouthpiece can be wiped over to remove any taste of metal polish. Finish with a dry, clean cloth.

- Soap all leatherwork, paying particular attention to folds and bends, which are all too easily missed in the daily cleaning.

- If the bridle has been wet, it may require an oil or leather dressing. Having treated the straps, fasten them at the bottom holes and runners. When the dressing has been absorbed, the bridle should be soaped and then put up correctly.

- When cleaning the headpiece, hold it firmly with one hand and use the cloth or sponge in the other hand.

- When cleaning the reins, step back from the bridle and hold the reins taut. Wrap the cloth or sponge round each rein in turn, and clean them or apply soap by rubbing to and fro.

- When cleaning a bridle with stitched-on bits, do not put the bits in the water, to avoid wetting the sewn ends of the leatherwork. Where the leather is folded over the bit rings, insert neatsfoot oil or leather dressing to help preserve the leather and keep it supple.

Re-assemble the bridle in the following order:

1. Thread the short side of the headpiece through the off side of the browband and then the near side, making sure the throatlash is to the rear.

2. Place the headpiece on the hook.

3. For a double bridle, first thread the bridoon headpiece (sliphead) through the near side of the browband. It should go under the main headpiece and buckle on the off side.

4. Next thread the noseband headpiece through the off side of the browband and buckle it on the near

side. It should be on the underside of the headpiece.

5. Attach the two bit cheek pieces. On a double bridle, there should now be two buckles on each side of the headpiece.

6. Attach the bit and the bridoon to their respective headpieces. Make sure that the bit is not upside down. Attach the lip strap.

7. Place all straps in their keepers and runners.

8. Place the noseband round the bridle, and secure by a keeper and runner.

9. Replace the reins. On a double bridle they should be of different widths. The wider rein fixes on the bridoon. The narrower rein goes on the bit. The bridoon rein lies behind the bit rein.

10. The bridoon bit should lie a little above and over the front of the bit.

11. The lip strap on a double bridle should be pushed through the 'Ds' on the bit – from the inner side towards the outer side – and buckled on the near side. For neatness it should be threaded through the runner.

12. The curb chain can be hooked on, and should hang round the front of the bit with the lip strap ring hanging down.

13. Many people fasten the lip strap through the loose ring of the curb chain to avoid the latter being lost.

When putting a bridle together, buckle fastenings should be on the outside, and billeted fastenings on the inside of the headpiece and reins. The flesh side of the leather should be nearest to the horse's skin.

2. If Not Taking the Bridle to Pieces

- Hang the bridle on the cleaning hook. The buckle at the back of the noseband should be unfastened.

- Wash and dry the bit. A nickel bit needs polishing.

- Unfasten all buckles and put them in the bottom holes.

- Clean and soap all leatherwork, starting with the headpiece.

- Replace the buckles in the correct holes, with spare strap ends in their runners and keepers.

- Put the noseband round the outside of the bridle and thread it through the keepers.

A bridle in daily use can be cleaned as above. However, **at least once a week** it should be taken to pieces and given more thorough attention as already described.

Putting Up a Bridle

If the bridle holder is high enough, it is better for the leather if the reins are allowed to hang down. The noseband should be put round the outside of the bridle and secured by the keeper and runner. The throatlash can be put round in a figure-of-eight, secured by a keeper and runner, but not buckled as this causes delay in using the bridle.

Alternative Methods

- Double up the reins at the back of the bridle. Put the throatlash through them, and then buckle or thread it through its keepers at the back.

- Double up the reins and wind the throatlash round them in a figure-of-eight. Put the throatlash round the front of the bridle and through the reins at the back. Then buckle or thread it through the keepers at the front.

General Notes

- Never balance the saddle on a narrow or unsteady surface – e.g. a pole or the back of a chair. The gullet may get damaged, and if the saddle falls, a broken or cracked tree and/or marked leather may result. If greater pressure is needed, place the saddle pommel down on top of your feet – never on the ground or on a hard surface – and work downwards.

- Use as little water as possible, but regularly rinse the washing cloth or sponge.

- Keep the soap sponge only slightly damp. If too wet, the soap will make lather and will lose its effectiveness, so that the leather remains dull.

- Never use hot water.

- Never use detergent or soda, unless for disinfectant purposes – e.g. for ringworm. All three may make cleaning easier, but they are very bad for the leather.

- Holes in stirrup leathers and bridles can become filled with soap and grease. Remove this with a matchstick or round nail.

- 'Jockeys', which are accumulations of black dirt and grease on the surface of the leather, can be removed by a fingernail or hairs from a horse's tail tied in a knot. Moisten the knotted horse hairs, cover in glycerine soap, and rub them over the 'jockeys'. Never use a knife or any sharp instrument, as this will scratch and damage the surface of the leather.

- When undoing billet studs leave the end of the leather in the keeper and push it with the thumbs, so that the stud comes out of the leather. Then remove the leather from the keeper. When replacing it, put the end of the strap in the keepers and push the leather over the billet stud.

- If the buckles are stiff, it is easier to push the leather piece below the buckle upwards and through the buckle, rather than pulling the end of the strap up.

- During cleaning, inspect for wear all leatherwork, stitching and metalwork. The tongues of all buckles should be regularly greased; they are the most likely parts of the buckles to become worn and rusty.

CHECKING TACK FOR SOUNDNESS

The Saddle

The **tree** can be damaged or broken:
- If it is dropped on the ground.

- If it falls from a doorway or similarly unsafe position.

- If it slips off a horse's back before it is secured with a girth.

- If the girth is pulled to reach the girth straps, and the saddle is not held with the left hand, it can slip over the off side.

- If it is left on the ground in the stable, and the horse steps on it.

- If a horse in the stable gets free, or is left free when saddled up, and rolls.

- If the horse falls and rolls over.

- If a heavy rider catches hold of the cantle of a spring-tree saddle when mounting and frequently exerts a twist and strain on the tree.

A saddle with a narrow or medium tree can be strained by being used on a horse with a flat or wide wither.

Damage can occur in four places: the pommel or front arch, the waist, and at one or both points of the continuations of the front arch.

If damage to the tree is suspected or confirmed, the saddle should not be used but should be sent for immediate repair. A damaged tree can seriously injure a horse's back. A fracture of the front arch or to the points can be mended. Damage to the waist or cantle is more serious.

Other Areas Which Need Regular Checks

- The stitching of the girth straps.

- The stitching round the pocket.

- The holes in the girth straps.

- The girth straps.

- The saddle lining. If this is made of leather and is well looked after it should wear and stay smooth for many years. Linen lasts longer than serge, but both wear out in time. In all cases, the stitching requires checking as does the padding, which must be kept even so that there are no lumps to give the horse a sore back.

The stitching on other parts of the saddle may give way, and look untidy, but the safety of the rider and the comfort of the horse should not be affected.

Parts of the Girth Which Need Regular Checks

• The suppleness and strength of the leather holding the buckles.

• The stitching of the leather round the buckles.

• The buckles and the wear of the buckle tongues. These must close firmly on the top part of the buckle. The buckles should be made of stainless steel.

• The strength of fabrics such as webbing, string and nylon, particularly in the buckle area.

• The front surface of a leather three-fold. This can crack and split, causing discomfort to the horse.

• The elastic inserts. These can become slack and unsafe.

Parts of the Leathers Which Need Checking

• The stitching round the buckle.

• The buckle itself, and the buckle tongues. Steel buckles are unlikely to break. Nickel is suspect and should not be used.

• The strength of the leather, and any stretching or signs of wear in the buckle holes. To avoid continual wear to the same areas, leathers can be shortened. This will bring the pressure on to a fresh piece of leather.

Stirrup Irons

STAINLESS-STEEL IRONS are preferable. They are strong and sufficiently heavy, so rarely cause trouble – although if treads are not used, they will in time wear smooth on the base. All irons require careful checking for cracks.

NICKEL IRONS are not recommended, as they: (1) crack and pull apart and (2) bend easily, lose shape and then break. When subjected to pressure, they can close and imprison the rider's foot – e.g. if a horse falls on its side, or if they are banged against a jump or gate.

The Bridle

• All stitching should be regularly examined and tested by stretching and pulling the leather.

• Buckles and buckle tongues should be checked for wear. Buckles should be of stainless steel.

• Reins should be checked for wear. Reins which have been broken and then repaired should be closely examined. Unless the break was in the buckle area, the reins are only suitable for light work, as the stitching will be subjected to friction and strain.

• Re-covered rubber reins are also suspect, as the re-stitching of the new rubber can weaken the rein.

• Webbing, plaited cotton and linen-corded reins all wear more quickly than leather. They should not be patched, but must be replaced.

Bits

Particular points of wear are:

• The centre joints or joints of the mouthpiece.

• Where the rings or cheeks join the mouthpiece.

Nickel mouthpieces, rings and cheeks can become rough and/or sharp. Jointed snaffles can come apart at the joint. Rubber and leather mouthpieces must be regularly checked for wear, as they have a limited life.

Other Equipment

MARTINGALES, BREASTPLATES, and other leather or webbing equipment should all be checked for wear. Particular attention should be paid to any buckle areas.

LUNGEING TACK The centre ring of the cavesson receives constant pressure, and is subject to strain. It should be regularly checked. The lunge rein can deteriorate, and should be checked for weak places.

CHAPTER

32 SADDLING UP AND UNSADDLING

SADDLING-UP PROCEDURE

1. Collect the Bridle (which is assumed to be a snaffle)

- Undo the throatlash and noseband. If necessary, double up the reins to keep them clear of the ground. Carry the bridle on the shoulder. Collect the martingale (if used) and the boots.

2. Collect the Saddle

- Check that it is complete, including buckle guards. If a numnah is to be worn, it should be put under the saddle and the straps fastened. Either:

- Place it on the lower arm with the pommel towards the elbow, or

- Place the cantle under the upper arm and hold the pommel with the same hand. Take the required tack to the stable. On a wet day, cover the top of the saddle with a stable rubber. Place the saddle and bridle in a safe place.

3. Tie up the Horse

- Check that he is clean, his feet picked out, etc.

- Put on the boots. It is usual to put the saddle on before the bridle.

4. Put on the Saddle

- Speak to the horse. Pat him on the neck and check that he is comfortably tied up, with the rope neither too tight nor too loose.

- Undo the rugs and then the roller or surcingles. Place the roller tidily in the corner or over the door. According to the temperature, either turn back the rugs or remove them. Fold them and put them with the roller.

- Put the saddle on the left arm, with the girth attached on the off side and folded over the saddle. Approach the near side of the horse. Pat the saddle area and then, using both hands, place the saddle up on the withers, and slide it back to its correct position. Never move a saddle against the lay of the coat. Check the saddle numnah to ensure that the surface next to the horse's skin is smooth. Pull the numnah well up into the arch of the saddle. Move round quietly under the horse's neck. Check the saddle on the off side and unfold the girth. Return to the near side and fasten the girth. This should be secure, but not tight. Make sure that skin is not wrinkled or caught up under the girth. Replace the rugs, but do not buckle them up, as if the rugs slip, it is better they fall clear than hang suspended round the horse's neck where they can frighten him and get torn. If restless horses are left, the rugs should be fastened and the roller replaced.

- If a martingale or breastplate is used, it should be put on before the saddle. Untie the horse. Put the neck strap over his head with the buckle on the near side. Re-tie the horse. Alternatively, unbuckle the martingale. Adjust it to the approximate length, and when doing up the girth, put it through the loop of the martingale. If two girths are used, fasten the one nearest the horse's elbow first. This is the one which should take the martingale. The second girth is placed on top and then buckled.

By putting the saddle on first, the horse's back has time to warm up before being ridden.

It is convenient and saves time if horses become accustomed to being saddled from either side.

After girthing up before mounting, pull the horse's foot and leg forward to stretch the skin and ease any wrinkling.

When saddling up a young or nervous horse, it is advisable to untie the headcollar rope. But leave it through the loop. Should the horse resent or play up, there will be less chance of his pulling back and getting frightened. It is advisable that such an animal, once he is saddled, is not left unattended.

5. Put on the Bridle (for fitting, see page 196)

- Take the bridle and reins over the left arm, with the browband nearest to the elbow.

- Go up to the horse, talk to him, and stand behind his eye on the near side. Unfasten the headcollar and put it round his neck. Put the reins over the horse's head. At this stage, either of the two following methods can be used:

1. Take hold of the headpiece of the bridle with the right hand. Put the left hand under the bit and move the first finger around to the off side of the horse's mouth. Persuade him to open his mouth by inserting your first finger in the gap behind his front teeth. As the horse opens his mouth, guide the bit into the mouth with your left hand and at the same time lift the bridle up towards the horse's ears with your right hand. Once the bit is in his mouth, use your left hand to help the right hand move the bridle carefully over the horse's ears. Ease out the mane and make it tidy, and put the forelock over the browband. Check that the browband, noseband and bit are level.

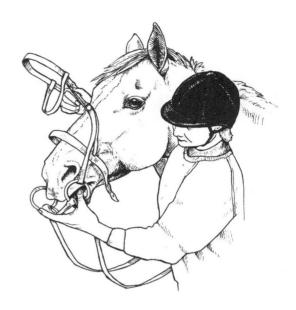

Putting on a snaffle bridle.

Adjust if necessary. Fasten the throatlash and noseband. Put all the keepers into place and replace the headcollar.

2. This method can be used with horses who resist or raise their head when being bridled. Hold the top of the bridle in the left hand. Place the right hand under the horse's jaw and round to the front of his face. Take hold of both cheek pieces of the bridle just below the browband. Move the left hand to hold the bit. Open the mouth with the left thumb, and continue the procedure as for 1.

To Put on a Double Bridle

Undo the throatlash, noseband, curb chain and lip strap. Make sure the bridoon is on top and in front of the bit. Proceed as when putting on a snaffle bridle. When checking the bridle for fit, ensure that any alteration to the noseband or the bridoon does not make the bridle crooked. If this happens, hold the browband and pull the leather through it. Fasten the curb chain and lip strap.

See also page 196, Fitting a Double Bridle.

Leaving a Saddled Horse

- If a saddled and bridled horse is to be left put the headcollar back over the top of the bridle. Double the reins twice round his neck, tie them in a knot on his withers or twist and loop them through the throatlash. If the reins are put behind the stirrups they may upset a sensitive horse and cause an accident.

- A saddled horse should always be left tied up. If loose in the stable, he may get down and roll.

- It is inadvisable to leave a horse tied up in the stable wearing a double bridle. The bridle should be left outside until just before the horse is required. The bridle should then be put on and the horse taken straight out to work.

Carrying and Putting on a Side-Saddle

The saddle should be carried with the seat under the right arm, and with the right hand holding the top fixed pommel. Attach the girth and balance strap to the near side. Hold them with the flap strap – if there is one – and the stirrup iron in the left hand. Do not put them over a doeskin seat, or the stirrup iron over the top of the pommel, as this marks and wears the leather. The girth should be fastened high on the near side, and to the front two straps; any adjustments should be made on the off side. Fasten the balance strap to the angled strap. Do not buckle it too high, as it lies under the right calf of the rider.

   Place the saddle on the horse's back from the near side, as for a normal saddle. Do up the girth and then the balance girth on the off side. The balance girth lies over the centre of the girth, and passes through a loop (if there is one); it attaches to the strap at the back of the saddle. It must not be too tight. Then fasten the flap strap, which comes on top. It must never be tighter than the girth and balance strap. When moving round the horse, try to ensure the saddle is held firm by a second person.

If a martingale is used, all three straps should go through it.

   When checking a saddle for fit, make sure that it sits evenly on the horse's back and does not sit too far up on the withers. The points of the tree must stay clear of the shoulder, or the saddle will rock about. For a saddle to sit well, the horse must have a good wither. Note that a horse's back requires some time to become hardened to the unaccustomed weight and shape of the side-saddle.

- Before mounting, the rider should check the girth. The balance strap should not be too tight nor too far back, as this may cause the horse to buck.

- When the rider is mounted, the girth and balance strap should be adjusted from the off side.

Care of the Side-Saddle

- If possible, bring the horse out of the stable before saddling. If he is saddled in the stable, ensure that the lower pommel does not catch on the door as the horse is led out, and also when he is led back in.

- Once the saddle is on and girthed up, avoid turning the horse sharply, as this can crack the tree.

- The saddle must never be left on the horse with the girths unfastened, unless he is held by another person.

- Never transport the horse in a horse box or trailer wearing a side-saddle.

- When unsaddling, do not put the girth and back strap across the seat as the doeskin is easily marked.

- Never rest a saddle on a door or insecure surface.

- If it has to be put on the ground, place it down pommel first.

UNSADDLING

There are several accepted methods of unsaddling. The following is the normal procedure:

1. On dismounting, run up the irons on the leathers and ease the girth one or two holes.
2. Take the reins over the horse's head, lead him into his stable, and close the door.
3. Take up the headcollar and place it round the horse's neck with the rope untied.
4. Unfasten the throatlash, the noseband and – if there is one – the curb chain. If there is a standing martingale, release it. If there is a running martingale, unfasten the reins, and release it.
5. With your right hand, ease the bridle over the horse's head, at the same time steadying his face and then taking the bridle with your left hand. The horse must be allowed time to quietly let go of the bit.
6. Place the bridle over your left arm so that both hands are free to put on the headcollar and to tie up the horse.
7. Place the bridle over your shoulder.
8. Unfasten the girth. If there is a martingale, take it out of the loop. Lift the saddle, draw it over the horse's shoulder and place it on your left arm. The girth can be put over the top of the saddle, but if the girth is wet or muddy, leave it down. The martingale neck strap should be undone, and the martingale should be placed, with the bridle, on your shoulder.
9. Put the saddle and bridle outside the stable. If you put the saddle on the ground, place the pommel down first, with the girth between the wall and the cantle; this prevents the leather from being marked or scratched. Avoid putting the saddle on the door where it could be knocked off.
10. Return to the horse, and briskly pat the saddle area to restore circulation.
11. Replace the rugs.
12. If the horse is wet, or is likely to break out, put either a sweat rug or straw underneath the rugs.
13. Take the saddle and bridle to the tack room ready for cleaning. Mud will come off easily if it is sponged when still wet.

An alternative procedure is to take the saddle off first, as detailed in 8, 9 and 10 above. Then remove the bridle as in 3, 4, 5 and 6 above. This method can be useful for restless horses, who are better controlled in a bridle, and obviates the necessity of separating a martingale.

Procedure after Unsaddling

1. Pick out the feet.
2. If the feet are muddy, wash them. Take care not to wet the heels.
3. Check the legs.
4. Sponge off any sweat marks. On a hot day, the horse can be washed off out doors.
5. Check for any signs of rubbing in the saddle, girth and mouth areas.
6. If the horse is wet put on a sweat sheet and walk him dry.
7. Tie up the haynet.
8. Groom horse – or at least brush off sweat marks.
9. Replace all the rugs.
10. Refill the water bucket.

SPECIALIST CARE OF THE COMPETITION HORSE

GUIDELINES FOR COMPETITION HORSES

The success and well-being of a horse who is in training for a stressful activity is very much dependent on his stable environment and the care and expert attention that he receives from his groom. As well as being given the specialist care outlined in the following chapters, the horse should be accorded the normal day-to-day care already detailed in Chapters 9–23.

Some General Points

Worming

An appropriate worming programme should be planned. Worming should be carried out when the horse comes in from grass, and every six to eight weeks thereafter, but avoiding major competition dates.

Flu/tetanus Vaccinations

These should be given as necessary, starting when the horse is first brought in, then according to current veterinary practice and as required by competition rules.

Teeth

Teeth should be inspected every six months, and attended to as necessary; some horses require attention more often.

Passport

This must be completed by a veterinary surgeon who is not the owner of the horse. It is a proof of identity. It must be correctly completed according to the regulations laid down for the current year. It must give an exact description of the horse, with details and dates of equine influenza and tetanus inoculations. For horses travelling abroad, it should contain details of immunisation programmes and blood tests.

In order to enter horses in affiliated competitions, county shows, etc., or to stable them on a racecourse, those who do not have passports must arrange a full and up-to-date inoculation programme. This must be carried out by a veterinary surgeon, and must be certified on a combined identification and vaccination certificate. The completed certificate must comply with the regulations laid down for the current year and must accompany the horse.

In the UK all horses competing in any affiliated discipline need to be registered with the British Horse Database.

General Notes on Fittening the Horse

These will obviously vary according to the competition for which the horse is being prepared, but many will apply to all disciplines.

- It is important to ensure that horses are given plenty of variety in their work. If hacking out, the route should be varied as far as possible. Loose jumping in the school may be of benefit. Drag hunting can provide a good opportunity for schooling over fences. A half-day out with hounds can add extra interest for an older horse who has become stale, or to educate a young horse.

Some horses find fast work more interesting if accompanied. This often improves the stride of the stuffy horse, but will not benefit an excitable one.

- Fitness work should aim to improve the overall performance of the horse through correct work and feed. It is essential to start with a healthy, sound horse. Fitter horses are less liable to suffer lameness problems. Fatigue is a common cause of breakdowns.

- Make the horse work into the bridle; he should not slop along on a loose rein.

- Discretion must be used as to the type of terrain. For example, sand is very heavy going, and artificial surfaces often slide under the horse. Grass may be uneven or slippery. Roads are hard but level.

- Strenuous work should not be undertaken the day before or the day after a competition or fast work.

- Fast work must be supported by rest periods, to allow the heart rate to return to normal.

- Extra care must be taken in humid conditions.

- If the horse is balanced he will suffer less strain: therefore, training should be over a large area and should avoid sharp turns.

- Fast work must not be in sudden bursts, as this puts strain on tendons and lungs. Horses must be well warmed up before starting any strenuous work, and speed should be decreased gradually to avoid strain to the limbs.

- In fittening work, go faster uphill to increase the length of stride. Take it easy downhill. In a competition, allow the horse to flow on downhill. Do not put on pressure uphill.

- After fast work the horse should be kept walking to allow the build up of lactic acid in the muscles to disperse.

- In bad weather, slow canters on a prepared surface

make it possible to continue fittening work. This also applies when there is hard ground in summer. If a horse is not used to the prepared, artificial surface of a gallop, this work should be carefully introduced, as he may well feel insecure and might slip.

- If canter work has to be in a school, the rein should be changed, through trot, every three circuits. Continual cornering makes for increased effort, and can also put greater strain on the legs. It is important to ensure that the surface is not dusty.

- It is important to remember that the rider should also be fit.

Checks on the Horse During Training

Check the horse's legs after work and again in the evening, to ensure that any signs of strain are noted in good time. This is particularly important after fast work. If there is any heat in one or more legs work should be reduced and cantering should be discontinued until the legs are normal. Concentrate diet should also be reduced.

Any change in the horse's attitude to work, appetite or stable habits may be a sign of stress. The workload may have to be reduced and possibly the competition schedule altered. It is better to under-train rather than to over-train. Once the horse attains peak fitness he will then stay fit longer, with less risk of being overstressed, or, as it is more generally described, 'going over the top'.

Blood Tests

Many owners arrange to have their horses blood-tested when first brought up from grass, and repeat the testing after eight to ten weeks. Should there be any incipient problems they can be recognised and dealt with in good time. It is sensible to have a blood test done when the horse is well and healthy so that a comparison may be made if problems arise. Blood tests for monitoring condition should be made when the horse is in regular work and eating his normal diet.

Feeding

It is important to feed the horse so that he is fit for the

work required. This will vary between different disciplines but the principles remain the same. In the case of the event horse his diet should be adjusted. When starting work the forage ration may well be 90% and concentrates 10%. By week 6 the ratio might be 60% forage, 40% concentrates; by week 8: 50% forage, 50% concentrates; and by week 12: 40% forage, 60% concentrates. To avoid digestive upsets, the forage ration should rarely be below 40% of the whole and then only for short periods of time for a specific reason (e.g. racehorse prior to a demanding race).

Daily consumption of feed varies enormously between different horses. Some horses get and stay fit on as little as 4.5kg (10lbs) of concentrates, while others may eat 8–9kg (18–20lbs). Many horses will eat their greatest quantity of food before they get fit and start competing. However, so long as they stay healthy and eager for work, there is no need for anxiety. The diet of 'good doers' may have to be restricted: unnecessary weight puts additional strain on heart, lungs and legs. Daily in-hand grazing, or a short daily spell in a suitable small paddock, helps to keep a horse happy and relaxed.

For more detailed instructions, see Chapters 19–23.

Equipment to be Taken to a Competition when Staying Overnight

- Stable tools, skip and muck sack.
- Sacks of shavings or paper bedding (if required).
- Two haynets.
- Two water buckets and water carrier.
- Feed bowl.
- Concentrate feed, which should be pre-packed and labelled.
- Hay or hay alternative, e.g. haylage.
- Supplements and electrolytes.
- Grooming kit, including extra sponges, towels, sweat scraper, hoof oil.
- Plaiting kit, fly spray.
- Two spare sets of shoes, studs and fitting kit.
- Tack-cleaning kit.
- Spare rugs and blankets.

- Two sweat sheets or coolers.
- Waterproof rug.
- Stable bandages, Gamgee and/or liners.
- Passport/vaccination certificate.

Other Essential Items

- Tack – depending on horse and competition.
- Bandages and/or brushing boots, over-reach boots, plus spares.
- Spare girths, stirrup leathers, irons, rubber-covered reins.
- Spare headcollar and rope.
- Lungeing equipment.
- Travelling equipment for the horse.
- First-aid kit for rider.
- First-aid kit for horse: to include scissors, cotton wool, crepe bandages, Gamgee, salt, wound dressings and wound spray, kaolin and/or Animalintex, ice-packs.

NB: No forbidden substances may be taken to a competition. Even being in possession of illegal drugs or syringes is an offence.

Packing the Lorry (see also Chapter 43)

- Equipment should be listed and ticked off as it is loaded. In this way vital equipment will not be overlooked.

- Use containers for horse equipment. This will help efficiency in loading and unloading. They should not be too large or too heavy as this will make handling difficult.

- Feed is best packed in plastic sacks. If the journey is long, the horse will need to be fed en route. These feeds should be individually packed and labelled: e.g. Monday am, Monday pm, etc.

- Filled water containers should be available for en route refreshment.

- Plan to arrive at your destination with plenty of time to spare so that the horse can settle in and recover from the stress of the journey.

34 THE EVENT HORSE

FITNESS

The following guidelines should help to plan a fittening programme, but schedules need to be adjusted to meet the requirements of individual horses. A chart of the intended work programme should be made and used to monitor the horse's progress. This should show diet, exercise (specifying fittening work, fast work and schooling), shoeing, worming, flu/tetanus inoculations, veterinary checks and competition dates. Ten to twelve weeks should be allowed for a horse to reach one-day event fitness, from grass.

The event season is now almost continuous from mid-March to mid-October. Schedules must be carefully planned, so that the horse can be given a rest. He cannot be expected to compete throughout the season. Bear in mind that it is not advisable for a horse to reach peak fitness by his first one-day event. If he is brought on steadily he will stay fit longer, with less risk of being overstressed and losing form.

Fittening must be a gradual process, so work must be increased progressively. Each horse must be treated as an individual and carefully observed, with note taken of his pulse and respiration rates and type of breathing. Body weight and condition of limbs must be carefully monitored and the programme adjusted accordingly.

Weeks 1–4: the horse requires walking exercise starting with ¾hr on the flat and gradually building up to 1½–2hrs with a progressive increase of hill work. Trotting can be introduced, initially on the flat, between the third and fourth week.

From **week 5, t**rotting can be gradually increased. The aim of this fitness programme is gradually to work the heart harder, to demand greater effort and thus to build up stamina. Hill work is highly recommended for this purpose, and as the horse's fitness improves the hill work can be develop from walk into trot.

From **week 6** dressage training may be included together with hacking; from **week 7**: jumping training can start. From **week 8** – depending on the individual horse and how long his lay-off has been – it should be possible to take part in small dressage and show-jumping competitions.

By **week 9** slow twice-weekly canters can be introduced. Begin with a 3–minute steady canter (15mph/400m/min), followed by a walking period during which the horse's breathing returns to normal. Follow this with another 3–minute canter. This canter work can be gradually increased in duration, building up to 15 minutes in total (5 minutes x 3) during weeks 9 to 12. After a 5–minute half-speed canter (15mph/400m/min) the breathing rate of a fit horse should return to normal in 5 minutes.

In **weeks 10 and 11**, the horse can do some cross-country schooling to give him additional fast work and jumping practice before his first event in week 12. The cross-country schooling replaces the canter work.

Extra Fittening Programme for the Two-or Three-Day Event Horse

A horse can be expected to tackle a three-day event four months after coming up from grass. The first part of the programme is the same as when preparing for a one-day event.

During **weeks 9–12** canter work can be built up to 20 minutes in total (5 minutes x 4), thus developing the horse's stamina. Two sessions a week is sufficient for this, each session followed by an easy day for recovery. If a good all-weather gallop is available on a hill, then the length of the canter work can be decreased.

To avoid pounding the legs, increase the workload without increasing the speed. If any hills can be used, the heart will be worked much harder.

In **weeks 12–14,** build up the half-speed cantering at this pace to 6–minute sessions by four repetitions; if the horse is recovering easily the rest periods can be reduced from 3 minutes to 2 minutes, and, in some cases, to 1 minute. The main criterion is the ease and speed of recovery.

During the weeks before a three-day event, one-day events provide additional fast work. If there are no such competitions, then Wednesday can continue as a cantering day and Saturday can become a day for faster work at cross-country speed, approximately 570m per minute for 1 mile, then rest, then repeat once more.

Some time during **weeks 12–14,** the horse can practise jumping steeplechase fences at the speed of 690m/min on two occasions. Then, at least ten days before the three-day event cross-country day, he can gallop at steeplechase speed over approximately the required distance for that phase of the competition. This is a fitness test and his last real workout before the three-day event.

All fast work must have a good warm-up period of at least 30 minutes, and also a working-down, cooling-off period.

During **week 15** work can be eased off. Keep the horse healthy and relaxed, and not overtaxed or overtired. The horse must stick to an exercise pattern and not lack schooling, but it should not be stressful. The horse will require a sharp burst on the Wednesday and Saturday.

Week 16 is the week of the three-day event. The horse should be given a short canter on the Tuesday, and another on the Friday after the dressage.

NB: Stuffy horses who take longer to get fit may require three short, sharp gallops a week at about 600m/min for ½mile in the last two weeks of fittening.

Interval Training

Interval training may be undertaken for fittening. This system puts limited stress on the heart, lungs and muscles, followed by short rest periods, then further stress. It involves careful checking of the horse's pulse, temperature, and respiration rate, which is not always within the capabilities of the average event rider.

A horse can start interval training if the following criteria have been met:

- he is 100% sound;
- he has undergone a basic fittening programme of approximately 6 weeks and is walk, trot and canter fit;
- he has been regularly wormed;
- he is in good health – a blood test at this stage would be useful;
- he is thriving.

A record should be kept of the horse's response to training, including temperature, pulse and respiration rates, work done, weather conditions (heat and humidity may have adverse effects) and any other relevant points.

Pulse is the most reliable gauge as respiration may vary more from one horse to another. During exercise the pulse and respiration rates will increase.

In interval training, the horse is given a set piece of work and the pulse and respiration are recorded before the work, immediately after it and then again 10 minutes later. This work will be repeated not more than once every four days. As the horse gets fitter, the recovery rate will improve. As the fitness develops, the periods of work are gradually increased, as are the number of repetitions; each time, the repetition of work is made just before the point at which he is almost recovered, to further develop the cardio-vascular system.

Interval training programmes vary from horse to horse and rider to rider. Some programmes involve three repetitions, some four or five. Initial rest periods are usually 3 minutes, but may reduce to 2 minutes and 1 minute as fitness develops.

Fitness pushes back the anaerobic threshold and increases the horse's capacity for using oxygen. Anaerobic respiration causes a build-up of lactic acid and this is the point when the horse becomes fatigued. Work must then stop to allow the system to disperse the lactic acid; prolonging work could cause muscle damage and other strain or injury.

A normal, 'warmed-up', starting pulse rate would be 50–60 beats per minute. During work, and for the work to be effective, the pulse rate should be raised to around 180. When the pulse rate rises over 200 then the anaerobic threshold is being reached. A pulse rate of 220 could not be sustained and is potentially dangerous.

In the first minute after exercise the pulse rate should fall to about 120. A pulse rate of 150 indicates over-stress, whilst a rate of 100 shows that the horse has not worked hard enough.

A 25–30% recovery should be evident within 10 minutes. If the pulse has not returned to normal within 30 minutes after a work-out, the programmes should be adapted as the horse has been over-stressed.

The respiration rate should never exceed the heart rate and should fall comparatively with heart rate.

Factors such as the feel of the horse to his work (how easily or willingly he works), how much he blows and sweats, and his temperament (e.g. sharpness, cheekiness, fractiousness when tacking up) must be prime pointers to his well-being.

Speeds of work will relate to type of horse and terrain used, but in training the event horse speeds ranging from 200m/min (trot) to 450m/min (canter) will be used for steady stamina work. Faster work will involve speeds required for cross-country and steeplechase, ranging from 520m/min for Novice cross-country to 690m/min for Advanced steeplechase speed.

TACK

A **SNAFFLE BRIDLE** should be used in basic training but later the horse may require a **DOUBLE BRIDLE**.

An **ALL-PURPOSE SADDLE**, suitable for all disciplines, is the basic necessity. A straighter-cut saddle for dressage is useful, but not essential. A forward-cut saddle for the cross-country is vital, as it enables the rider to get his weight forward over the centre of gravity.

A **BREASTPLATE** and a **SURCINGLE** are necessary for the cross-country. Holding straps on the girth prevent the surcingle from sliding back. Rawhide or buffalo hide leathers and stainless steel irons are also necessary.

A **WEIGHT CLOTH** is no longer required in horse trials competitions.

REINS for the cross-country must prevent the hands from slipping, so rubber ones are best. For the cross-country a bootlace can be used to secure the headpiece of the bridle to a plait just behind the horse's ears.

BOOTS or **BANDAGES** are advisable for protection in the cross-country phases. Tendon and over-reach boots are often used for the show-jumping phase. Boots must be strong enough to protect the legs, and supple enough to prevent sores, but not too heavy and absorbent, particularly on a wet day or if there are water obstacles. Care must be taken that the boots allow for movement of the joints and do not only fit when the horse is standing still.

Bandages must be applied with great care and with even tension. They must always be worn with Gamgee or other similar protection underneath. If Porter boots are used under bandages, the boots must be carefully trimmed to fit the horse when the leg is bent. Bandages must be secured by sewing or tape, as cotton tie-tapes have no give and may cause problems.

Protective boots should be worn when schooling, both on the flat and jumping, to prevent injury. For hacking on roads, it is sensible to fit exercise knee caps.

CARE OF THE HORSE AT A THREE-DAY EVENT

Much of the following advice also applies to one-day event horses.

Care Before Cross-Country Day

- Clipping will help the horse in hot weather.

- The night before the cross-country, give the horse his usual evening meal, but reduce the hay ration.

- It is preferable not to bed the horse on straw. If there is no choice, then use a muzzle.

Care on the Morning of Cross-Country Day

- For a one- or two-day event feed the horse at least four hours before the start of the cross-country. For a three-day event feed him four hours before Phase A. A small concentrate feed is necessary to keep the digestive juices working, and thus to prevent the possibility of colic. Horses going late in the day may have their normal early morning feed, plus 0.5–0.9kg (1–2lbs) of hay.

- Do not feed hay unless the horse is going very late in the day, as the blood is required to take oxygen to the muscles, not to digest food.

- Water should be available for the horse at all times. However, do not suddenly offer water in the last hour before the start.

- All necessary equipment should be taken to the 'Box' for use in the 10–minute halt between Phases C and D.

- Everything should be left ready for the return of the horse after Phase D.

Care of the Horse at the End of Phase B (Steeplechase)

- There is a designated area at the end of Phase B, where attention may be given to the horse. It is useful to have a helper at hand with the following equipment: bucket, sponge, water, spanner for studs, tools for removing a shoe.

- If possible, without stopping the horse as he starts walking on Phase C, the helper should check his soundness, shoes and studs. If the horse should need

the farrier's attention before Phase D, arrangements for this can be made while the horse is completing Phase C.

- In very hot conditions, the horse will benefit from a quick, cool sponge-down.

Care of the Horse in the 'Box'

- The 'Box' is a fenced-in area at the end of Phase C, where competitors can attend to the horse during the compulsory 10–minute break between Phases C and D.

- The groom must be fully aware of the areas where the horse finishes phase C, starts on Phase D, and finishes after D.

- The groom should know where to find water, vet, farrier, first aid, starter, timer and lavatory.

- All the equipment required in the Box should be taken there (packed in a trunk) before the start of Phase A. If possible, the groom should try and lay out the equipment in the Box, somewhere quiet and sheltered from wind, rain or sun.

List of Equipment

- Water containers.
- Two water buckets.
- Two sponges – one for washing down, one for rinsing the horse's mouth.
- Sweat scraper.
- Towels.
- Waterproof rug.
- Sweat rug.
- Thick rug.
- Spare set of shoes with studs already in place.
- Stud equipment and hoof pick.
- Scissors.
- Spare bridle, reins, saddle, stirrups and girth.
- Spare boots.
- Hole puncher.
- Grease.
- Jacket, gloves, whip and drink for rider.

- As the horse completes Phase C he goes through a compulsory veterinary check. This varies from a brief look to determine soundness and well-being, to a more thorough check of pulse, temperature and respiration. As soon as it is completed, the groom should get to work on the horse.

- The rider must now be free to talk to the trainer, and to prepare mentally for Phase D.

- It is useful to have someone to hold the horse and one or two workers to prepare him for Phase D.

- Put a headcollar over the horse's bridle, slacken the girths, move the saddle forward, and loosen the noseband. Do not take the saddle off.

- Check shoes, boots/bandages.

- Wash the areas where the blood vessels are closest to the surface, i.e. the head, the insides of the legs, and the neck.

- In hot weather, plenty of water helps to reduce the temperature. In cold weather, just use sufficient water to sponge the sweaty areas.

- Sponge out the mouth with clean water.

- Scrape off excess water. In cold conditions, towel the horse dry.

- Cover the horse with rug suitable for the conditions. However, if the weather is very hot do not use a rug as it will trap air and slow cooling. Lead the horse around. All this should take approximately three minutes.

- Keep an eye on the horse.

- Start to prepare the horse for Phase D four minutes before he is due to leave the Box.

- Check the saddle, tighten the girths, refasten the noseband, remove the headcollar.

- Help the rider to mount. It may be necessary to alter the stirrups.

- Wipe off any excess water on the reins. Using rubber gloves, grease the fronts of all four legs to protect knees and stifles from bangs and bruises. (See also notes on grease application in hot weather, page 252.)

- Check the surcingle.

- Use a piece of rope, minus the clip, to lead the horse into the starting box, if necessary.

Care of the Horse After the Cross-Country in Normal Weather Conditions

- In two- or three-day events, the groom must wait for permission from the steward at the scales before touching the horse. The horse will be checked by the vet.

- Remove the tack. Cover the horse with a sweat rug, and a light blanket if the weather is cold; do not use any rugs if the weather is hot. Offer the horse a small drink of about a litre (¼ gallon) of water with chill taken off. Walk the horse slowly for about five minutes, or until the horse appears to be breathing normally. This allows the build up of lactic acid in the muscles formed during work to disperse.

- The horse, particularly in hot or very humid weather, recovers more quickly if electrolyte salts are added to his drinking water. If he is used to being given these after strenuous exertion at home, he will be more likely to accept them at a competition. For advice on type and quantity, consult your veterinary surgeon.

- Take the horse to the stable area or to his box and remove the boots and bandages.

- Check the horse for injury.

- Remove the studs.

- Wash off the horse. Remove any surplus water with a sweat scraper. Cover the horse with a sweat rug and a light blanket if the weather is cool (in hot or humid weather dispense with the rugs), and walk the horse again.

- Give the horse another drink of about 2 litres (½ gallon). From this time on he may have 2 litres

(½ gallon) of water every 15 minutes.

- Put the horse in the stable and encourage him to stale.

- If all is well, bandage the horse with dry bandages.

- Give the horse a small mash to get his digestive juices working. This may be bran and oats or sugar-beet feed and grass nuts, according to the horse's normal diet.

- Tie up a small haynet, give the horse a bucket of water and leave him to rest.

- After two hours, remove the bandages and check the legs for signs of injury.

- Treat any injury as necessary, and re-bandage with dry bandages over cooling lotion or cold kaolin. Do not use any cooling preparations that you have not used before as the horse may be allergic to them.

- Lead the horse around to loosen his muscles.

- Return to give the horse another feed and more hay. At this stage, the horse should have water ad lib.

- Do not give the horse more feed than he is used to eating.

- Make a late-night check to see that the horse is resting and comfortable.

Care in Extreme Heat

- Immediately after the horse has finished the cross-country lead him around for a few minutes.

- Wash the horse down, scrape off the surplus water and lead him around again. The cooling process is brought about by evaporation. Repeat until the horse appears to be less distressed. He must continue walking until his temperature, pulse and respiration are back to normal.

- Keep a check on the horse's temperature.

- Wash the horse down with iced water all over.

- Small drinks of water can be given every fifteen minutes, to prevent dehydration.

- Add electrolytes to the water, but offer plain water as well.

Cold-Water Cooling for Hot Horses by Dr David Marlin

The following guidelines are reproduced by kind permission of the Animal Health Trust.

Horses that are hot (above 40°C/104°F) and competing in hot environments (above 26.5°C/80°F) and that are cooled quickly during or after competition are less likely to suffer heat stress, will recover more quickly, will not become as dehydrated and are almost certain to perform better.

Technique

All that is needed are:

- Some large buckets to hold 40–50 litres (9–11 gallons) of water and ice.
- Smaller buckets.
- Giant sponges.
- Three assistants – one to hold the horse and one person to cool each side. (It is not necessary to remove the tack.)

Start to cool the horse whilst taking the horse's rectal temperature. Liberally apply cold water to all parts of the body including the quarters, as this is where most of the large muscles used for movement are located and so it is an area that gets particularly hot.

It is not necessary to scrape off excess water after each application; it is more important to continue to apply cold water. If you wish to scrape off the excess water, do so quickly at the end of each 30-second cooling period and while the horse is being walked between cooling periods.

Carry on cooling the horse for 20–30 seconds, walk the horse for 20–30 seconds and cool again. The walking and cooling sequence is **IMPORTANT**. The walking promotes skin blood flow and the movement of air aids evaporation. If possible, carry out the cooling and walking in the shade.

Check the horse's rectal temperature at intervals. It should be possible to reduce rectal temperature by around 1°C in 10 minutes.

There is no evidence to suggest that there is any harm in letting the horse drink small amounts of water (half a bucket) during a competition (e.g. whilst in the 10–minute box in three-day eventing), between rounds (e.g. show jumping) and during long warm-up periods (e.g. dressage), which will also help to cool the horse down and reduce the effects of dehydration.

When to stop

- When the horse's rectal temperature is less than 38°C–39°C.

- When the horse's skin feels cool to touch (over the quarters) after a walking period.

- If the respiratory rate is less than 30 breaths per minute

- If the horse begins to shiver.

What not to do

1. Ice in the rectum does very little to lower body temperature. It makes it hard to assess body temperature and can hide a high temperature. Masking a high temperature from vets at events is unwise as it will prevent a horse receiving appropriate cooling and other necessary treatments, which may result in the development of heat exhaustion and death. The chances of injury will also increase if the horse is allowed to continue when overheated and dehydrated.

2. Don't hold small bags of ice over the head, neck, under the tail, on the quarters, etc. Instead, concentrate on cooling as much of the body surface as possible. Holding bags of ice is likely to reduce cooling by stopping skin blood flow to the area under the pack.

3. Do not place wet towels on the neck or quarters. Although at first the towel may be wet and cold, it soon warms, and hinders the loss of heat, acting as an insulator.

4. Excessive application of grease prior to the cross-country acts as an insulator, prevents sweating and limits sweat evaporation.

5. Do not let horses stand still for prolonged periods. If cold-water cooling is adopted, do so completely and not tentatively. The cold water on the skin will reduce the horse's sweating rate. This has the advantage that because the horse sweats less, it becomes less dehydrated.

6. As stated, there is no harm in allowing horses to drink small amounts (half a bucket) during competition. Water should also be left in the stable until 15–30 minutes before exercise. Water is emptied very rapidly from the stomach. Do NOT give the horse ice-cold water to drink. Recent research has also shown that it is important to feed hard feed and some hay together, at least four hours prior to exercise.

There is **no** evidence to suggest that cold-water cooling causes other problems such as 'tying-up'.

Care Before the Veterinary Inspection on the Final Day of a Three-Day Event

- Start early.

- Feed and water the horse.

- Remove bandages, wash off poultices, and hose legs.

- Lead the horse out at the walk, then jog him to check the state of soundness. If the horse is in good order, lead him out for half an hour to remove stiffness from the muscles.

- On return to the stables, groom the horse and plait him.

- Thirty minutes before the veterinary inspection, take the horse out again for further walking. In some cases the rider may prefer to give the horse a hack or a light school to loosen the muscles.

CHAPTER 35

THE DRESSAGE HORSE

The dressage horse is the gymnast of the equine world. His training is a continuing process. For this reason, as soon as he is fit and working he is rarely given long rest periods. Any prolonged time without work allows his muscles to slacken, which will necessitate weeks of preparation before he is once more in hard condition and ready to continue his training. If it is decided that a horse needs a rest from intensive work, he can be turned out or hacked out daily.

The basic fittening work is described on pages 243 onwards. Once the horse is fit, his exercise programme will usually consist of lunge work, ridden work in the school, and hacking out. Most horses will benefit from a daily spell in the field to allow them to relax mentally. Jumping makes a refreshing change from work on the flat, and gymnastic jumping exercises in particular also improve athleticism. The horse's mental attitude to his work is of prime importance, and this is affected by his stable environment and daily care, as well as by the work itself. Every horse must be treated as an individual.

Feeding

The dressage horse needs to be fit, but not in the same way as the event horse. He must be muscle fit, but does not need the same reserves for endurance and galloping as an eventer or racehorse. He should look well covered, but should not be overweight, as this makes his work unnecessarily stressful.

Feeding must balance the energy requirements of his work, and he must approach this work with the right mental attitude, especially when at a competition. Time must be spent working out a suitable diet for the individual horse.

Grooming and Clipping

The dressage horse requires thorough daily grooming to keep him healthy. In warm weather, a wash-down after work ensures a clean coat.

He should be clipped as necessary. Most horses do not require clipping in summer, but those who grow a heavy summer coat should have it removed by a total clip. In winter, when competition is more limited, most horses are comfortable if given a blanket or chaser clip, but by February they could have a whole clip, so that they will grow an even summer coat. Horses should be kept neatly trimmed through the year. Tails are usually left long, well below the hock, unlike those of the hunter or event horse.

Plaiting

For competition, manes should be plaited. Some riders use Continental-type plaits, with white tape wound around the top of the plait. Tails can be either plaited or left full.

Shoeing

An experienced farrier should advise on the best type of shoes for the individual horse. It is usual to put on fullered and concave shoes of light iron, with stud holes in front and behind. Cushion pads may be used in front to lessen concussion. As slipping is less of a problem in dressage, flat, Continental-type shoes are now often used for dressage horses.

When competing on grass, the choice of stud depends on local conditions. It is usual to put studs in the hind shoes. Studs in front shoes can shorten a horse's action and sometimes they can cause unlevel steps. On the other hand, slippery conditions are just as likely to affect the stride. Studs are not necessary for competing on artificial arenas.

Tack

- A dressage saddle is an essential item.

- Use of a numnah is general practice. The square-shaped Continental ones are the most fashionable, but those shaped to fit the saddle are neater and less conspicuous.

- To stop the saddle slipping forward a foregirth can be worn.

- A snaffle or double bridle, depending on the test.

- For permitted saddlery and bits, see the current British Dressage rules.

- Boots or bandages are usually worn when working at home and riding in before a test. Sandown or similar bandages, which can be put on without Gamgee, are often used to protect the leg from below the knee to the middle of the fetlock.

Cooling Down

In very hot weather, follow the guidelines set down for event horses (see page 251).

36 THE SHOW JUMPER

Show jumping is an all-year-round sport, but a horse cannot be expected to compete indefinitely. Careful forward planning is therefore necessary to ensure that there are rest periods so that the horse will be fresh and keen and will give of his best.

Fittening

- After a rest period of two or three months, a horse requires an eight-week fittening programme before competing in a major competition. The initial exercise should be organised as outlined in Chapter 33.

- Roadwork and hacking out should continue throughout the fittening period, but after the third or fourth week, lungeing, school work and gymnastic jumping may be introduced.

- A show jumper needs to be fit to compete over present-day courses, and slow cantering, as detailed for one-day event horses, is needed.

- During the summer, if the ground becomes hard, it is essential to avoid unnecessary pounding of the legs.

- Road exercise should be restricted to active walking, preferably including some hills. School work and jumping should be carried out in a field with a thick, matt surface of grass, or on a non-skid prepared surface

- After a three- or four-day show, it is sensible to arrange for a few days rest and daily grazing.

Feeding

The show jumper must feel and look well in himself, but should not carry any excess weight.

He requires plenty of energy food, but high protein feeds should be avoided. It is important to adjust feeding so that the horse will keep his condition. A daily spell at grass helps to ease any stress.

A horse may have a normal feed two hours before jumping. When at a show, arrange his feeds according to the time of his classes, but as close to his normal feed times as possible. If necessary, give a smaller feed during the day and make up for this at night.

He may have a small amount of hay (1kg/2lbs) after jumping, but he should not be allowed more if he is expected to jump again within three hours.

Grooming and General Care

For grooming details see Chapter 10. A horse should always be well turned out for a show.

Clipping

In winter, a blanket or chaser clip is recommended, so that the horse will not catch a chill when standing about at a competition.

In summer, a well-bred horse should only need tidying up. A more common-bred horse with a thicker coat may benefit from a full clip.

The show jumper should be kept well trimmed, with mane pulled short, and a tidy or pulled tail.

Plaiting

This improves the turnout of most horses. Although it is not a requirement, it is expected at the bigger shows. Pulled tails are quicker and easier to keep tidy.

Shoeing

Normal fullered and concave shoes are used. They can be fitted with road studs behind. The inner web of the hind shoes may be bevelled smooth to minimise the risk of an over-reach. Stud holes should be put in all four shoes, so that suitable studs can be used according to local conditions. A spare set of shoes should be taken to the show.

Tack

- Jumping saddle.
- Breastplate.
- Surcingle.
- Bridle and martingale to suit.
- Tendon boots and over-reach boots in front.
- Hind leg protection may be necessary for some horses.
- Horses who catch their girth area with their front feet when jumping big fences may need a leather guard fitted to the girth.

Care at the Show

It is important to warm up the horse before jumping. There will then be less risk of strain to muscles and tendons. After jumping, the horse should be kept warm and walking. Warm rugs should be put over him in winter, and a summer sheet and/or sweat sheet in summer.

When competing in very hot conditions, follow the cooling guidelines on page 251.

THE ENDURANCE HORSE

Types of Competition/Ride

Set-Speed Rides

These are usually between 20–40 miles (32–64km), and must be ridden at or above a certain speed according to the classification of the ride. Training rides are usually 15–20 miles (32km). Rides for novice horses are usually at 6½mph (10km/h) for 30 miles (48km). Qualifiers for the Golden Horseshoe Ride Final are at 7½mph (12km/h) for 40 miles (64km). The Golden Horseshoe Final is a 100–mile (120km) ride over two days, with the distance divided 50/50. At the Final, 8mph (13km/h) is required for a gold, 7mph (11km/h) for a silver and 6½mph (10km/h) for a bronze. All horses must have a clean veterinary sheet. Any penalties incur a drop in grade. Speeds could be liable to alteration and should be checked.

Vet Gate Rides

These are usually 50 miles (80km) and over. After each 15 miles or 24km (approximately) they pass through 'gates'. Here the horses are examined by the vets and, having passed this examination, may continue. The clock is not stopped during the examination. The horse completing the course in the least time is the winner.

To be successful in this type of ride, riders must be capable of assessing the condition of their horse as they approach the 'gate', and should adjust their speed so that time will be not be wasted if the horse fails to pass

the first examination. If they fail, they are held and may re-present for examination any time within the next 30 minutes. Failure to pass the second examination results in elimination. The required pulse rate to be achieved is below 64.

Race Rides

These are races held in major competitions. All horses start together at the drop of a flag. There are compulsory rest periods or vet gates every 15½ miles or 25km (approximately). At these halts horses are re-examined by the vets and may only proceed if they pass the examination. The first horse to finish is the winner.

Distances vary, but all are over 50 miles (80km), with a maximum within 24 hours of 100 miles (160km). Rides can continue for several days.

In any race ride, elimination is incurred if a minimum average speed of 6½mph (10km/h) is not maintained.

The British Endurance Riding Association and the Endurance Horse and Pony Society are the main organising bodies for endurance rides in Great Britain.

Fittening

Basic work for the first weeks after a rest is as for any discipline, with long periods of walking followed by slow trotting. Most of it is carried out on roads or other hard, level surfaces. Later, faster trotting, particularly up hills, helps to develop the muscles, especially those of the heart and lungs. The horse must learn to trot down hills, but once he has learned to do so, downhill trotting is not recommended for fittening purposes as it is a strain on the tendons.

After the first month, training can be varied by work on the lunge, in the school, over jumps, etc. The type of competition will determine how fit the horse should be. For a 20–mile (32km) ride, an hour daily is sufficient with a three-hour ride once a week. For 50 miles (80km) or further in faster competitions, more hours of work are needed. These may include interval-training type fittening and, at least once a week, a long ride of about five hours covering 30 miles (48km) or more at a speed in excess of 6mph (9.6km/h). This accustoms the horse to carrying a rider for long periods at the required speed, and hardens his back in the saddle area.

Whenever the horse is being exercised, it is important to make sure that he is 'working'. If he is just slopping along on a loose rein, he does not use his muscles, lungs or heart and cannot get fitter.

If the ground is very hard or deep and unsuitable for interval training at the canter, work up long hills at a strong trot. This will raise the heart and breathing rates as in cantering, and will cause less strain on the legs.

Interval Training

This is a method of increasing a horse's fitness by using varying intervals of cantering and walking. For a horse competing in 25–40 mile rides, the interval training system would be very similar to that outlined for novice event horses (see page 247). For more demanding, longer rides, e.g. the Golden Horseshoe Ride, more specialist stamina work will need to be undertaken to extend the endurance horse's fitness level.

The horse should be basic 'walk, trot and canter fit' (4–6 weeks) before commencing.

Before the interval training it is important to warm up the horse at the trot and afterwards to walk him until he is cool.

If the heart rate is checked immediately after cantering, and again after the 3 minutes' walking, a fair idea of the horse's fitness can be gained from noting the speed with which the heart rate drops back to normal. When the heart rate fails to return to normal in less time than on the previous occasion after doing the same work in the same weather conditions, the horse is as fit as possible. The trainer should aim to achieve this condition shortly before the day of a competition.

A stethoscope can be used to check the heart rate at rest in the stable before starting, and also during and after interval training.

The resting pulse rate of a fit horse, standing in the stable, can be as low as 25 up to 40. The warm-up pulse rate would be in the range of 40–50. During trot work the pulse would rise to 80–100, and in canter would be between 100 and 150. Recovery to below 64 is required before presentation to the vet in competition.

Feeding and General Management

This is the same as for any horse in hard work. All feedstuffs should be of the best quality obtainable. As training progresses, bulk feeds should be cut down and replaced by a series of short feeds – oats, barley, proprietary nuts etc., supplemented by anything that the horse likes – e.g. soaked sugar-beet pulp, carrots, boiled linseed – to encourage his appetite. When the horse is in hard work, salt should be added to the feeds, as this is lost in sweat and lack of it may cause muscle cramps. Most trainers also like to add a mineral supplement to supply any trace elements which may be lacking in the diet. (See also page 154.)

There are no hard-and-fast rules as to the quantity of feed required. It varies according to the horse's build and temperament. If, however, a large quantity of short feed is given, this should be divided into four feeds during the 24 hours to avoid putting the horse off his feed altogether.

The horse can be kept in a fairly bare paddock for part of the time, as long as there is shelter from cold winds and flies. This helps to keep him relaxed and to give him exercise.

Water should always be available for him to drink. On rides, he should be encouraged to drink freely from any source, whether a bucket, stream, cattle trough, or even a clean puddle. This is particularly important in hot weather as, because of sweating, distance horses soon become dehydrated. No harm will come to the horse if he drinks when he is hot as long as he moves off immediately. He should not be allowed to drink too much water at compulsory halts or just after finishing. At these times he will be standing about, so only a little chilled water should be allowed, or colic might develop. When competing it is advantageous to add electrolytes to water given from buckets. These replace the salts, lost in sweat. If the horse will not drink water with electrolytes, plain water should be offered. At no time should he be put off drinking.

Grooming

The horse should be thoroughly groomed each day. Any sweat should be removed and particular attention should be paid to the saddle area. Take special care of his feet, and have him shod regularly by a skilled farrier.

Worming and Vaccination

As well as regular worming all distance horses should be immunised against influenza and tetanus, as required by Jockey Club regulations. Many rides are based at racecourses, and any horses not immunised will not be allowed on racecourse premises.

Saddlery

SADDLE English general-purpose, Western or special endurance type, worn with or without numnah or saddle blanket as required.

BRIDLE Snaffle, pelham or double bridle, or any form of hackamore or bitless bridle. If a gag bit is used, there must be two reins, one of which should be a snaffle rein.

Preparation for the Competition

Each horse and rider should have a back-up crew, consisting of one person, or more usually two, in a vehicle, one of whom should be a good navigator. This crew follows the horse, meeting up with him at points en route and at the veterinary checks.

Before the day of the competition, the rider and crew should discuss the plan of campaign. They should outline the route on a local Ordnance Survey map, study it, and decide upon suitable meeting places. They must calculate the times the rider expects to pass each point, and check whether it is possible for the helper to get there by car in time to meet him. The helper often has to take a longer route, for example driving around a mountain and up the next valley, while the competitor travels straight over the top, so timing is vital. Having settled the route to be taken, with meeting places and arrival times, the crew's next duty is to load the vehicle with all the equipment needed during the ride.

Equipment

- 25-litre (5-gallon) water container (full).

- Two buckets: one for drinking and one for washing down.

- Electrolytes, and measure, to add to drinking water if required.

- Funnel to pour unused clean water back into container.

- Sponges and 'slosh-bottles' for washing down. Sweat scraper.

- Towels.

- Headcollar and rope.

- Rugs, including sweat rug and waterproof sheet.

- Surcingle.

- Spare set of shoes, ready shaped to fit the horse, plus protective hoof boot.

- Spare reins, leathers, girth and numnah (if worn), plus a leather bootlace or string for emergency tack repairs.

- Crepe bandages and basic first aid kit for both horse and rider. Veterinary surgeons are always present on rides and can deal with more serious problems.

- Drinks for rider, with plastic cup.

- Jacket or anorak for rider at compulsory halts.

- Pocket knife.

- Grooming kit.

- Haynet if halts are not back at base.

The above should be carefully packed in the vehicle, making sure that essential items, such as water, are readily available.

When competing, the rider should carry:

- Folding hoof pick.

- Map of route, together with 'talk round' if supplied. Both should be in a plastic case to keep them dry.

- Compass and whistle in case of fog or losing way.

- Card showing times of meeting up with helper. This acts as a guide as to whether the rider is keeping up the correct speed.

- Safety pin and piece of string or leather bootlace (for emergency repairs).

- Small crepe bandage and wound powder for first aid en route.

- Coins for telephone box and note of the 'base' telephone number.

- Energy-giving snack bar.

Veterinary Inspection

On the day of the ride, the helper takes the horse to the preliminary vetting, and holds him while the vet checks his pulse and respiratory rates and examines him for 'lumps and bumps'. The helper will then be expected to run the horse up for soundness. This should have been practised beforehand, so that he runs up freely and obeys the helper's spoken commands to trot, walk, etc. The rider should attend the vetting in order to hold rugs etc. and to answer any queries about the horse that the veterinary surgeon may ask.

When the vetting has been completed, the rider then saddles up and the helper takes the horse to have his tack and shoes inspected, if required by the ride rules.

The Start

The helper assists the rider to mount, checks the girth etc., and ensures that the rider has everything he needs. As soon as the horse has left the start, the helper makes sure that all the necessary equipment is in the car. He can then set off for the first meeting place.

Meeting Places

As the horse approaches, the helper should have ready a bucket of water for the horse and a drink for the rider At later points on the ride, a bucket of water for spongeing down the horse should also be set aside:

- First, the horse should be offered a drink, then a quick check should be made to see that all shoes are in order and that there are no cuts or scratches in need of attention.

- If the horse is very hot, a quick spongeing of his face and neck will freshen him. (See also the guidelines on cooling horses on page 251.)

Compulsory Halt

On arrival at compulsory halts, the helper should take the horse while the rider has something to eat and drink and takes a rest:

- The bridle should be changed for a headcollar, and a little water should be offered.

- The girth should be loosened but the saddle should not be removed. Time must be allowed for the blood vessels of the back to refill after taking the rider's weight for a long period. Removing the saddle before this has taken place can cause lumps to form under the skin.

- The legs may be sponged down and any cuts dressed. The head and neck may also be sponged.

- The reins should be washed free of sweat Later, when the saddle is removed, the girth and numnah should be changed and the girth area cleaned.

While this is being done the horse should be allowed to graze or nibble a little hay, as this will help him to relax and to lower the pulse rate before vetting takes place.

After 20 minutes the horse is called to the vetting area for his check. Again, the helper leads him, accompanied by the rider:

- Vetting completed, the rider re-saddles and re-mounts for the next stage of the ride

- The helper re-packs the vehicle.

The Finish

At the finish the procedure for the first 30 minutes is the same as at the halts:

- Surplus mud and sweat should be removed, any cuts should be dressed and his legs should be well bandaged with Gamgee tissue under the bandages.

- He may then be offered more water and a small feed and some hay, or allowed to graze and possibly have a roll.

- If rugs are worn, frequent checks should be made to see that he has not broken out.

- He should not be loaded for the journey home until perfectly dry and cool.

CHAPTER 38

THE DRIVING HORSE OR PONY

Driving can be an all-year-round pursuit, although the competition season starts in April and ends in October. Activities range from showing classes and driving club meets, to horse driving trials and marathons.

Fittening

This depends on the type of activity, as driving trials horses require a much more intense fittening programme than show horses. The programme starts according to the date of the first competition; six to eight weeks should be sufficient fittening for show classes, but a minimum of sixteen weeks would be required for driving trials and marathons.

When first starting work, horses may be lunged or long-reined, beginning with 5 minutes on either rein, building up to a total of 30–45 minutes according to the required degree of fitness. The work is carried out in trot, with boots or bandages on all four legs. In some yards the horses are not ridden, but in many cases – particularly with older horses already broken to harness – it is more practical to hack them out. If hill work can be included, it helps the fittening process.

Horses due to compete in marathons should, by the sixth week of their basic fittening programme, be doing at least 2 hours work a day. From the twelfth week onwards they can mostly be driven, and they should be covering 20 miles a day. The week before a competition their work may then be eased off, so that they are fresh and eager for work on the competition day.

The suppling work on the lunge and long-reins is a very important part of the preparation of any driving horse, especially for the dressage section of a three-day event.

In the first few weeks of driven work, care must be taken to see that the horse's shoulders and chest do not become tender from pressure by the collar or breast collar. The skin can be hardened with methylated spirits.

Horses working in double harness, as a tandem or particularly to a coach, require extra time to get used to working together as a team.

Feeding

The show horse must look good and must be well covered; but the driving trials horse must be – and must look – very fit, so he needs more energy feed than the show horse.

Health

Horses competing in driving trials in very hot weather may suffer from dehydration, and precautions must be taken against this.

Grooming

Details are given in Chapter 10.

Clipping

This should be carried out as required. See Chapter 11. A horse taking part in driving trials may benefit from a full clip in summer.

Trimming

No part of the mane should be clipped, nor should it be hogged: it is a protection against collar sores on the withers. For presentation showing and turnout classes manes should be pulled short and plaited. Tails if not pulled may be plaited. Legs should be tidied. Mountain and moorland ponies should have their manes and tails tidied, but not pulled or plaited.

Shoeing

Shoes can be fullered or of plain iron. The custom of putting very heavy shoes on a show horse to improve his action is now less popular. The weight and type of shoe varies between different horses and according to intended work. If possible, obtain advice from a farrier experienced in shoeing horses for driving.

Road studs are used behind for roadwork. Depending on conditions on the day, studs may also be used in the ring.

An alternative is the use of borium let into the steel. This provides a good non-slip surface, and unlike studs, it does not jar the legs.

Horses destined for driving trials should preferably be shod several days before a competition, so that the shoes are well bedded and sure. A spare set of shoes should be taken to the competition or show.

Harness

Driving harness is more complex than riding tack, and takes much longer to clean and maintain. Collars, breast collars and cruppers should be kept clean and the leather supple, so that the risk of rubbing is minimised. All harness should be closely inspected for worn material, leather and stitching. The underside of show harness should be treated with saddle soap, and the top surface with boot polish or patent leather cream. Harness for presentation and turnout classes should be carefully adjusted to ensure a correct fit.

Vehicles

These must be kept in immaculate condition and must be well serviced. Wheels may need professional attention. All parts of the vehicle should have regular safety checks.

In presentation, show, and turnout classes, it is essential for the turnout of horse, vehicle and groom to be correct and according to the requirements of the competition.

CHAPTER

39 THE HUNTER

Fittening Work

Heavyweight hunters, particularly if in gross condition, should be brought in to work at the beginning of August. This gives them time to get fit before hunting starts in November. Thoroughbreds and lightly built horses require less time to get fit, and their fittening programme may be left to the end of August. The value of plenty of walking exercise cannot be over-emphasised. It helps to harden the legs and to develop muscles without strain.

During September, the hunter is generally given some quiet cubhunting, and more energetic cubhunting in October. This should look after the canter side of the fittening. Horses not going out with hounds until November require canter work twice a week in October. The actual speed and distance of the cantering depends on the individual horse and whether suitable going is available. Trotting up hills can be a good substitute if canter work is difficult to organise. The aim is to have the horse fit to gallop by 1 November.

Work During Hunting

A fit horse hunting three full days a fortnight or two short days a week should only require about 1 to 1¼ hours' exercise a day to maintain his fitness. Should there be no hunting, he should have a minimum 1½ hours each day.

- During snow or freezing weather, if normal exercise is not possible, and if there is no indoor school, a large ring of shavings or mucked-out straw can be laid

down. The horse should either be ridden, lunged, or led round on this surface.

- If the snow is dry and powdery it will not ball in the feet so riding in fields or woods is usually quite safe Roads must be avoided because of the danger from and to traffic. Wet snow does ball in the feet, and may not be safe to ride on unless the feet are packed with grease or fitted with pads.

Feeding

For suitable diet, see Chapter 21.

The night before the horse's rest day he should be given less concentrates.

- Horses subject to azoturia require special treatment, i.e. on the day off reduced concentrates and walking out.

- All horses benefit from some in-hand grazing or regular turn out to grass for up to an hour a day.

- On September cubhunting mornings, horses need not be fed beforehand, and can be given their breakfast on return to the stable. For October cubhunting, horses can be given a small feed at 6 am, followed by normal daily rations on their return.

- For November hunting onwards, horses should have their normal hard feed at least two hours before leaving the stable. They can be given a small amount of hay at 7 am, despite the fact that this takes longer to eat and to digest. Assuming that the meet is not before 11 am, strenuous effort would not be required before 11.30 am, allowing four hours for digestion.

On return from hunting the horse should be watered according to the method below, then given a small net of hay, and half an hour later a small feed. He requires a nourishing but an easily digested feed. Three hours later he should be able to digest his normal evening ration of a concentrate feed and hay.

The next day should normally be a rest day, and as long as the horse is well and sound, he can be given normal rations.

Preparing for Hunting

The Day Before

- Check the horse, particularly his shoes.

- He should be given a thorough grooming.

- Tack should be checked for stitching and put ready for the morning.

- Bits and stirrup irons should be of stainless steel.

- Reins should be rubber-covered to give control when the horse sweats and/or when the weather is wet.

- Double bridles may have a laced or rubber-covered bridoon rein.

- The headcollar and roller should be cleaned.

- The day rug, sweat sheet, leg protectors or bandages with Gamgee and knee caps should be put ready.

- Plaiting should be left until the next morning.

Equipment for Hunting Day

- Grooming kit for tidying horse on arrival.

- Haynet for the return journey.

- Water carrier and bucket for giving the horse a small drink on his return to the box and also to enable any cuts to be washed clean.

- First-aid kit consisting of small bowl, cotton wool, packet of salt, antiseptic powder or spray, oiled lint dressing and crepe bandages.

The Morning of the Hunt

- The horse should be allowed to eat his feed undisturbed. If time allows, he can then be mucked out. If not, this can be left until later, providing that someone is available to do it. Otherwise, the box must be mucked out and bedded down before leaving.

- The horse should be groomed, tail bandage put on, mane plaited and feet oiled. If the bed is of shavings, his feet should not be oiled until he is ready to leave. If the going is likely to be very muddy, it is sensible to plait the bottom of the tail, turn it up, and stitch it. This makes it more comfortable for the horse. For horses who kick out when their tails are wet and muddy this should definitely be done. An unpulled tail looks smarter if plaited.

- Any small cuts or cracks should be protected from infection with antiseptic cream or spray. The horse's legs can be oiled (with neatsfoot, olive oil or liquid paraffin) to prevent mud getting into the pores of the skin and to make cleaning easier.

- Horses are often transported to the meet saddled and bridled and without protective clothing, except for a tail bandage or tail guard. As long as the journey is short and the driver careful, this arrangement is quite satisfactory. Horses travelling on longer journeys should wear protective clothing and be saddled up on arrival.

- Rugging should be according to temperature. A sweat rug, though not worn on the way to the meet, should be taken and put on for the return journey. Roller or surcingle must be long enough to buckle over the saddle.

- Ideally, transport should be parked a minimum of one mile from the meet, and out of the way of passing traffic. A short hack to the meet helps to settle both horse and rider.

- Transport should be left locked and any equipment or valuables put out of sight and made secure. Trailers can be padlocked to the towing vehicle to deter thieves, or a wheel clamp can be employed.

Return Journey by Horsebox or Trailer

On a dry day, horses should arrive back at the box cool and dry, having walked the last mile. On a wet day, they are better trotted for the last mile, so that they are warm before being loaded.

Procedure

- Loosen girths and ease the saddle. Put on sweat and top rug and fasten roller or surcingle. Alternatively, the saddle can be taken off and the back of the horse slapped to restore circulation. Considerate riders dismount and lead their horse for the last half mile.

- Take off the bridle and put on a headcollar. Tie the horse to a safety loop on the box. Anxious or excited horses should not have their bridle removed until they are safely loaded.

- Offer a small drink of water. Check and deal with any obvious injuries.

- Put on travelling gear, in particular a tail bandage or tail guard to prevent rubbing. Horses facing a long journey (over 1 hour) benefit from well applied stable bandages over Gamgee or similar, as these give support to a tired horse and keep him warm. Knee caps are advisable.

- Load the horse. Tie up the haynet.

- If transport is parked on busy road or it is raining, it may be safer to load the horse first before putting on the travelling gear.

Hacking Home

On a **dry day,** walk for ½ mile, then intersperse slow, steady trotting with short walk periods. Always walk the last mile home and lead the horse for the last ½ mile. He will then return cool and dry.

On a **wet day,** proceed as above, but trot the last mile home so that the horse returns warm to his stable.

On Return to the Stable

- Unload the horse. His feet can either be washed and checked, or done later. Take the horse to his stable. Hay should be available. Remove the saddle. Check over the back and girth area and replace the rugs. Stand back, shake up the bed and whistle to encourage the horse to stale. Should he roll, watch to make sure that he does not get cast.

- Offer the horse a drink. In cold weather, take off the chill. Very tired and dehydrated horses may be offered water plus electrolytes, but plain water should be offered as well, as he may prefer this and it is important that he should drink. If he has had a drink before travelling home, he can be allowed 4.5 litres (1 gallon) of water at this stage, and then allowed his fill 15 minutes later. If he has not had any water, he should be rationed to 2.5 litres (½ gallon) every 15 minutes. This avoids too much chill to the stomach.

- Tie him up and allow the horse to eat his hay.

- Remove the travelling gear. Pick out and check feet. Check shoes, heels and legs for cuts and thorns. Treat any injuries. The knee, upper arm and front and inner thigh are likely places for thorns, which can usually be felt under the skin, and the horse will flinch under pressure. Thorns can sometimes be removed with tweezers. If they are deeply embedded, the area has to be poulticed or fomented to soften the skin. Blackthorns, in particular, often cause infection and should always be treated with care. Sponge the eyes, nostrils, dock and sweat marks behind the ears and around the nose. If dry, the legs should be brushed off and bandaged with fresh lining put on.

- Turn back the rugs: first from the front and then from the quarters. Clean off sweat and mud. Sticky sweat in the saddle and girth area is often better removed with warm water and a sponge. Rub with a towel to dry. Replace the rugs. Make sure that they are dry. Wash the tail and swish it dry. Undo the mane.

- Give the horse a small feed and haynet.

- The tack can now be cleaned. It is best to do this the same evening.

- After an hour, return and check that the horse has not broken out, i.e. in a cold sweat. Should this happen, first rub or towel-dry the ears until they are warm and dry. If the evening is suitable, walk the horse out. If it is not suitable he should be rubbed with straw or towel-dried – particularly in the loin area; or stand him under an infra-red lamp (if you have one) to warm and dry him. Fresh rugs should now be put on. When attending to the horse, shut the top doors so that he does not catch cold.

- He may then be left, but should be checked every half hour until he remains dry and is settled.

- Water buckets should be filled and more hay given.

- In all cases, hunters should be visited later in the evening to check that all is well. They can then have water buckets refilled and be given their late feed.

- Very tired horses should not be worried by too much cleaning. They should be dried, kept warm, fed and left to rest.

For horses who return wet and muddy

METHOD 1

- Place straw under a fresh sweat rug. Over these put dry top rugs.

- Thatch the legs with hay or straw and loosely bandage them.

- Rub or towel-dry the ears.

- Feed the horse.

- Clean the tack.

- The bandages can either be left on until morning or taken off later that evening. In either case, after removal the legs should be brushed down and the bandages replaced over dry lining.

- The rugs should be turned back and the horse cleaned, as already detailed. Replaced rugs must be dry.

METHOD 2

- In some areas, particularly if the mud is likely to blister, it is better to wash the horse's legs, belly and inner thighs with cool water (not hot) and a horse shampoo.

- Remove all mud.

- Use a sweat scraper to remove any surplus water from the belly and upper arm. These areas can be towel-dried or rubbed with straw.

- Dry the heels and bandage the legs over a dry lining, or thatch them.

The Day After

- If the horse appears well and has eaten up, he should be given his normal feeds and hay.

- Remove the bandages and check the legs for heat and/or swelling. Check the shoes.

- Weather permitting, remove the rugs. Check the saddle and girth areas.

- Check for soundness by having the horse walked and then trotted up. He may at first appear stiff, but this should walk off.

- Replace the rugs.

- The horse should be thoroughly cleaned, and again inspected for injuries and sores.

- The horse should be walked out for 45 minutes at a brisk pace to ease any stiffness. An exercise blanket (with a fillet string) could be used on a cold day. To check on his soundness, trot him up again.

- He should be left to rest.

40 THE POINT-TO-POINTER

The point-to-point season begins in early February and continues until late May. Many owners aim to have their horses fit for particular races in their own and neighbouring counties, and many do not wish to race throughout the season. When choosing meetings, take into account the soundness of the horse and the going. Some horses race better on soft going and cannot cope with hard ground, whereas others prefer firm ground. Fittening work is therefore geared to the date of the first race at which it is planned to run the horse.

In order to qualify for a point-to-point and to obtain the necessary certificate, a horse must have had a minimum of eight days' hunting, although this may vary according to the rules of a particular hunt. The programme, therefore, will begin with a horse who is already hunting fit. Depending on the horse, it will take six to eight weeks to get him racing fit.

Point-to-pointers are regularly hunted until Christmas. Most horses are then given a short break, approximately 10 to 14 days, to freshen up before their more specific fitness programme begins. Exercise during this break can be reduced to one hour's walking. Amounts of concentrates are reduced and hay is increased. The horse should then return to full work with increased zest and a good appetite. Daily in-hand grazing should be organised if weather permits and if grass is available.

Certain horses take a long time to become racing fit, e.g. those who are naturally lazy, carry too much weight or have wind problems. With these horses, it may be advisable to continue hunting for several more weeks and to forego the Christmas break.

Fittening Work

Fittening work for the first 3 to 4 weeks consists of long, slow cantering every third day. The horse is worked at half-speed, as described for the event horse. He must canter on the bit, and in balance, over approximately 3.2km (2 miles). As the horse's fitness improves, the speed can be very gradually increased, until in the last 3 to 4 weeks before his first race he works at threequarter-speed over a reduced maximum distance of 2.4km (1½ miles). As a general principle, it is better to gallop too little rather than too much.

During the last 4 to 5 canter days, the horse should be given a pipe-opener over the final 0.4km (¼ mile), at near maximum speed. This helps to clear his wind. Horses who are thick in the wind require sharper, faster work.

A horse should be given his last gallop 3 to 4 days before a race, and then a pipe-opener the day before, or on the morning of the race.

To help in training, horses should be worked in company, and preferably on ground with a slight uphill incline. It is important for the cantering and galloping terrain to consist of good going.

To avoid strain on muscles and joints, horses must be warmed up before starting their canter work. After finishing they should be walked for at least half an hour. This allows for the dispersal of the lactic acid in the muscles, which builds up during fast work.

Between canter days, work should consist of 1½ to 2 hours' walking exercise, with some trotting up hills. If no hill work is possible, a longer time has to be spent on daily exercise. If canter work is difficult to organise, more fittening hill work should be done. An occasional day's flat work or gymnastic jumping in a school helps to keep a horse supple, and adds variety to his routine work.

It is essential to give horses the opportunity to jump fences at racing speed with one or two others, as it is a skill which they need to learn. Experienced horses require only one or two schools over fences. Novice horses must be given as many as are necessary to get them jumping well. Failure to provide thorough schooling can result in a dangerous fall.

During the week before the first race, the horse's road exercise can be reduced to approximately 1 hour so that he is fresh for the race. He should still have two short, sharp threequarter-speed canter sessions. Care must be taken to ensure that reduced exercise does not result in health problems, or in a horse getting too wound up. If fit, highly strung horses become subject to excess stress, races can be lost. Some trainers with a suitable surface canter a horse on the lunge. This allows him a chance to get the freshness out of his system and to relax mentally.

The Day of the Race

The horse should be transported to the race meeting in good time, preferably arriving 1½– 2 hours before his race is due to start (see Transportation, Chapter 43). After a long journey, he may profit from a walk out in hand to unstiffen. He should be allowed access to water but no food. He must be declared as a runner 45 minutes before the start.

He should be prepared in time to walk down to the paddock for the parade, which will take place 20–30 minutes before the race is due to start. He should wear a paddock sheet over his saddle; in cold weather extra blankets; and on a wet day a waterproof sheet. Once the jockey is mounted and has walked round, it is important to re-check the girths.

After the race, if the horse has been placed in the first four he has to go to the winners' enclosure and the jockey has to weigh-in with the saddle, which he should remove himself. An unplaced horse can be walked straight back to his box. In each case a sweat sheet and possibly a light top rug should be put on, and the horse should be walked about until he is cool and his breathing normal. He may now be offered 2 litres (¼ gallon) of water. To assist recovery, this may contain electrolytes. His legs and heels should be checked for injuries. If any are found, they should be attended to. Weather permitting, the horse should now be washed down with warm water, scraped off, rugged up (including a sweat sheet), and walked about until dry. In cold or wet weather, the horse should be boxed up and then the worst of the sweat should be sponged off and the eyes, nose and dock cleaned. He should then be rugged up and watered. Before leaving for home, his legs should be bandaged over Gamgee or thick leg-wraps, and his rugs, if damp, should be changed. He can be given a haynet. He should be encouraged to stale, although many horses will not do so until back in their own stable.

On arrival at home the horse should be checked over in case he has broken out (see The Hunter, Chapter 39). If the sweat sheet is damp it should be changed. The horse should be given water, then a small feed and his normal haynet. Later, if he has eaten up, he can have his normal hard feed.

Post-race Treatment

The next day the horse should be trotted up and, if all is well, put on normal rations. Horses are usually expected to race six to eight times a season, and on occasions they may have to race on succeeding Saturdays. It is essential for them to be kept fresh and keen to run. If a horse blows up in his first race, or if he finishes very tired, he should be given time to recover (about 14 days) before racing again.

In between races, he should have steady exercise of 1½ hours a day, and two 1.2km (¾ mile) canters a week. Canter distances and speed vary according to the horse: some require more fast work than others. At this stage, it is a mistake to do more fast work than necessary, as he may well get stale and go off his feed. Should this happen, he must be given an easier time and not raced again until he is back to normal.

After the season is finished, the horse should be roughed off (see Chapter 46). As it will then be summer, he should be turned out for a few weeks, but from July it is better to bring him in by day to save him from attacks by flies and also to prevent him from getting too fat. It is sensible to start quiet walking exercises in August, so that the horse can be brought back into work with the minimum of strain. A long period at grass is not recommended – unless it is necessary because of injury – as muscles get slack and there will be more risk of strain when the horse is put back to work.

Feeding (see also Chapter 21)

This varies according to the individual horse. In all cases the horse should look fit and well covered when he starts serious training in January. If at this stage he is in light condition he is unlikely to last the season. Unless he is a very good 'doer' and over-plump, his hay should not be rationed; usually extra concentrate feed limits the desire for hay. If a horse is not clear in his wind from a fitness point of view, hay may need to be reduced, but not to below 40% of the diet, as some bulk is essential for proper digestion A special diet may be necessary, see page 167.

Daily hay should be withheld until after exercise – assuming that exercise will be in the morning. The night before a race, the hay ration should be reduced to 1.8kg (4lbs) and hay should not be given on the morning of the race. The horse should be given his normal feed and then no more until after the race, unless he is racing late in the day, when he may have a small haynet. Normal access to water should be allowed, but he should not be permitted to drink more than half a bucket within 45 minutes of the race.

Bedding

Horses should preferably be bedded on shavings or paper. If bedded on straw they are likely to eat a certain amount, which may affect their wind and certainly increases their intake of bulk.

Grooming

Thorough daily strapping helps to get the horse fit and to keep him healthy. It is essential to clean him thoroughly after any fast work. In suitable weather conditions he can be washed down.

Clipping

Horses should be clipped as considered necessary. Early in the year they are likely to have a chaser or blanket clip, but by April they are more likely to be clipped right out to avoid unnecessary sweating.

Shoeing

Most horses are shod with lightweight steel shoes, which stand up to normal exercise but are also suitable for racing. Some horses are plated, which entails fitting racing plates the day before the race and replacing the normal shoes a day or so later. This practice puts considerable strain on a horse's feet and is not advisable unless the horse has a strong growth of horn. Studs are not allowed because of the risk to horses and jockeys in the case of a fall. Screw-in road studs can be used for exercise but must be removed before a race.

Racing Tack

Saddlery Requirements

- A medium-weight steeplechasing saddle, i.e. 4.5–5.4kg (10–12lbs).

- A chamois leather under the saddle prevents it slipping.

- Stainless-steel irons.

- Rawhide or buffalo-hide leathers. As these can stretch, they should be well used before being put on for racing.

- Two strong webbing girths or leather girths with elastic inserts. These are preferable to plain leather as they are less restrictive when a horse gallops and expands his lungs to the limit.

- Weight cloth and sufficient lead. This may or may not be necessary, depending on the weight of the jockey and the required weight for the particular race.

- Surcingle.

- Aintree breastplate.

- Snaffle bridle with a stainless-steel bit to suit the horse. Cavesson noseband and rubber-covered reins.

- Irish martingale.

During a race, considerable strain is put on all tack. It must be thoroughly checked beforehand.

Jockeys riding horses for other owners may wish to use their own saddles. Though most saddles will usually fit more than one horse, it is a sensible precaution to take a spare racing saddle in case of problems.

Turnout

- Manes are best plaited, unless this causes the horse to sweat up. A loose mane is awkward for the jockey as it can get tangled in hands and reins. For safety, the bridle headpiece should be plaited into the top plait.

- Tails may be plaited, tied-up, and secured with three or more tapes.

- Boots or bandages may be worn. Bandages should be put on over Gamgee and securely stitched. Over-reach boots are not usually worn.

Clothing for the Race Day

- Normal day rugs.
- Two sweat sheets.
- Spare blanket.
- Paddock sheet.
- Waterproof sheet.

Paperwork

Schedules must be applied for in good time. Entries should be made to an agreed plan, so that if for any reason the horse should have to be withdrawn for one day there will be another suitable engagement the following week.

Hunters' certificates signed by the MFH should be registered at Weatherbys as soon as received. Except for members' races, a horse is not eligible to run unless registered.

The horse must have a passport, which must be taken with him to the race meeting. It shows his breeding and provides an exact description, with proof of up-to-date vaccinations. It must be completed by a qualified veterinary surgeon.

41 THE SHOW HORSE OR PONY

The show horse is often rested during the winter months. He is usually stabled at night and turned out during the day. Preparation for the next season begins in early February, to give a clear eight weeks before the early shows. Young horses are given a rest after the summer shows, and then continue their education and training throughout the winter. For animals who do not appear in the ring until May or June, preparation can begin later.

Some people prefer to give their horses a little hunting, to provide a change, but there is a risk of injury or acquiring a blemish.

Well-bred ponies who winter out need extra feed to ensure that they are in good condition. They are usually brought in during February, when they require careful rugging to encourage the summer coat to come through. Ponies who are known to be slow in losing their winter coats may need to be brought in by mid-January. They should then be clipped out so that their summer coat is well established by the early shows. Mountain and moorland ponies should not be clipped unless they are to be entered in show pony classes.

All animals should be regularly wormed, their teeth checked and their inoculation programme completed.

Feeding

- Feeding a show horse or pony needs thought and care. Show animals must look and feel well, but must accept discipline, and must display good manners. The show hack and pony must look well covered, and have an elegant movement, great presence and impeccable manners.

- A show hunter should carry more flesh than a competition horse, but must not be overweight or heavy-topped. Overweight animals – particularly young growing stock – suffer strain on joints, tendons and muscles, which can lead to soundness problems in later life. An experienced judge is quick to recognise any weak points in conformation, even if the owner has hoped to disguise them with excess flesh. The show hunter must be sufficiently fit to gallop without undue strain on heart, legs and wind. His diet should have more energy content than that of the show pony and hack.

- The feeding of leading rein, first and small ponies, requires extra care. The pony must look well, but must be safe for a small child to ride. Such ponies require very little energy food in comparison to horses. They are often better if kept in a small paddock, and only stabled the night before a show.

- Food must be nourishing but not overheating. Cooked, flaked barley, boiled oats, sugar beet and horse and pony cubes provide a suitable diet. Pre-cooked food, as marketed by several firms, is also suitable. Good-quality meadow hay is generally preferable to seed hay, as it has more taste, and the variety of herbs and grasses is important for horses stabled throughout the summer. A period of daily grazing gives a change of diet and helps to keep the horse happy and relaxed. If grazing is not possible, fresh-cut grass or sliced carrots make a welcome addition to the diet.

Grooming and General Care

Thorough daily grooming is essential for the horse's appearance. See also Chapter 10. If the horse is kept well rugged and well fed, his summer coat should come through in time for the spring shows. He should not be clipped after mid-January, or if this proves necessary he should be given a total clip so that his coat grows in complete without untidy saddle, leg or blanket lines.

Trimming

See also Chapter 11. The correct trimming of show ponies is of the utmost importance: well done it can enhance their appearance; badly done it detracts from their looks.

Shoeing

- Regular attention from the farrier is essential. Show hacks and ponies are often shod with lightweight steel shoes or aluminium plates to encourage the sought-after floating and extravagant action. Toe-in horses should have the clip of the shoe to the outside of centre; toe-out horses clip to the inside.

- Rolled toes in front or behind should not be used, as they may be taken as a sign of foot or joint problems.

- Screw-in road studs can be used on hind shoes. Front shoes should have prepared stud holes. The type of stud, and whether to use them all round or only behind, depends on conditions on the day. If the ground is hard, front studs can shorten the action.

Exercise

A show animal requires sufficient work to keep him cheerful, happy and well mannered, and to get him supple, obedient and well balanced. Daily hacking, with as much variety as possible, should be combined with regular lungeing and schooling. Systematic and skilful training enhances the muscular development of the horse and improves any weaknesses in conformation. At the same time, it is essential to keep the horse bright and cheerful in himself, so that he will give his best in the ring. Towards the end of the showing season, many horses become stale and bored, which affects their performance.

- Horses unaccustomed to a double bridle often accept this more happily if it is put on for hacking.

- Over-fat ponies when brought up in the spring may need the fat sweating off, if they are to be ready for the early shows. They can be exercised in rugs and hoods, but on returning to the stable care must be taken that they do not catch a chill.

- All show animals must be trained to stand correctly when they are being looked at in hand. They must be obedient and active when trotted up. This should be practised at home.

- All young animals should be trained to load, and should be accustomed to travelling before the show season starts. This helps them to cope with the added stress of a show without undue worry. See Chapter 43, Transportation.

Tack

All tack should be of best-quality leather and must be looked after well.

Ridden Classes

SADDLE

This must be comfortable for both horse and rider, must sit well on the horse and must show off the shoulder. A showing saddle with too straight a flap can be uncomfortable, both for rider and judge. A modified version of the general-purpose saddle with a less forward-cut flap and a longer and less deep seat is the most practical and does not spoil the line of the horse's back. A dressage-type saddle can also be suitable for hack classes. Stirrup irons and leathers must suit both the rider and the judge. A spare pair can be taken.

Double bridle suitable for showing and its component parts.

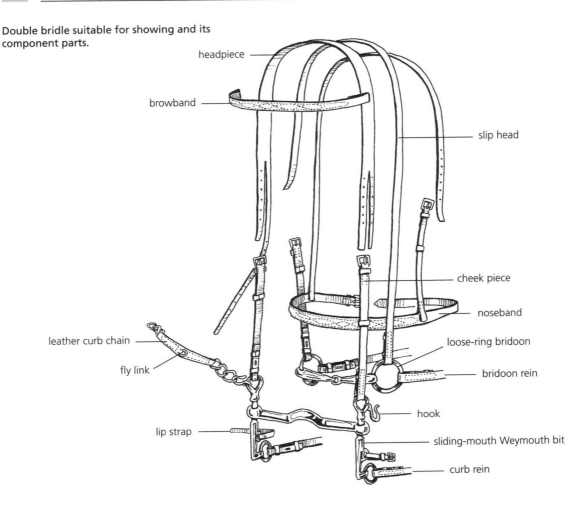

headpiece

browband

slip head

cheek piece

noseband

loose-ring bridoon

bridoon rein

leather curb chain

fly link

hook

lip strap

sliding-mouth Weymouth bit

curb rein

Girths in hunter classes are usually leather; in hack and pony classes either leather, or white lampwick or Cottage Craft type. Bright colours should be avoided. Saddles should fit without the use of a numnah, but horses with sensitive backs may go better if a numnah is worn. Linen-lined saddles can be more comfortable for the horse.

BRIDLE

- Hunters, hacks and cobs are normally shown in double bridles, but pelhams are acceptable.

- Working hunters can be shown in snaffles and may wear a standing or running martingale.

- The leather in hunter and cob bridles should not be too light. The bridoon rein can be plaited for more control in wet weather. Browbands should be plain.

- Hack and pony bridles should be of lightweight leather with stitching on the noseband and browband. Coloured browbands can be used, except for working pony classes. They should preferably be of plaited satin or velvet ribbon, not plastic.

- Novice ponies can be shown in snaffles.

- Mountain and moorland ponies are shown in doubles or pelhams. They should have plain browbands and no stitching on the noseband.

In-hand Classes

BRIDLE

- All leatherwork should be of the best quality, and finely stitched.

- Buckles should be of brass, and neat in shape, unless a riding bridle is used. The buckles should then be of stainless steel and the browband plain.

- Fancy browbands can be worn, except for hunters and mountain and moorland ponies. Brass link on leather, braided ribbon over leather, or plaited leather are the most acceptable. If in doubt as to suitable equipment, ask the advice of an experienced exhibitor.

Hunter Classes

BROOD MARES – Double bridle or snaffle bridle.

FOALS – Foal slip or brass-mounted foal headcollar with white lead rope or leather rein.

YEARLING, TWO- OR THREE-YEAR-OLDS – Young horses are more easily and more safely controlled if shown with a bit in their mouths. This is usually a mullen-mouth snaffle and is buckled on to a brass-mounted headcollar. It may have a white browband. Well-behaved young horses or ponies may be shown without a bit in their mouths. A 3.6m (12ft) leather or white lead rein with a coupling is fastened to the headcollar or bit. Alternatively, a brass-mounted bridle with a snaffle bit can be used. A ring bit is sometimes put on strong yearlings, as it gives more control.

STALLIONS

- A brass-mounted stallion bridle with a mullen-mouth snaffle bit should be used. It may have couplings to which the lead rein is fastened.

- The lead rein should be at least 3.6m (12ft) long. It may have a chain end to go through the bit rings or couplings. This will give greater control and safety to the handler if the horse should play up and rear.

Attendants leading in-hand animals should be suitably dressed. Leather lace-up shoes make running easier.

The Day Before the Show

- After exercise, the horse should be thoroughly groomed. He can be washed all over with a suitable shampoo. Grey horses may need the addition of 'blue bag' or a chalk block to remove yellow stains, particularly on the tail. The tail should be bandaged to the bottom or alternatively put in a bag. White socks should be washed and bandaged.

- Tack must be taken to pieces and thoroughly cleaned. It should be dressed with glycerine saddle soap. The outside surfaces of bridles can be treated with cream or tan boot polish to give a bright, hard sheen. The underside of the leather should be soaped to keep it supple. Saddles should be rubbed over with a duster to avoid the possibility of marking light breeches.

- Day rugs and protective clothing should be brushed and put ready.

- A list of necessary equipment should be drawn up. Some of this may be ticked off and loaded the day before.

EQUIPMENT

- Clean grooming kit. Two clean stable rubbers.
- Tack cleaning kit.
- Plaiting equipment: needles, thread, scissors and comb.
- Studs and fitting tools.
- Hoof oil, hoof blacking or substitute.
- Chalk block.
- Waterproof rug.
- Exercise rug for use in the ring.
- Spare rug and sweat sheet.
- Headcollar and clean rope.
- Protective travelling gear.
- Show saddle and bridle complete with girths, irons and leathers.
- Exercise saddle complete with numnah and bridle.

- Lungeing cavesson and rein; side reins and whip.
- Exercise bandages.
- Water container.
- Two buckets.
- Hard feed.
- Two filled haynets.
- First-aid kits for horses and humans.
- Spare set of shoes.
- Spare headcollar and rope.
- Spare lengths of binder twine.
- Suitable boxes and chests, with a printed list of contents, so that the above items can be easily packed and located.

The Day of the Show

- On some occasions it is necessary to plait before leaving for the show. If possible, leave the plait on the withers and on the forelock, and do them at the show.

    Unpulled tails can be plaited and then bandaged. To keep the bottom of the tail clean it can be bandaged all the way down or put in a stocking or bag. The bottom of a tail should not be plaited or turned up, as it will be difficult to straighten out.

    When undoing plaits, be sure that hairs in the mane are not snipped off with the thread, as these will give an untidy appearance when the horse is next plaited. Such hairs have to be continually trimmed back, resulting eventually in a thin mane and very small plaits.

dragon's teeth   chequer board

Quarter marks.

- Travel arrangements should ensure that there is plenty of time to see to the horse on arrival at the showground. Allow extra time for horses who have to have more exercise than others.

- On arrival, saddle up the horse with the spare tack. Then exercise him. A tail bandage should be worn, with either travelling bandages or fresh bandages put on to cover most of the pastern to help keep him clean. In some cases the horse will be lunged.

- After exercise, box him or tie him up outside the horsebox. He must then be groomed, have his mane plaited (if not already done) and his tail bandaged to produce an immaculate turnout. White socks may require re-washing. Any whitening powder used must be well brushed out, so that it does not mark the opposite leg or hoof.

- Use the body brush or tail comb to draw lines on the quarters against the lie of the hair. Then draw it down the quarters in the lie of the hair to form a pattern. The patterning shows off the quarters and proves that the coat is clean. Hair lacquer can be used to set the hair. 'Dragon's teeth' can be drawn on the flanks by brushing the hair with alternate angled strokes.

- Studs should be fitted if required.

- On whole-coloured horses with dark feet, hoof blacking can be used. Horses with white feet and/or legs should have their feet dressed with colourless oil. If the show ring is not on grass, hooves are better polished with shoe cream rather than oiled, as most arena surface materials stick to oil.

- In hot weather, anti-fly dressing may be necessary. Some horses will not put up with being sprayed, in which case put the spray on a cloth and wipe it over the horse. Take care when doing this, as some fly dressings leave oily marks.

- Vaseline or olive oil can be applied around the eyes and nose to give a gloss.

- Saddle the horse up and walk him about to keep him

warm. In wet weather, he can wear a waterproof sheet over his saddle and back while he is waiting to enter the ring. This can then be removed quickly.

- Immediately before entering the ring, wipe the horse over with a cloth. His eyes, nose and mouth may need spongeing clean. Hooves may need attention. The rider's boots may need a dust.

- During the class, an attendant is required to strip the horse and tidy him before the rider shows him in hand to the judge. Attendants should be clean and neatly dressed. Girls should tie their hair back tidily, and men should wear a cap. Bright colours should be avoided. At county shows, girl attendants should wear jodhpurs or riding trousers, jodhpur boots, a tweed coat and a hunting cap. Blue or black coats should not be worn. The assistant should carry a body brush and two clean damp sponges in a clean stable rubber. The attendant should take off the saddle and put it behind the horse, standing it on its pommel. He should then remove any sweat marks, and sponge the horse's nose and mouth.

CHAPTER 42

THE POLO PONY

The polo season lasts from May to September. Ponies usually winter out, but in a very cold or wet season they may be stabled or yarded at night. They require hay throughout the winter, and when the weather gets cold they should be fed concentrates to ensure that they will be in good condition when they are brought in to start work in early March. During the winter they should be regularly wormed and their feet should be trimmed as required.

Condition

Many ponies start the playing season carrying too much weight. This puts unnecessary strain on tendons and joints and limits their capacity for fast work. Ponies should look fit and well, but more like racehorses than show horses. They should be well muscled and round, with ribs covered but no excess tummy weight. This hard condition must not be mistaken for that of an animal receiving insufficient feed with poor muscle development and therefore no condition.

Fittening Work

Ponies need approximately eight weeks to get fit before playing their first match. A few practice sessions during the ten days before the first match provide the chance to check on their fitness and also to re-introduce them to stick and ball.

A suitable programme is:

Week 1: Half an hour's walking exercise, increasing to 1 hour by the end of the week. Ponies may be lunged or ridden in an arena for a few days if not suffficiently settled to hack out on the roads.

Weeks 2 and 3: Begin schooling in walk and trot to make the pony supple and obedient. This can be ridden work in an enclosed arena, lungeing and/or long-reining. Sessions should start with 5 minutes and increase over two weeks to 30 minutes. Cavaletti work and gymnastic jumping also develop balance and agility. Schooling sessions should be interspersed every other day with 1 hour of road exercise, but this will depend on individual ponies and circumstances.

After schooling, ponies should still have at least half an hour per day active walking exercise, or if weather is suitable they may be turned out in a field for a short time.

Week 4: Active walking exercise, together with some jogging. At first these slow trot periods should be of 10 minutes' duration, working up to 20 minutes, e.g. walk 10 minutes, jog 10 minutes, walk 3 minutes, jog 10 minutes, walk 3 minutes, jog 10–20 minutes, walk 20 minutes to finish. Total time: 1¼ hours. If it is possible, some hill work will improve fitness.

Weeks 5 and 6: Some cantering can be started with 3-minute sessions on alternate days. Canter work should develop progressively over the next two weeks, with the periods gradually extending to 4- and 5-minute canters, repeated twice and up to three times, to give a total canter work-out of 12–15 minutes. Ponies should have at least 15 minutes warming up before starting canter work. When cantering, they should be relaxed and on a very light contact. They must not be pressurised.

Weeks 7 and 8: To clear the wind in the last 10 days before a match, cantering can be done at half speed and then threequarter speed. This can be every other day and for short spells, i.e. 5 minutes followed by short road exercise. On alternate days, either schooling or road exercise may continue.

When hacking out, it is usual for one groom to ride one pony and lead another, but the pony to be ridden should be changed each day. If ponies are well behaved and traffic-proof, and if the hacking circuit is suitable, two ponies may be led out. Care should be taken to ensure that other road users are not hindered.

Playing Tack and Turnout

BRIDLES The most popular bits are doubles, pelhams, or varying types of gags. Running/draw reins are often used with a gag bit. These keep the head straight when riding with the reins in one hand, thus assisting control of the quarters and quick turns. Rubber bit-guards can be used and may save a sore mouth, particularly with a pelham. Curb chains should be adjusted with particular care. On a pelham it is preferable to fit them through the ring, thus keeping the hook away from the face. A lip strap helps to keep the curb chain in the chin groove.

SADDLES Fixed-tree half-panel saddles with reinforced plates at the pommel are usual. They have no knee or thigh rolls. The girths should be of leather or webbing, the irons of stainless steel, and the leathers of buffalo hide. Rawhide leathers tend to stretch. A thin blanket or a lightweight numnah helps to prevent scalding of the back.

MARTINGALES A standing martingale is used.

BOOTS It is essential to protect both front and hind legs. Polo boots extend from just below the knee and hock to the coronet. Shorter boots or bandages can be used, together with coronet boots and/or over-reach boots.

TAILS These are plaited up and tied up and secured either with Velcro straps or with a tail bandage.

MANES These should be hogged.

Match Days

- On match days, ponies should be warmed up before starting to play. Older ponies, or those likely to hot up, can either be walked or trotted about. Others may be ridden by their rider, who perhaps needs to get his eye in and have some practice with stick and ball.

- A polo match is divided into four or six 7-minute chukkas. Ponies usually begin by playing one 7-minute chukka an afternoon. When fully fit they may play two chukkas in an afternoon, with a rest in between.

- At the end of his first chukka, the pony should be unsaddled and sponged down, particularly around the eye, nose, dock, saddle and girth areas. He should be walked about, and in cool weather needs a sweat sheet and light rug.

  After the second chukka, or if not to play again, he should be unsaddled and his boots and/or bandages be taken off. He should then be washed down and surplus water scraped off. He may need a sweat sheet and light rug. He should be checked for injury. His tail should be unplaited, studs removed and he should be walked about to dry off.

Between Matches

- Ponies may be expected to play in two matches and one practice session a week. The day after a match, they should be trotted up to check on soundness. They should then be walked out for 15–30 minutes to relieve stiffness. Weather permitting, it is then usual to turn them out in a field for 2 hours or to graze them in hand.

- Between matches, exercise consists of 1 hour's active walking and jogging exercise. Because of the concussion which may be incurred from galloping on hard ground during matches, hard trotting on roads must be avoided. If there are enough spare days between matches, ponies may be schooled.

TYPICAL ROUTINE

Day 1: 15 minutes' walk-out and/or spell in field.
Day 2: Road exercise.
Day 3: Gymnastic exercises.
Day 4: Work with stick and ball.

Roughing Off

This is done at the end of the season in September. If the pony has had daily grazing it takes up to two weeks.

- If the weather is warm and dry – particularly at night – it may be possible to turn the pony out after five or six days. Concentrate food is reduced, hay is increased and the pony is turned out to grass for increasingly longer periods each day.

- Rugs are gradually removed until none are needed. Grooming is restricted to spongeing eyes, nose, dock and attending to feet.

- Shoes should be removed and the pony may then be left out all the time.

- He must be well fed during the winter, so that he will be in good condition in the spring and ready for work.

Feeding and Watering

- Concentrate feed is built up steadily to a total of 4.5–6.4kg (10–14lbs), according to the requirements of the individual pony.

- It is important to ensure that a pony does not hot up. Cubes or coarse mix are often found more suitable than oats, particularly for Thoroughbred ponies.

- Amounts of hay should range from 3.6–5.4kg (8–12lbs) a day.

- Because of their playing timetable it has been found that ponies do better if fed their concentrate feed twice a day: in the morning and (the larger amount) at night. There is then the minimum alteration to feeding routine on match days. During tournaments, ponies may play several days in succession. On a match day, the normal feed is given in the morning.

- Hay, if given, should not exceed 0.9kg (2lbs). Food should be withheld for four hours before the match. Water should always be freely available but a large intake of water should be avoided within 30 minutes of the match. After the match, the pony may be allowed a quarter of a bucket of water and then his fill half an hour later. On a cold day he should be preferably kept walking for 10 minutes after watering.

On return to the stable, he may be given a small feed followed later by his normal feed and hay. During hot weather, it is essential to check late in the evening that the pony has access to plenty of water.

Bedding

Ponies who are bedded on straw and who eat their bedding must be muzzled after allowing them time to eat their feed and hay, e.g. at midday on non-polo days. The muzzle should be of the lightweight wire mesh type. This particular problem can be avoided if ponies are bedded on shavings or paper.

Health

Apart from specialised treatment, the health of polo ponies should be looked after in the same way as for other horses.

- It is preferable for all necessary flu and tetanus inoculations to be completed before the ponies start work in March.

- During the season, they should be wormed every 4–6 weeks according to the agreed programme.

- After any fast work – whether when fittening or playing – their legs should be checked for any indications of strain or bruising. If there is the slightest sign of heat or swelling, work should be discontinued until the leg is back to normal.

- Ice is the best immediate first-aid for a bruised or

strained leg. Cold water bandages may also be put on after work, should there be any question of strain.

- After a match, the pony's mouth should be carefully checked. If it is sore, it should be bathed twice daily with salt water. To allow his mouth to heal, the pony should be led out in a cavesson or ridden in a bitless bridle.

Teeth

A polo pony's teeth require attention from a 'polo dentist'. Teeth should be examined in the fifth or sixth week of the fittening work, and rasped down as necessary.

Many mouth injuries can be avoided by attention to teeth and by an examination of the mouth and shape of the jaw before deciding on which type of bit to use. Teeth should again be checked halfway through the season.

Grooming

A thorough daily groom helps to get the pony fit. After fast work or a match, it is essential for the pony to be thoroughly cleaned.

Clipping and Trimming

Usually it is unnecessary to clip ponies when they come in during March, as late clipping spoils their summer coat. The exception is a pony recently imported from the southern hemisphere, such as South America, who will not be accustomed to the change of season and will grow a thick coat.

Normal trimming of heels and chins can be carried out as required. Manes should be hogged, which may need to be done every week.

Shoeing

During fittening, fullered or concave shoes of normal weight should be put on. When fast work starts, lighter shoes should be used. These must be securely fixed, with pencilled heels in front and hind shoes well set back.

To save strain on joints when turning, feet should be kept as short as possible.

Ponies should be shod at least every four weeks and their shoes should be carefully checked before a match.

Screw-in road studs should be used for exercise in the outer web of both hind shoes. The rules allow two studs in each hind shoe, if required. Studs are not allowed in front shoes.

43 TRANSPORTATION

LEGAL REQUIREMENTS FOR ROAD TRANSPORT

There are strict regulations regarding the construction of vehicles intended for the transportation of horses. They have been drawn up to protect the horse's welfare, and can be found in the *Transit of Animals (Road and Rail)*, current edition available from HMSO. They apply to all horse transport, whether it is used for business or pleasure.

When first constructed, all chassis are plated and issued with a plating certificate. The plate gives the weights which each axle can carry, and above which it should not be loaded. To exceed these limits can be dangerous, and is an offence. The vehicles also have to have an annual Ministry of Transport test to make sure that the plating certificate is still correct.

Before deciding to buy a horsebox or trailer, especially if being ordered to personal specifications, it is advisable to check that its measurements and its weight-carrying capacity conform with EEC regulations. All vehicles with an unladen weight of 1525kg (30cwt) must have an annual test at an HGV Ministry of Transport station. An MOT certificate, as issued by garages, is not valid for these heavier vehicles.

Drivers of vehicles over 7.5 metric tonnes gross weight must have an HGV (heavy goods vehicle driving) licence; they may also need an operator's licence.

The vehicle must be fitted with a tachograph.

There are regulations as to the weight of the towing vehicle in relation to the loaded trailer. These should be verified before you arrange to tow a horse trailer.

CONSTRUCTION

Horse transport must be strongly built and must fulfil legal requirements. Before buying your vehicle check with your local Department of Transport office for current regulations.

BODY WORK is usually steel framed, and may be constructed from aluminium, fibre glass, glassonite, steel, hardwood or wood. Wood is the heaviest and requires most maintenance.

FLOORS must be safe, sound, non-slip and easy to clean. Wood, rubber composition on wood, or rubber granulastic floors are all suitable, the last-named being the easiest to clean. Wood floors should have gripper treads, as they can be slippery. Horses travel better – particularly over long distances – if the floor of the box is low to the ground.

RAMPS are made of wood or metal. They should be covered with a non-slip surface or heavy-duty matting. Gripper treads can be used. Ramps should all be fitted with efficient control springs to facilitate lowering and raising. Some vehicles have both side and rear ramps; others only one of these. Some modern lorries have hydraulically operated ramps.

PARTITIONS should be strong, well padded, and fitted so that they can be easily moved over or removed if a larger space is required.

THE GROOM'S DOOR is a small side door which enables an attendant to reach the horses without opening the ramp. Many boxes have an integral cab and therefore direct access to the horses.

VENTILATION is very important. It should be arranged by means of sliding windows (which should be of toughened glass and protected by bars) adjustable vents or roof ventilators.

LIGHTING Skylights provide natural light and a more pleasant travelling environment, although in hot sunny weather the box can become very warm. Internal lighting is required by law.

LIGHTWEIGHT GATES, when opened, act as sides to the ramp. They help to guide the horse up the ramp, and prevent him from slipping or stepping off the side. They can be quickly closed if the horse should try to back out.

All horse transport should display clear **warning notices** front and back, stating that horses are being carried. All vehicles and trailers should be given regular servicing and safety checks, both from the maintenance aspect and for the safety of the animal.

TYPES OF TRANSPORT

There are three main types of road transport:

1. The Horsebox

The modern purpose-built horsebox varies in quality from functional to very luxurious. Boxes for competition horses can be fitted out according to the customer's requirements. They can be made to take from two to eight horses, as well as providing storage areas, sleeping and sitting accommodation, and washing and cooking facilities. Even the least elaborate boxes are expensive, but they offer the safest and most comfortable transport for the horse, giving a smoother, quieter ride with less vibration than a lorry or trailer.

Partitions can be designed so that horses are divided off from each other to face forwards, backwards or diagonally.

Ventilation arrangements should always be checked. Whilst a stuffy atmosphere makes the horses more liable to respiratory problems, draughts increase the risk of chill.

The design of the box should allow horses to lower their heads. This is particularly important for horses with respiratory problems.

2. The Cattle Lorry

These lorries, if high enough, make suitable horse transport, but they are understandably less comfortable than a purpose-built vehicle. They lack padding and insulation, and can be draughty and noisy. Extra clothing may be needed in cold weather. Usually there is only a back ramp.

Horses may be divided by wood partitions or slotted-in rails. It is inadvisable to stand them parallel to the front of the box, as this makes balance difficult when the vehicle is moving. They should stand either facing forwards, backwards or diagonally.

The floor should be bedded down with clean straw or shavings. The ramp may be spread with straw to make it less slippery and to assist loading. This is particularly helpful when the ramp is uncovered metal.

NB: If a caravan or similar vehicle is towed behind a box it will then become a long vehicle, subject to HGV regulations.

3. Horse Trailers

These are built to carry one to four horses.

Most trailers are now built on a double axle, and have four wheels which makes for greater stability and a better allocation of weight.

Two-wheel trailers, though they are light and manoeuvrable and can be towed by a smaller car, are far less stable. They are suitable only for small ponies.

The great advantage of a trailer is that it is low-loading, unlike a horsebox or lorry with its steep ramp, which can be off-putting for a nervous horse. Conversely, the roof of a trailer is lower than that of a horsebox, and the space is more confined. This can be inhibiting for a tall horse, both when loading and when travelling, and can often turn him into a bad traveller and make him difficult to load.

Trailers can be towed by Land Rovers or similar 4 wheel drive vehicles, or by a suitably powerful motor car. The 4-wheel drive type is preferable, as it tows a heavy load with less strain and copes better with rough or muddy going.

It is essential for the towing-bar on the towing vehicle to be at the correct height, so that the trailer rides level and in line.

The joint lighting system has to be professionally installed, and the towing arrangements should be checked at the same time. Trailers have been known to break loose from their towing vehicle. The latest models have an automatic brake which prevents this from happening. A strong extra chain between vehicle and trailer is a useful safeguard.

THREE- AND FOUR-HORSE TRAILERS are heavy to handle and to tow. With the three-horse type, two horses travel facing the front; the third – and preferably smallest horse – is loaded from the front and travels in the middle, facing backwards. The horses are separated by partitions which swing across for easier loading and unloading.

TWO-HORSE TRAILERS are the most popular type. They have one partition, but if a horse travels better with more space, the partition may be removed (see Transit of Animals). There are two models: a front unload, which has an angled front, a back ramp for loading and a front side ramp for unloading; and the shorter model which loads and unloads from the back. The front unload trailers are more expensive, but are worth the extra cost if young or valuable horses are to be carried. The horses are less worried when being unloaded from a front ramp, as they can see where they are going. Horses unloaded from the rear are more likely to rush backwards, and may be injured. Making some horses

step back can be difficult; others learn to pull back when they are tied up in the trailer, although the back strap should control this. Difficult loaders may be more willing to enter the trailer if the front ramp is down and they can see out.

ONE-HORSE TRAILERS should be considered only if the horse is a good loader and travels well. The small space can be very off-putting, and many horses understandably refuse to go inside.

PONY TRAILERS are smaller in height and width, and are also cheaper.

Breeching straps should be fitted at the rear of each partition. When one horse has been loaded, the strap should be hooked on to the centre partition. It will help to keep him in place while the second horse is loaded. Remember that if a small horse or pony fights to get out he may end up under the breeching strap. If partitions are not used, the straps can be hooked together behind both horses.

All trailers are equipped with **breast bars** which fit across the front of the partitions. If the partitions are removed, a single long bar should be fixed in the same place. The bars give the horse support if there should be an abrupt stop, and prevent him from standing too close to the front of the trailer.

Adjustable legs should be positioned at all four corners of a two-wheel trailer. These should be lowered to help to steady the two-wheel trailer during loading and unloading. They are rarely needed with a four-wheel trailer. If it is necessary to remove the towing vehicle while the horses are on board (in an emergency only) the legs can be lowered. But note that it is very easy to forget to raise them again and to drive off with them dragging along the ground.

A **jockey wheel** on the front of the trailer is a worthwhile extra. It enables the front of the trailer to be raised or lowered, thus making for easy hitching on and off. It also makes the trailer easier to move when not attached. It is important to check before moving off that the wheel is well clear of the road surface.

For **security**, it is advisable to have a postcode or

address painted clearly on the trailer roof and to have a wheel clamp or tow-lock in place when parked at home. This will reduce the trailer's vulnerability to thieves.

PROCEDURE BEFORE TRAVELLING

Home Transport

Regular servicing of all vehicles prevents unnecessary problems. All ramp and door fastenings should be kept well-oiled, so that they can be easily undone and done up securely. It is possible for the back ramp of a horsebox to come undone when travelling, and for it to fall on top of a following car.

Horsebox

Check petrol/diesel, oil, water, battery, lights, indicators and tyre pressures.

Towing Vehicle and Trailer

- Check as for horsebox. Also, when the trailer is hitched on, check the coupling hitch and safety chain. If the towing attachment has been adjusted or moved for use on another vehicle, make sure that it is securely bolted and at the correct height. The trailer should ride level, so that the tow bar is in line with the ball hitch on the back of the towing vehicle. Check that the following are working: trailer brakes, brake lights, indicators, side and back lights, internal light.

- Before moving off, make sure that the groom's door is closed, the jockey wheel and light cables clear of the ground, and the stability legs raised. If the top doors at the back of the trailer are to be left open, make sure that they are securely fastened back.

- To avoid theft, it is often advisable to carry the spare wheel of the trailer in the towing vehicle.

- The floor of all horse transport should be regularly checked for wear. This is very important. Wooden floors often get damp and in time will deteriorate and

may break under the weight of a horse. When travelling the box floor should be bedded down with straw or shavings to give protection from vibration and jar. It also encourages a horse to stale on a long journey. The floor should be swept clean immediately after each journey and allowed to dry out.

- If you are planning a long journey, particularly if using a trailer and towing vehicle, it is advisable to have them both checked over by a garage experienced in this type of transport. Make sure that the trailer springs and tyre pressures are correct, as these affect the ease of towing and the horse's comfort.

Hired Transport

Before hiring a horsebox it is advisable to check its condition and to make sure that the driver is experienced.

As there is always a risk of infection and contagion, the box should have been washed out, and clean bedding should have been put down.

The Driver

- Anyone new to driving a horsebox or trailer should have a test drive, with an unloaded vehicle.

- Drivers must always take extra care to avoid sudden breaking, rapid acceleration and cornering too fast. They should pay particular attention when travelling over a rough or uneven surface. Though horseboxes can be driven safely at a greater speed than horse trailers, it should be remembered that horses may still become frightened and liable to injury when their driver is cornering at speed or driving too fast.

- Overhanging boughs and hedgerows along narrow lanes can hit the top and sides of the box and make frightening noises. Drivers should be aware that this often upsets horses.

- The driver of a horse trailer must be aware of the stability of the towed vehicle. Trailers when towed too fast soon start to sway and can easily turn over or jack-knife, with disastrous consequences. Drivers should also take extra care in high winds or on a slippery surface, as the vehicle and trailer can be difficult to control. No exact limits can be set under normal road conditions. A speed of 30–35mph (48–56km) is acceptable.

- In hilly districts drivers can expect excessive wear to the clutch and gear system, particularly on a car.

- Reversing or turning a horsebox or a towed vehicle takes considerable skill and practice. It is recommended that drivers practise with the vehicle unloaded. On some trailers, before reversing, the brake pin has to be put in to prevent operation of the automatic trailer brake. Before moving off, the pin must be taken out so that the brake functions correctly. Many modern trailers have an automatic reversing system.

PROCEDURE AFTER TRAVELLING

- The box or trailer should be skipped out and all damp bedding removed.

- The remaining bed should be placed at the front of the vehicle, and the floor should be thoroughly swept and allowed to dry.

- If necessary, remove and store dry bedding.

- Scrub down and hose out the vehicle regularly. Removable rubber matting should be lifted so that the floor can be aired and also checked for wear.

LOADING

- Loading creates few problems if the handler is experienced and knows what he is doing. He will then give the horse confidence.

- Site the vehicle alongside a wall or solid fence. Avoid a slippery surface. If loading from a field or on a road, try and park by a fence or gateway. It is often helpful to site the vehicle so that a gate can be closed behind the horse so that he is surrounded on three sides. If in a yard the gate should be closed.

- Inside the box or trailer, a short loop of string should be attached to each tie ring. The headcollar rope should be passed through the string and not tied directly onto the ring. This makes for greater safety – if the horse pulls back the string should break.

- Wooden flooring should be bedded down.

- To make loading easier, partitions may be swung across, secured, and then replaced when the horse is on board.

- A horse in a box without partitions can travel loose. A mare and foal are best travelled loose, with the partitions taken out.

- The ramp should be level and firm. To achieve this it may be necessary to use wedges.

- Some yards have a loading bay, i.e. a built-up bank with high boarded sides. Park the box alongside the bank and let the ramp down on to it.

- If there is no unloading bay the ramp of the horsebox will be less steep and more inviting if the box is parked on a slight downward slope with the ramp opening up the slope.

- A covered school is a safe and convenient loading area. Lead the horses into the school and leave the doors open. Park the vehicle so that the ramp opens into the school. Fill any gaps with straw bales. Horses should associate the school with discipline, and the enclosed space encourages them to load without trouble. If a scuffle should develop, there is less risk of injury to the horse. Alternatively the vehicle can be backed into a corner of the school along one wall. The doors of the school can then be shut.

Training Horses to Load

Horses should be trained to load at an early age, there will then rarely be problems later. As long as the mare is a good loader, a foal will usually follow her. Young or inexperienced horses can be encouraged by loading an older stable companion first.

- A trailer should have the partition removed so that there is a visibly larger space.

- Front-unload trailers may have the front ramp let down or just the front top opened.

- A bowl of food should be at hand to encourage a shy loader and as a reward after loading. A worried animal can be given a feed in the horsebox or trailer, and then unloaded. This procedure should be followed several times before any travelling is done. For the first journey a short, smooth trip should be arranged. If you are using a box, an attendant should stay with the horse. This is illegal in a trailer.

Time spent training a horse to load is never wasted. A horse who is easy to load can save hours of frustration, worry and possible injury to both himself and his handlers.

Loading Procedure for Experienced Horse Handler with One Assistant

ASSISTANT
1. Check that everything is ready.
2. Be ready to carry out any instructions.
3. Stand quietly at the side of the ramp, making sure to be behind the horse's eye as he walks in.
4. When the horse is in the vehicle, secure the partitions or bar; in the case of a trailer, secure the breeching strap, then the ramp. When lifting a ramp, always stand at the side. If the breeching strap should break and the horse backs out, the ramp will drop down and can trap the lifter. A heavy ramp requires an assistant on either side to lift it.

HANDLER
1. Stand on the near side by the horse's shoulder. Look forward, not at the horse.
2. Start on a half-circle or well away from the vehicle and lead the horse straight up the ramp into the vehicle.

3. If the horse hangs back, do not pull him. Feed him a titbit, pat him on the neck, look ahead and persuade him forward. If he is inclined to pull back, allow him to do so, and move back with him. The horse will then not be upset and the loading process can be resumed without fuss.
4. When safe inside, as long as there is room, stand beside the horse without going under the front bar, as this is likely to make the horse pull back. Wait until the partition bar or back strap is in place.
5. Go under bar and tie the horse to a loop of string placed through the ring. A single horse in a trailer without partitions should be tied on both sides so that he cannot swing round. He must have room to move his head and should not feel too restrained.
6. Go out through the groom's door.

Loading Procedure without an Assistant

SIDE-LOADING HORSEBOX
Tie a haynet to the tie-ring. Shut the yard gate. Lead the horse in as described above. Manoeuvre him into position. Thread rope through the ring and hold the end while fixing the partition. Or tie the horse to the string loop and then fix the partition.

BACK-LOADING BOX OR TRAILER
Lead the horse in. Put the rope through the rings. Do not tie him up. Quietly move to the back of the vehicle and secure the bar, the breeching strap or the gates. If the horse has a tendency to run back, use a lunge rein or long rope with which to lead him. Put it through the ring and take it back; this way the horse can be encouraged to stand. Should he pull back you can retake the rein as he comes down the ramp.

A good horse who travels well can often be persuaded to enter the vehicle ahead of the handler. Many horses prefer this. As the horse moves forward, the end of the rope can be put quietly over his neck. The bar, breeching strap or gates can then be secured behind him before he is tied up to the string loop.

If the haynet is not required it should be removed. Horses on a restricted diet, or before hunting, or fast or concentrated work, should not be given a haynet on the outward journey. Worried travellers, however, often stand more quietly if hay is available. If a haynet is used it is essential to tie it as high as possible and to make it very secure. Serious accidents have occurred when horses in transit have become entangled in empty or loose haynets.

Many horses load and travel more quietly if the centre partition of the trailer is taken out, or if it is swung across and secured. Horses rarely kick when travelling, but there can be real risks of a tread, so precautions should be taken – i.e. use protectors and coronet boots. (See Chapter 29.) Some horses panic in a restricted space, as they like to spread their legs for balance themselves. The Transit of Animals Act stipulates that horses travelling without a partition should be unshod behind.

Loading a Reluctant Horse

Try to establish the cause of the problem.

- If the horse is unwilling to leave the others, a stable companion can be loaded first.
- If the horse is frightened the causes can be:
- Fear of the unknown.
- Fear of a small enclosed space.
- The memory of a frightening journey.
- Fear of a steep ramp.

Some of these problems can be resolved by dealing with their causes and by not hurrying the horse. However, if he has had a fright when travelling, or has been upset by being loaded with unnecessary force, it is a more difficult problem to overcome.

Many horses and ponies are difficult and very obstinate to load because they have been handled by inexperienced and often nervous people who have given in to them. Such animals handled by experienced people often load with little argument.

PROCEDURE

- Make suitable arrangements for loading, as above. The handler should be experienced, strong enough to

prevent the horse from whipping round and running off, and able to hold him straight, facing the ramp.

- The handler should wear suitable clothing – i.e. non-slip boots or shoes, gloves and a hard hat. Assistants should also be suitably dressed.
- Start loading in plenty of time so that there is no feeling of haste or worry. Enlist experienced helpers, and if possible choose a quiet time and place so that outside noises, unwanted onlookers, cars, horses, etc., do not prove a distraction.
- Put plenty of straw on the ramp and under the bottom edge. The ramp must be firm.
- Make sure that the horse is well protected by suitable clothing, bandages, knee caps, and hock caps. Make sure that he is not worried by the extra clothing. However, young horses may be better loaded without anything on their hind legs unless they are used to it.
- Have food available to encourage and to reward.
- Have at hand two spare lunge lines and lunge whip, in case you may need them. The horse should not wear a saddle.
- A cavesson with lunge line attached to the centre ring should be put on over the headcollar.
- If the horse is very headstrong, a snaffle bridle can be put on. The cavesson and lunge rein can be placed above it. The bridle gives greater control and ensures that the horse can be kept straight and looking where he is going.
- Remember that a firm attitude and a quiet, encouraging voice are the most important aids.

Other Loading Methods for Reluctant Horses

Handler with Two or Three Assistants

- Stand the horse by the ramp, facing forwards. Feed him from a bowl. Pat and encourage him to sniff the

Loading a reluctant horse, with two assistants.

ramp. Encourage the horse to go forward. An assistant on either side can lift first one front leg and then the other, followed by the hind legs. Do not use force. If the horse draws back, follow him and start again.

- Buckle lunge lines on either side of the vehicle. Have two assistants holding the ends to make a passageway. As the horse walks forward, the assistants cross over and bring pressure to bear on the horse's quarters. A sensitive horse can rear up and may slip or fall backwards. The handler must instruct the assistants and be quick to forestall any such problem.

- If the horse is not a kicker, two assistants can link fingers (not hands) behind his quarters and push him up the ramp. It is safer to do this with a short rope or strap. If a fourth assistant can be found he can lift one foot at a time. This is probably the most successful method of all, and most horses will accept and give in to it. Though there is some risk, an experienced person can usually assess the horse and his reactions. The closer the two assistants are to the hind legs, the less is the danger of injury should the horse try to kick. Alternatively, a short rope can be held between two assistants and forced up against the horse's quarters. They may hold the side of the trailer or horsebox gate to apply more pressure. This is not as effective as linked fingers at close range; however, by linking fingers it is possible for one assistant to be pulled behind the horse because he is unable to detach his fingers.

Handler with One Assistant

- Lead the horse towards the ramp. An assistant walking behind with a lunge whip can swish or crack it, or touch the horse on the quarters, to encourage him forward. He can also flick the horse on the hocks, first on one side then the other. He must stand out of kicking distance.

- Make a sudden, unexpected noise, such as a bang or swishing straw immediately behind the horse.

- Apply a wet yard broom (gently) to the quarters.

- A blindfold can be used, but this may cause panic; some horses will not accept it.

The successful persuasion of an unwilling horse to enter a horsebox or trailer depends on the handler's

(a) experience and firmness;

(b) horse sense; quickness of reaction; skill in forestalling trouble; and ability to foresee if strong methods will succeed or are more likely to cause increased resistance and possible injury.

It can be comparatively easy to load an awkward horse if experienced help is available and if firm methods are used. If it is not, then inexperienced handlers must rely on kindness and patience. In time these should result in a more amenable and obedient horse.

UNLOADING

If you are away from home, park in a safe and suitable place. Make sure that there is enough room for the horse to move safely down and off the ramp, and to keep straight for a few strides to avoid strain. Avoid unloading a side ramp horsebox or front-unload trailer towards the road unless help is available to warn traffic.

When unloading horses, the safest method is always to untie the horse before removing the holding barrier, whether it is a partition, front bar or breeching strap.

Horsebox, Side Unload, Handler with One Assistant

- Handler enters by small door or through the cab, speaks to the horse, pats him and unties him.

- Assistant lets down the ramp and, when told, opens the gates and swings the partition across. Handler turns horse and leads him quietly down the ramp.

Trailer, Front Unload, Handler with One Assistant

- Begin as for horsebox.

- Assistant opens top door of front trailer ramp and, when told, lets down the ramp. Assistant and handler remove front bar from horse nearest to the door. Handler then leads horse out.

- Any attempt on the horse's part to hurry or jump off the bottom of the ramp should be restrained. If a second horse is being unloaded, and if help is available, keep the first horse close by so that the horses can see each other. When unloading the second horse after undoing the bar, move the partition across so that there is more room.

Horsebox, Side Unload, Handler without Assistant

- Speak to horse.

- Let down ramp.

- Untie horse but leave rope through the loop.

- Open gates and/or partition.

- Free rope and lead horse out as before.

Trailer, Front Unload, Handler without Assistant

- Speak to horse.

- Open top door and let down front ramp.

- Untie horse nearest to door and remove bar.

- Lead horse out. Put the horse in a loose box. If away from home, tie him to a string loop on the back of the trailer.

- The second horse can then be unloaded in a similar manner.

Trailer, Rear Unload, Handler with One Assistant

- Handler speaks to horse and enters by small door.

- Handler unties horse.

- Assistant lowers ramp and, standing to the side, removes bar or breeching strap.

- Handler standing at first to side and then in front of horse allows him to step back. Assistant stands beside ramp and helps to keep the horse straight with a hand on his quarters.

- If quarters swing to the right, the horse's head should be moved to the right, which helps to straighten him. If the horse rushes back, he should not be restrained but followed, keeping hold of the rope.

Trailer, Rear Unload, Handler without Assistant

- Speak to horse and enter by small door.

- Untie horse.

- Climb out of the small door. Move round to the rear. Let down the ramp and unfasten the breeching strap.

- Allow the horse to step back. Take hold of the rope as he moves back. Alternatively, if there is room, move in alongside the horse and persuade him back. A lunge line or longer rope put through the tie-ring and attached to his headcollar can be of help should he rush back.

Horses travelling on their own in a trailer without partitions should not be allowed to turn round. Ponies and animals under 15hh are usually able to turn and walk out forwards without strain.

IMPORTANT: When unloading two horses it is advisable to have someone to help.

CARE OF THE HORSE WHEN TRAVELLING

The requirements of the horse when travelling are similar to those in his normal loosebox. He should be warm in winter, cool in summer, with a plentiful supply of fresh air, but no draughts; have a resilient, non-skid surface to stand on; and the minimum amount of noise and disturbance.

Most horses travel well, providing that they are driven with consideration, that they have sufficient space, and that the vehicle has suitable suspension.

Many of the problems and ills associated with travelling can be traced to overheated or draughty transport and inappropriate clothing. These can be avoided if the person in charge of the horses is thoughtful and conscientious about their welfare.

Many horses sweat when travelling. Common sense should be used as to suitable clothing.

Regular checks should be made, especially on long journeys. Bear in mind the type of vehicle that you are using. A fully enclosed horsebox containing three or more horses is very much warmer than a cattle lorry containing one horse. Check the front and roof ventilators, as these can cause draughts.

Two horses in a trailer may generate considerable heat. The back top doors of trailers can be left open except in very cold weather or if on motorways, in tunnels, or in heavy traffic when horses become disturbed by vehicles coming up close behind them. In these conditions they travel better if the top doors are closed.

Food and water are not usually required on a short journey of under 2 hours. On a long journey of 6–8 hours, regular watering, feeding and hay are necessary. If feasible, the opportunity to stretch legs adds to their comfort.

Clothing For Travelling

The following factors should be taken into consideration:

- Time of year.
- Temperature on the day of travelling.
- Length of journey. Whether travelling alone or in company.
- Whether a good traveller, or likely to get upset and sweat.
- Type of horse.
- Type of vehicle.

Stabled Horse in Winter – Journey 1–4 Hours

- Headcollar and clip-on rope (poll guard is optional).

- Sweat rug plus normal stable rugs or their equivalent in day rugs.

- Bandages with Gamgee, or leg protectors that cover the knees and hocks.

- Knee caps and hock boots if no leg protectors.

- Tail bandage.

- Tail guard – preferable to a tail bandage for a long journey.

- Surcingle or roller with pad and breast girth.

- If thought advisable: coronet boots and hock boots.

- In very cold weather, a hood is often put on a clipped Thoroughbred.

Spare rugs should be available during and at the end of the journey, in case the travelling rugs get wet or the horse has sweated.

A sweat rug helps to ensure an even temperature, and keeps the horse from getting a chill should he sweat. If needed, the sweat rug can be fastened round the horse's chest, the other rugs may be folded back and buckled over the withers. They should be secured by the roller.

Thermatex rugs are light but warm and make good travel rugs.

Stabled Horse in Summer

Summer sheet with sweat rug or sweat rug only. Protective clothing as detailed above.

Long Journeys

Horses should be regularly checked to see that they are comfortable. They may require an extra rug, or if too hot, a blanket removed or turned back at the shoulders.

Short Journeys

When travelling hunters to a meet, or horses and ponies to a rally, they can often be fully tacked up. This can save worry and hassle on arrival, and is a safe procedure if the animals are good travellers. Protective clothing is often not used except for a tail bandage and knee caps.

The bridle reins can be twisted up as for lungeing, or crossed and doubled round the neck, or tied in a knot. The headcollar should be put on top of the bridle. Running martingales should not be attached to the reins. The ends of the martingale should be tied up and then fitted on arrival.

Rugs should be used as already described. The roller or surcingle must be sufficiently long to fit over the saddle. Rugs must always be secure. Horses may be wet and/or hot when boxed up and their rugs can get damp. A sweat rug should be put on and dry rugs should be available when they arrive home.

It is advisable to carry an extra blanket or sweat rug.

If protective clothing is required, leg protectors are easier to put on.

Rugs are not necessary for grass-kept ponies, although in summer a sweat rug or summer sheet will help to keep them clean.

An unclipped pony in winter is best left without rugs, unless he is wet and cold, in which case a sweat rug and light blanket, with straw underneath, is recommended. This 'thatching' allows more air to circulate.

Make sure that the pony does not become overheated.

NB: It is dangerous for horses and ponies to travel with jump studs in their shoes.

Equipment for One or Two Horses

SHORT JOURNEY
- 13.6-litre (3-gallon) water-carrier (filled).
- Bucket. Haynet for return journey.
- Suitable grooming kit and hoof pick.
- Horse first-aid kit, e.g. wound powder etc.
- Spare rope.

LONG JOURNEY
- Spare headcollar and rope.
- 13.6-litre (3-gallon) water-carrier (filled).
- 2 buckets.
- 2 haynets.
- Feeds.
- Removable manger.
- Grooming kit.
- Horse first-aid kit.
- Spare rugs.
- Skip and tools.

TRAVELLING ABROAD

- Research is necessary to find out about the various injections and blood tests necessary for foreign travel. These must be carried out in good time. The bloodstock agency arranging your transport will provide all necessary information. If the horses are travelling as part of a team, this is the responsibility of the team manager.

- The necessary documents for the countries to be visited or travelled through must be obtained. This applies to horses, people and vehicles.

- Check that all dates and descriptions on the horse's passport are correct.

- Have all the necessary documents for returning home.

- Check that any medicines carried are permitted in the countries which you are visiting.

- Arrange for the horsebox to be serviced and supplied with any small spare parts which might be difficult to obtain overseas and which could cause delay.

- Careful driving will diminish the stress of travelling.

- Air conditioning in long-distance horseboxes is a great advantage, as humidity in the box can cause respiratory problems. When a horse becomes affected in this way it lowers his resistance to infection.

- When loading the horsebox it is healthier for the horses if hay and straw are stored separately and partitioned off.

Feeding

Before a long journey, give a laxative, low-energy diet and continue this throughout the journey to avoid the risk of azoturia and/or colic. Check if it is possible to import horse feed to your destination. If so, carry as much as space allows in the horsebox. If there are restrictions, it will be necessary to find out what feed is available.

Keep a supply of apples and carrots to help encourage the horses to eat up and to calm and distract them during moments of stress.

Watering

Horses should be watered every 2 hours. Filled water containers should be carried.

Horse Clothing

Use rugs which are easy to alter, so that layers can be removed or added as the temperature and climate dictate. Tail bandages can tighten and rub, so it is better to use a tail guard. Knee caps and hock boot straps should be protected with Gamgee. Leg protectors are easier to fit and more comfortable. If necessary bandages and Gamgee may be fitted under them. With leg protectors, knee and hock boots are unnecessary. A poll guard is advisable.

Keep a simple bridle within reach at all times. It may be necessary to restrain the horse, and a bridle will give greater control. This applies particularly when loading and unloading.

Regular exercise is essential. Try to walk the horse in hand in a bridle for at least 30 minutes at every morning and evening stopover. If possible, a short trot on the lunge allows the horse to lower his head and encourages him to cough up any mucus. It is important to allow him to drop his head.

Bandages or leg protectors should be removed twice daily, and the horse's legs should be given a brisk rub by hand, upwards towards the heart. Bandages may become tight on long journeys, so it is preferable to wear leg protectors.

After a long journey, the horse requires a day to recover – with very quiet exercise – and then a gradual build-up to the competition. It is therefore very important to leave plenty of time, both for the journey and for recovery.

AIR TRAVEL

- As well as being the speediest method of transport, this has the least effect on the horse's fitness.

- There is usually a baggage restriction, which should be checked with the agency or airline.

- Various methods are used for travelling horses by air. The most usual, and the best, method is the loading of two horses into a crate (similar to a horse trailer) which is fork-lifted into the aircraft and then secured.

The actual loading of the horses is carried out by expert handlers, who are employed either by the airport authority or by the company arranging the transport. The groom or owner should always be at hand so that the horse can be reassured.

During the inevitable wait before customs clearance, and during loading and unloading, the unaccustomed activity and noise may well alarm the horse. A supply of carrots will help to keep him occupied and to allay fear.

At no time on the journey – whether it is by air or by ship – should the horse be left unattended. It may be possible to arrange a rota system with other grooms, so that a rest period is possible. Once the aircraft is airborne, the journey is usually smooth, and most horses relax and travel well.

Sedating a horse, either before or during transit, is not usually advisable. He needs to be fully conscious in order to balance himself. Most airlines prefer a horse to travel in a bridle.

On rare occasions, it has been known for a horse to react adversely when travelling by air, by trying to kick himself out of his crate. For the safety of all concerned, the captain of the aircraft may demand, under such circumstances, that the horse should be destroyed.

SEA TRAVEL

It is now customary for horses to travel in their horsebox, which is driven into the hold of the ship.

Try to organise sufficient space for the ramp to be let down or opened slightly at the top. This allows the horses more air in the hold, which can be very stuffy and hot. All vents in the horsebox should be open and clothing should be adjusted as necessary. Someone should stay with the horses throughout the journey.

Partitions in the horsebox should be arranged to allow horses to spread their feet and thus help their balance.

Horses can suffer badly from seasickness – although, of course, they are unable to be sick. The effects can last for several days.

When passing through customs and frontier posts there may be considerable delays. Make sure to allow for this when planning the journey.

THE HORSE AT GRASS

CHAPTER 44

GRASSLAND MANAGEMENT

Constant use of the land for grazing by horses will always present problems.

Horses are very selective in their grazing habits. Unlike ruminants, who spend much of their day resting and chewing their cud, horses graze more or less continuously – except for short periods of rest or when sheltering from hot sun and flies. Fields grazed only by horses, even with a low number on a large acreage, develop 'roughs' where they dung (also called 'camp areas') and 'lawns' where they graze. Only if very short of food will horses graze on these 'roughs', and unless good management is practised, the 'roughs' increase in size and so reduce the grazing area. They also become reservoirs of infection for internal parasites and a home for infective larvae. In wet weather, the latter are able to migrate to the 'lawns', with a corresponding greater risk of infection to the grazing horse. Thus fields grazed entirely by horses are likely to deteriorate, unless given knowledgeable and experienced management.

Objectives of Good Management

1. To provide grazing over a long period of time by the establishment and/or maintenance of a dense vigorous sward of suitable variety.

2. To minimise infection from worm larvae, although in comparison to anthelmintic strategy this plays a minor role.

3. To provide suitable working areas: e.g. for jumping and schooling.

4. To maintain the land as a visual amenity, thus avoiding complaints from neighbours and possible action by local authorities. Well-kept fields and fencing are a good advertisement for any commercial establishment, and a source of satisfaction to the private owner.

Management must be geared to provide conditions which encourage the most productive and palatable grasses and discourage weeds and less productive species. Palatability tends to be a reflection of good management rather than of species .

The following procedures are recommended:

Soil Analysis

This should be carried out every four or five years, and when moving into a new establishment, particularly if the grassland appears to have been neglected.

The Ministry of Agriculture through their local Agricultural Development Advisory Service Office (ADAS) advise on local seed and fertiliser merchants qualified to carry out soil analysis. Results are in accordance with the ADAS index. In fields grazed only by horses, soil samples should be taken both from the 'lawn' and the 'rough' areas. The latter are likely to be high in potash as a result of horses urinating in the area. According to the results of the analysis, lime, phosphate and potash can then be applied.

Beware of expensive 'trace element' fertilisers and sprays. If horses are found to be deficient in particular trace elements, it is usually more effective to feed or inject these, rather than to make up the deficiency in the soil. Breeding establishments may need to take specialist advice.

Drainage

Signs of poor drainage are:

- Surface water.
- Plants such as rushes, tussock grass, buttercups and couch grass, though the appearance of these plants may also indicate poor management.

Any drainage scheme is expensive and may be beyond the finances of a small establishment, but if the field is part of an agricultural holding it may be possible to obtain a grant.

Before embarking on a drainage scheme, make sure that all boundary ditches and outlets are cleared, as these can be one of the main causes of surface water in a field. Drainage problems can also result from damage to soil structure, by the use of heavy machinery when the ground is wet, or by 'poaching' from animals left out in the winter. If damage to soil structure is suspected, advice should be sought from ADAS. Improvement can usually be achieved by subsoiling and pan bursting (to break up the top layer of heavily compacted soil). Such work has to be carried out by a specialist firm.

If a drainage system has to be installed, its design depends on soil structure, the lie of the land, and other natural features. In areas of clay soil, underground channels – preferably in a herringbone design and connected to piped major drains – can be successful. It is possible to install drainage pipes by machines without disturbing the top surface. In all cases, professional advice from a reputable firm should be sought.

Harrowing

Before allowing tractors on to a field, the surface should be sufficiently firm to take the weight without damaging soil structure.

Harrowing should begin as soon as the land is dry enough. Depending on the season and the soil, this may range from late February to early April. A spiked or pitch-pole type harrow is the most efficient, as it removes the dead grass and moss, aerates the soil, and encourages new growth. Chain harrowing to spread the dung and tidy the paddock may be carried out through the summer, when paddocks are rested, and preferably during hot, dry weather. In a dry season, the hot sun helps to desiccate the manure and kill off worm eggs and larvae. In a wet season, both eggs and larvae tend to thrive, so there is a danger in spreading the droppings over a wide area. They could increase the area of tainted pasture, especially in fields which do not drain well.

Rolling

Rolling is beneficial:

- If a severely poached field is to be used for riding. Harrowing followed by rolling will provide a level surface. To achieve a good result the soil must be dry enough not to stick, but not so dry that the lumps of earth will not break up.

- If the field is to be cut for hay. Rolling can reduce the risk of damage by hay-making machinery when it is travelling over a rough surface.

- In early spring, when it can encourage early growth by raising soil temperature.

A light-ribbed Cambridge roller is most suitable. With certain soils – particularly clay and heavy loam – there is a risk that rolling may damage the soil structure and assist in the formation of an impervious pan. This inhibits growth and causes drainage problems. Such areas should not be rolled unless this is absolutely essential.

Fertilisers

In areas where soil fertility is good (usually a sign of careful management over many years), the application of fertiliser may not be essential. Lush grass is not required for horses, and unless other stock is available to graze off the first growth, the application of fertiliser can be unnecessary and indeed unwise. Ponies are quick to gorge on new grass, with resultant digestive upsets and risk of laminitis. Establishments with a limited acreage find that the application of a suitable fertiliser is essential if grass is to be encouraged and weeds kept in check.

The four major elements necessary for healthy plant growth are lime, phosphate, potash and nitrogen. Magnesium is included in the soil analysis, but appears to be of no importance to the health or growth of a horse. Nitrogen is not included, as amounts vary according to the season and rainfall.

Following the soil analysis, local fertiliser firms who know about the soil features in the district can be consulted. Representatives may need to be reminded that horses have different grazing requirements from milking cows, beef cattle and sheep. A steady growth over a longer period is needed, not a lush growth of grass.

LIME Levels are measured on a 'pH' scale, 7.0 being neutral. A 'pH' level of 6.5 is suitable for grass – below this level the soil is acid, leading to the growth of poorer grass – above pH7 the soil is alkaline and the large amount of calcium results in the locking up of certain minerals. A high pH (excess lime) can seriously affect the growth of young stock. A level above pH7 is inevitable on chalk soil.

A deficiency of lime is usually adjusted by a dressing of limestone or chalk, both of which contain the calcium which is so important for healthy bone growth in the young horse. This dressing can be applied at any time of the year, but preferably in autumn or spring and on a calm day to avoid drift loss by the wind. In the past, basic slag was used, but it is now difficult to obtain.

PHOSPHORUS (P_2O_5 is phosphate) The Ministry of Agriculture have drawn up an ADAS index from 0–8. For horses, a suitable index should be 1–2. This allows for the phosphate level to be below the calcium level. This correct balance is essential for good bone

development in young stock and general health. A low phosphate level, e.g. index 0, hinders grass growth.

ROCK PHOSPHATE This should not be applied unless the soil is very acid (below pH5), but experiments are continuing to make rock phosphate more soluble and thus more readily available to plants. Super-phosphate can cause sudden bone failure and should not be used.

POTASSIUM (K_2O_1 is potash) Potash is necessary for plant growth. An index level of ADAS 1–2 is suitable for grazing land. If hay is to be cut, then a dressing of potash after the hay has been taken corrects any loss. In general, if land is grazed by horses, potash levels once established remain constant. Little is known about the ingestion of this mineral by horses.

NITROGEN (N) Nitrogen is the key to grass growth, although success also depends on the correct balance of the lime, phosphate and potash. Nitrogen is leaked out of the soil by rain, so it is best applied during the spring and summer months. Swards containing large amounts of clover manufacture their own organic nitrogen from bacteria on the clover root. It can make them self-supporting in nitrogen, but the amount of clover needed to achieve this is too much for horses. It can lead to digestive upsets and laminitis.

Too much nitrogen on grazing land, with resultant over fast growth, can produce unsuitable grasses and a lack of balanced nutrients. This can affect the bone growth of young horses.

Application of Fertilisers
These are best applied in the spring when the field is not being grazed. If the grasses are to take full advantage of the treatment, a period of rest is advisable. Should this not be possible, the fields should be safe to graze once a liquid fertiliser has been washed in by a light shower or heavy dew, or the dry granules are not visible on the surface.

To save expense, a suitable mix of required fertilisers in either granulated or liquid form may be put on in one operation as a compound. Since grass growth for horses is required over a long period, it may be advisable to apply a small amount of nitrogen in the spring and some more later when the fields or paddocks are being rested.

Some people object to the continual use of chemical (inorganic) fertilisers. Owners of breeding stock often prefer to use organic fertilisers (such as farmyard manure or seaweed). These are slower-acting and more expensive but provide minerals which may otherwise be lacking. Adult horses, other than broodmares, are unlikely to be affected by the use of chemical fertilisers. There is normally a second natural growth of grass in September, and fertiliser applied in mid-August promotes this.

Manures

FARMYARD MANURE (FYM) If available, FYM is a complete fertiliser. It also sweetens those parts of a field which have been soured and made 'horse sick'. It provides material to improve soil texture, corrects any tendency for the soil to be alkaline, and supplies the main plant nutrients of phosphate, potash and nitrogen. It is best used on fields intended for hay and should be applied in the autumn when the weather and earthworm population can work it into the soil. If applied on grazing fields it should be stored for at least six weeks before use and the field should not be grazed for six weeks after spreading, to ensure that any infection is eliminated.

Unless linked to a farm or small animal enterprise, the average horse establishment is unlikely to have any opportunity of obtaining FYM. It is sometimes possible to obtain cattle or pig slurry and also human sewage sludge from council sewage farms. The latter is treated and sterilised. All of these make excellent fertilisers, but are strong and therefore they should only be used in small quantities and not on grazing fields. They are suitable for hay fields, encouraging strong growth.

Once spread, the strong smell of any FYM or slurry can linger for days, which is a consideration for establishments dealing with the general public.

HORSE MANURE This is not as valuable as FYM and is generally inadvisable for fields grazed by horses, as it

may increase the area of tainted grass. There is also a risk of worm infection, as it requires considerable heat over several weeks to kill worm eggs and larvae. Manure must be left to heat and rot well before applying.

Early Grazing

Grass starts to grow as soon as (a) soil temperature rises and (b) there is sufficient rainfall. If the sward is to profit from the treatment already outlined, it must not be grazed until the grasses are well established, i.e. approximately 10cm (4ins) tall. Earlier grazing is to the detriment of the better-quality grasses, which need to develop strong roots and invariably grow more slowly than the poorer grasses and weeds.

Rotation

Where there is sufficient acreage, a system of rotational grazing should be practised. This involves dividing large fields into smaller paddocks. An area of 3 acres is a useful grazing size. The divisions can be achieved either with permanent fencing or by the use of electric fencing (see Fences, page 311). In early summer – i.e. May or June – the stocking rate could be as high as ten horses or fifteen ponies per paddock for periods from ten to fourteen days. The paddock may then need top-cutting (see below), and should be rested for three weeks before being grazed again. In July and August grass growth slows down and the paddocks take longer to recover. If more growth is required, they may benefit from some nitrogen. Growth accelerates again in September, but the first frosts check it until the following spring. In June, the grass which is surplus to grazing requirements may be cut for hay. An alternative is to cut grass daily and use it for fresh green feed for stabled horses. If possible, the paddocks chosen to be cut for hay or green feed should be rotated each year.

Top-Cutting

From June onwards, it may be necessary to mow over lightly or top-cut all paddocks. If horses stay in the field and if the weather is fine, much of this taller grass may be eaten as the sun helps to sweeten it. Any heavy patches of grass which remain should be forked up and burned. If left, they check the re-growth of the underneath grasses.

This top-cutting is necessary to prevent weeds and uneaten coarse grasses from seeding and to encourage growth from the better grasses.

When you are top-cutting, a careful watch should be kept for ragwort. It is sensible to have the paddocks searched before cutting is started. Any ragwort plants should be pulled up and burned. This task is most easily carried out after rain, when the ground is soft.

Over-Grazing

Fields which are over-grazed, and the sward not allowed time to recover and establish new growth, are bound to deteriorate. Such treatment can result only in the encouragement of weeds and coarse unpalatable grasses, an increase in worm infestation, and the loss of what could be valuable grazing. Where the amount of land is limited, it is sensible to try to obtain alternative grazing, and certainly to avoid grazing any damp fields between October and May.

It should also be noted that the sward deteriorates if under-grazed and the grass allowed to seed.

Cattle as Alternative Grazers

Cattle make excellent followers-on for paddocks grazed by horses. They eat the coarse grasses rejected by horses, and ingest and kill the worm larvae deposited on the soil from the horse's dung. Sheep and goats can also help to free grass from worm larvae, but may require special fencing. Sheep in particular prefer short, sweet grass, similar to that enjoyed by horses. To buy cattle or sheep for the above purpose can be a considerable capital expense with no guarantee of a cash profit at the end of the summer. As an alternative, it may be possible to arrange for a neighbouring farmer to graze his cattle in return for hay or help with upkeep and field cultivation.

Picking up of Droppings

If time and labour are available, the daily picking up of droppings in small paddocks is extremely worthwhile. It decreases the worm infection, improves the appearance of the field, and encourages the growth of more palatable grass. A vacuum machine, which collects droppings and can be attached to a tractor, is now on the market.

Poaching

If animals are kept out during the winter months considerable damage can be done to a sward. Low-lying or clay soil areas are likely to be worse affected. Damage is caused by deep hoof prints in the sward, which in a wet season often remain full of water, adversely affecting the root system of the grasses. Exposure to frost causes further damage. If wintering out animals, it is sensible to use only one or two paddocks, preferably the best drained. Shelter, water and access must also be considered. In low-lying areas, in a wet season it is sensible for fields to be shut up in October and not grazed again until April or May. The decision to close the field depends on the season. Well-drained and/or hill fields may be safely grazed through the winter.

Stocking Rate

This varies according to the following factors:

- Soil type and fertility.
- Length of grazing season.
- Rainfall.
- Period of daily grazing.
- Area of land available and whether divided into separate paddocks.
- Parasite control.
- Quality of management.

Given good soil fertility, an approximate stocking rate throughout the year would be one horse or two ponies per acre, or two horses or five ponies per hectare (possibly more during the summer months if grazing is very good). From October until the following May, supplementary feeding will be necessary. On land with poor fertility, this rate of stocking could be halved: one horse or two ponies per two acres. Supplementary feeding may be necessary from July onwards, particularly if the animals are working.

If rotational grazing is practised from May to October, three 3–acre paddocks with periods of rest should accommodate ten horses or fifteen ponies. The period of daily grazing affects the consumption of grass. Many private and riding school animals stand in during the day. Others, if working, may need to have their grazing time limited so that they will not become overweight.

IMPROVEMENT OF GRASSLAND

Reduction of 'Camp' Areas or 'Roughs'

These are areas of the field contaminated by the horse passing dung and urinating. The grass becomes coarse, horses refuse to graze it, and the sward rapidly deteriorates. The situation is worse on poorly drained soil.

Procedure

- Daily removal of droppings.

- Alternate grazing with cattle. Cattle eat down the coarse grass and also ingest and kill the worm larvae and eggs. Special attention to the fencing may be necessary to contain the cattle.

- Regular top-cutting and resting of the field. This allows the grasses and plants to establish a healthy root system, which leads to increased growth.

- Use of farmyard manure.

Control of Weeds

- Good management encourages the better grasses, and discourages weeds: buttercups in particular.

- Some weeds, such as nettles, thistles and bracken, can be controlled or eradicated by regular cutting, i.e. six times in the growing season. This method may have to be repeated for a second year before success is achieved.

- Clumps of weeds, i.e. nettles, thistles and docks, can be treated by using a suitable herbicide applied directly on the clumps by a knapsack sprayer. Strict regulations exist concerning the use of sprays by untrained personnel. Unless used by the land owner a NPTC (National Proficiency Test Certificate) of competence must be held. This necessitates attendance by the individual at a one-day training course to ensure competence and gain a certificate in use and management of sprays.

If the weeds are widespread, total spraying of the field may be necessary. A weed-killer can often be combined with a fertiliser to give a boost to the fine-leaved grasses. It should be remembered that some selective weed-killers are effective against all broad-leaved species, and thus many other beneficial and appetising herbs and grasses will be destroyed. More sophisticated weed-killers are now available which may avoid this problem. There is specific legislation covering the spraying of weed-killers. This should be checked before any spraying is done. Professional advice should be sought.

Weed-killers should be applied in late spring when the plants are growing vigorously. In some cases, a second application will be necessary later in the summer when regrowth has occurred. Spraying should take place in calm weather to avoid drift and when rain is not expected for 24 hours. Some sprays are toxic to stock, so fields should not be grazed for some time after spraying. If poisonous plants are present – e.g. ragwort – they must have withered and crumbled before stock can return to the field.

Agricultural merchants can usually be trusted to give some advice as to the necessary type and quantity of weed-killer to use. For some species – e.g. docks and bracken – this has to be applied later in the summer. They will also advise as to which weed-killers are toxic to stock. It is important to remember that horses are much more sensitive to weed-killers than other stock.

Direct Drilling

Small areas of badly poached ground should be harrowed or, if necessary, raked over by hand. Suitable horse paddock mixture should then be broadcast, and the area should be rolled. For small areas a grass 'hand seeder' (on sale in specialist shops) can be used to introduce herbs into a pasture which is deficient in them. It is then advisable to treat a narrow strip of the field.

Whole fields can be improved by direct drilling or slot seeding. This work is usually undertaken by contractors. Advice should be taken as to whether (a) the soil is suitable and (b) the process is likely to succeed. The best times of year are spring or late summer, when the soil is warm and there is sufficient rain to germinate the seed. It may be advisable to kill off the poor-quality turf and weeds with herbicide. This will not affect the new seed and will ensure that the new grass has a good start and is not smothered by already established seeds. A special drill is used which cuts into the ground, making channels into which the grass seed and fertiliser are directed. The field is then rolled.

This process does not disturb the top layer of soil and encourages the rapid establishment of a dense sward, which can be grazed by horses at a much earlier date. It is of particular value in fields where stones and flints are close to the surface. If such fields are ploughed, the surface can be unsuitable for riding and schooling horses for several years.

Ploughing and Re-Seeding

It is now considered more sensible to improve pasture by good management and perhaps by direct drilling, rather than by ploughing up and re-seeding. Ploughing should only be resorted to in urgent cases where the sward appears to have gone beyond reclaim. Advice should be sought from the local ADAS office.

Ploughing and re-seeding are usually undertaken by local contractors in late summer. The old poor-quality, weed-infested turf can be killed off by herbicide. The field is then ploughed, and then scarified to make a suitable seed bed. Seed and fertiliser are drilled together and the field is rolled. The following summer, the field is best grazed by sheep to consolidate and encourage a dense sward. In the first year it should not be ridden on, but a hay crop may be taken. It can be grazed by horses in late summer, but in a wet season care should be taken to ensure that it does not get poached. If the new grasses are to thrive it will require careful management. There will be competition from indigenous plants, which may have been germinated with the ploughing.

Special seed mixtures are now available for horse pasture. Advice should be taken as to whether they are suitable for the particular area. They contain appropriate quantities of grasses palatable to horses and some clover and selected herbs. Hardy grasses which will withstand being ridden on should also be a consideration when choosing the seed mix if the grassland is to be all-purpose.

GRASSES

The requirements for horse pasture are prostrate growing grasses with a good bottom growth and the ability to produce a dense sward. The following grasses are of value:

- Perennial rye grass. There are several varieties, which are all of value.
- Smooth meadow grass, known in the USA as 'Kentucky blue grass'.
- Creeping red fescue.
- Chewing fescue.
- Sheep's fescue.
- Tall fescue.
- White clover in small quantities. Too high a percentage of clover will make the grass too nitrogen rich. Clover can also be slippery to ride on.
- Timothy, cocksfoot and the tall, early rye grass

(Lolium spp) are all suitable for hay, but do not make good grazing for horses or a hard-wearing sward.

Some grasses, which on fertile soil would be classed as weeds and therefore undesirable, have advantage where the soil is poor in natural fertility, e.g. hill areas. In such areas, some of the better grasses will not flourish. Crested dog's tail, sweet vernal and common bent grass all flourish on poor soil, but they only provide third-rate pasture (often suitable for ponies).

Comprehensive research is being carried out on the development of new strains of grasses, which may in time supersede the strains recommended above.

HERBS

Horses relish herbs, which provide nutrients often lacking in more shallow-rooted species. The following are of value:

- Narrow-leaved plantain.
- Yarrow.
- Dandelion.
- Chicory.
- Comfrey.
- Ribwort.
- Burnet.

HAYMAKING

Aspects of haymaking which need consideration are:

- The sward should be of a reasonable quality. Fields infested with weeds and growing inferior grasses with only a small proportion of good grasses are not suitable.

- When the fields are shut off for hay, sufficient grazing should be available. Depending on soil fertility, sward and rainfall, hay fields will not be ready for grazing for at least four to six weeks after cutting.

- It is not economical to buy haymaking machinery for small acreages, so a local contractor or a neighbouring farmer should be hired to do the work. Such arrangements often result in small acreages being left until last, which will mean the risk of deteriorating weather, a crop of poor quality, and grass past its best.

- Fertilisers cost money, but they are necessary if the crop is to be worth harvesting. Contractors charge per acre for cutting and turning. A poor crop of grass can often be just as costly to make as a heavy crop, although baling costs will be less.

The **advantages** of making your own hay are:

- The hay (subject to the factors mentioned above) will be of good quality and costs less to make and store than buying from a corn merchant.

- The fields should benefit from being cut for hay; good grasses will be encouraged and weeds and inferior grass cut before being allowed to seed. The sward, if well harrowed, improves by not being continually trodden.

- In May and June, establishments with ample grazing have an over-abundance of grass. Unless this is (a) set aside for hay, (b) cut as green fodder for stabled horses, or (c) efficiently grazed, the sward deteriorates. Grasses are likely to be swamped by stronger-growing grasses and weeds. Cattle can be brought in to eat off this grass. This is an expensive capital outlay and not necessarily profitable.

- Shutting up the field helps to ease the worm burden, which is likely to increase if fields are continually grazed by horses.

The **disadvantage** is that it is very likely that horses and ponies will dislike and not eat hay that has been made from pastures that have been grazed only by horses.

Preparation of Hay Fields

Meadow hay is made from fields of permanent grass specially 'shut up' (not grazed). Low-lying fields should not be grazed during the winter, in order to avoid poaching of the ground. In early spring, fields should be harrowed and fertiliser should be spread; rolling is not usually necessary, and can be detrimental by increasing panning. The grass can then be left to grow. Before cutting it must be checked for ragwort.

On dry, well-drained fields it is possible to allow grazing during the winter and not to 'shut up' until mid-April. Harrowing and fertilising can then be carried out. Though the resultant crop will not be as heavy and will be late growing, it can provide a useful supply of hay.

Seed hay is made from fields specially sown down. They may not have been grazed at all; the turf may not be established, and grazing – particularly if the ground is soft – would seriously damage the sward. The crop is much heavier than that from permanent pasture, and may take longer to 'make'. It should be of superior quality because of the specific choice of grasses sown, and free from weeds.

Cutting and Baling the Hay

Hay is made by cutting the grass preferably just before flowering, but if the weather is uncertain it may have to be delayed. The grass is left lying in rows. After a few hours of hot sun the top surface begins to wilt. The grass is scattered by a machine, a process known as 'teddering', which scatters the rows of grass and allows wind and sun to penetrate and dry the crop. A heavy crop will need teddering at least twice; after rain it may need to be done several times. When the hay feels dry and crisp it can be 'rowed up' (two rows put into one) ready for baling.

The hay should then be baled. If the baling is not completed in a day, the hay must be turned again the following morning so that it will be dry and crisp. If the baler is drawing a sledge, the bales can be left in convenient stands of eight to twelve. This makes for quicker and more convenient carting. If there is no sledge, the bales should be stood up in groups of four. The knots of the baling strings should be at the bottom of the bale nearest to the ground. Hay stacked in this manner 'shoots' any rain.

It is usual to cart the bales as soon as possible. However, if the weather is settled, the bales improve by being left out in the field for a few days before carting, so that they can sweat and dry naturally. There will then be no overheating.

The making of good hay is a skilled process. Fine weather is essential. In a wet summer, good hay can be difficult to make and find.

Silage

Silage is cut from young grass – (5–10cm/2–4ins) – and carted at once, with no wilting. It is stored in a 'clamp'. Here the grass is piled up and covered tightly with polythene sheeting. Air is excluded. The grass then 'pickles' and preserves in its own juices. This method of grass preservation is primarily used for feeding cattle. As ruminants, cattle are able to further ferment and digest any possible bacterial contamination possibly associated with the close cutting of the grass to the soil surface (e.g. botulism). While some use is made of silage for horses and they thrive on and enjoy the rich nutritional forage, there is an element of risk when feeding it.

Haylage/Big Bale Silage

This is harvested from grass older than that taken for silage but younger than grass to be taken for hay. The cut grass is left to wilt for 24 hours and then vacuum-packed in sealed polythene bags. Depending on the manufacturer the bags may be round or square and vary in size from 20kg (45lbs) up to large bales which can only be moved mechanically with a tractor or fork-lift truck.

Haylage is widely used as a dust-free forage for competition horses. It is nutritious but higher in water content than hay so it should be fed at least pound for pound as hay or in slightly greater quantities so that the fibre ration is maintained in the diet (see page 163). Due to the nutritious value of haylage, it is usually possible to reduce the concentrate intake for the horse thus making an economic saving.

Storage of haylage is a major consideration. While it can be safely stored outside it must not be vulnerable to accidental puncture by vermin or machinery. If the vacuum is broken, contact with air will cause spoiling of the contents and there will be a risk of bacterial contamination if fed to horses. Once a bale is opened it should be fed in three to five days (depending on the weather – it will keep longer in winter). Some smaller yards would be unable to utilise the big bales satisfactorily (see also page 149).

MANAGEMENT OF HORSES AND PONIES AT GRASS

In most parts of Britain, ponies and cobs can be kept out throughout the year. It is their natural lifestyle, especially if they are part of a herd. If suitably fed and properly looked after, they keep healthy and sufficiently fit for hacking and light to medium work. Most of them can withstand dry cold and freezing temperatures. The worst possible conditions are cold, driving rain, inadequate shelter and a wet, poorly drained field. In such weather conditions, horses benefit from New Zealand rugs.

The better-quality animals with lighter coats do less well in winter, and will benefit from being stabled at night. Those turned out by day may wear a New Zealand rug (see Clothing, Chapter 30). If stabling is not available, quality animals need extra concentrate food to maintain body heat. This should maintain their health, but if they are to keep their condition they must be very well looked after.

FIELDS

Appropriate Fields

Though your choice may be limited, before buying or renting a field to be grazed by horses you should first consider the location, the type of land and the keep. The following factors should be considered:

- The field must be easily accessible in winter, so that the horse can be fed and the water supply checked.

- Well-managed, old-established ley or meadow field is the best type of grazing. It provides a great variety of grasses and herbs, whereas a newly sown ley often lacks herbs in the sward. Old turf also provides the thick sward which is important for maintaining the soundness of young growing stock and will stand up to poaching better than newer leys.

- Rich pasture makes most animals fat and more liable to laminitis; sudden access to lush growth can cause laminitis, even in thin ponies. A strict system of controlled grazing may have to be practised.

- If the fields are wet and marshy animals will not thrive on them in winter. In summer, however, as the grass is often of poor quality, it suits ponies likely to get over-fat.

- High, exposed fields lacking shelter are very cold in winter and provide no escape from sun and flies in summer.

- Very steep fields are only suitable for small ponies. Larger ponies and horses – particularly young stock – may develop back, hock and/or stifle strains.

- Small paddocks can be invaluable in providing limited grazing for small ponies, or a daily exercise area for a stabled horse. They can, however, become heavily poached in winter. In general, horses are happier and thrive better in larger fields where there is more room.

- Low-lying fields or those with a clay-based soil are not suitable for winter grazing. The land becomes poached – i.e. cut up – with the roots of the grass exposed and often killed by frost. The quality of the pasture then deteriorates. Animals standing in mud are reluctant to lie down to rest. Much of the hay fed in the field is trampled in and lost.

WATER SUPPLY (SEE ALSO CHAPTER 17)

A regular water supply is essential for health. If piped water or a suitable unpolluted stream is not available, alternative arrangements have to be made. When grass is plentiful and rainfall normal, a horse may well drink less than 4.5 litres (1 gallon) a day, the extra water intake being provided in the grass. In hot weather, when the grass is dry, or in freezing weather when the horse's diet is hay or other dry food, his daily water consumption is much greater, up to 45–54 litres (10–12 gallons) a day.

Water Troughs

The best way to ensure a satisfactory supply of water is to pipe it to a carefully sited concrete or metal trough and control it with a covered ballcock system. As the horse drinks and the water level drops, the ballcock drops with it; this releases a valve and allows the water to flow. As the trough fills, the ballcock rises with the water level and the valve shuts off the supply. This system usually works without problems. However, on occasions, the valve can stick and restrict the water, or the ballcock can puncture, so that the water will not be cut off and will spill over, causing flooding. Daily inspection is advisable.

Troughs vary in size and can hold as much as 900 litres (200 gallons). They should relate to the size of field and the normal stocking rate, and should be large enough to allow several animals to drink at once.

Old baths and other metal and plastic containers are often used in small paddocks as a cheap alternative to purpose-built water troughs. These are suitable as long as there are no projecting edges likely to cause injury. Boxed-in baths are safer and less of an eyesore. If they are filled by piped water and a tap, the tap should be placed out of reach of the horses, so they cannot play with it or get caught up.

If they are close to a suitable supply, water troughs can be filled by means of a hosepipe. When the trough is full, the hose should be moved out of the way so that water will not syphon out. In freezing weather, the trough should be emptied. If water has to be transported, 25–45 litre (5–10 gallon) plastic water containers can be used. A strong wooden broom put through the hand grips enables two people to carry the container more easily.

Siting of a Water Trough

The cost of piping may influence the siting of your trough.

It should preferably be sited in a well-drained area of the field, away from the corner and near enough to the gate for easy checking. Do not put it near trees, as their roots make pipe-laying difficult and in autumn their leaves foul the water.

Troughs can be placed lengthways along a fence line, so that there is the minimum of projection on which a horse can knock his legs. In this position they can be made available to an adjoining field. Rails or boards built above the trough will prevent horses jumping out or fighting above it.

If sited away from the fence line, troughs should be at least 10m (33ft) into the field and well away from the

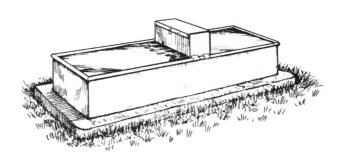

Metal water trough with a covered ballcock system.

corner, so that there is less risk of a horse being trapped and kicked.

All troughs should be set on solid brick or concrete supports.

The ground surrounding a water trough may become poached. Horses' legs can get very muddy, with the consequent risk of mud fever and cracked heels. Thick mud may also discourage regular drinking. To avoid these problems, loads of rubble, stone or rammed chalk can be put down, or a roughened concrete apron (approximately 10m² (33ft²) can be placed up to or around the trough (but note that in freezing weather the surface can become slippery, in which case straw should be put down).

Water troughs must be kept clean, so they should be regularly scrubbed out. During the process, the ballcock should be tied up. Troughs which do not have a base outlet must be emptied by bucket. A plastic dustpan is useful to finish the job.

Precautions in Freezing Weather

All exposed piping should be lagged and then boarded. The cavity should be filled with fibreglass. An alternative is to remove the bottom of an old dustbin or other container, put it over the pipe, and fill it with hot horse manure. It can be replaced as necessary. This system also helps to prevent the pipe freezing in the ground.

If metal pipes should freeze, they can be thawed out by covering them with straw and then setting light to it, or by using a blow lamp. Remember that as ballcocks are usually made of plastic they will puncture if exposed to flame.

With plastic piping, hot water can be poured over the pipe, which may also thaw the ballcock. Frozen plastic piping should not be subjected to direct flame, i.e. a blow lamp or burning straw.

When pipes freeze, they sometimes burst as they thaw. It is then necessary to turn off the water at the mains. The location of all held stop-cocks should be noted.

In freezing weather, ice on a water trough must be broken each morning. Horses drinking during the day help to keep the water from freezing, but in very cold weather it may be necessary to re-break the ice at night. These tasks must never be neglected.

Rivers and Streams

This water supply can be satisfactory as long as the watering place is safe and the water unpolluted. There must be a sound approach area, sufficiently wide for several animals to be able to drink at once. If the stream is a field boundary, fencing should be constructed to prevent horses crossing over or moving up or down the stream.

Streams with steep banks must have the approach area set well back so that the slope is less steep. Streams with a sandy bottom are not suitable, as a horse when drinking absorbs a certain amount of sand and this in time can cause sand colic. Streams adjoining or near buildings are often polluted by farm or stable drains, or by a neighbouring factory. If there is any doubt, a sample of water should be sent for testing.

Stagnant ponds should be fenced off, as they are not suitable drinking places.

POISONOUS PLANTS

Ragwort, buttercup, bracken, cowbane, foxglove, deadly nightshade, horsetails, meadow saffron, water dropwort and leaves of most evergreen plants are all poisonous to horses. Fortunately, their bitter flavour ensures that as long as other grazing is available, they are rarely eaten. Occasionally an animal develops a taste for one of them and will suffer 'colic' or may even die.

After the use of weed-killer many of the plants listed above become palatable, so stock should not be allowed into a field which has been sprayed until all such plants have rotted down.

When found in hay poisonous plants are particularly dangerous – except buttercup, which when cut and dried loses its toxic properties. Old pasture which is about to be cut for hay should be closely examined, especially for ragwort.

RAGWORT Of the above plants, ragwort is the most common, and the one most likely to be eaten. In winter and early spring, young plants can be recognised by their prostrate growth and distinctive leaf shape. They bloom in July, August and September, when the yellow flowers can be easily seen. Mature plants grow to a height of 90–120cm (3–4ft). Young plants may only be 10–15cm (4–6ins) in height. When cut, pulled, or poisoned by herbicides, they become palatable. If you are top-cutting fields in July and August, take particular care either to pull up the plants or to pick up the cut plants and burn them. This is even more important if the grass is cut for hay.

The poison from ragwort acts on the liver and has a cumulative and long-term effect, which is usually fatal. Grazing with sheep in the early spring will eliminate ragwort.

YEW All parts of a yew tree are poisonous, even when the tree is dead. It appears to be palatable. Twigs falling off a tree are easily eaten with a mouthful of grass. Even a small quantity can be lethal. If a horse snatches at yew tree branches when being ridden through woods, the rider should dismount and remove all of it from the animal's mouth.

DEADLY NIGHTSHADE The berries, although rarely eaten by horses, are poisonous. They are brown to purple in colour. The plant is found as a creeper on hedges only in very limited areas. The less toxic plant, woody nightshade, which has red berries and a purple flower, is more common. Hedges should be checked and any plants pulled out and burned.

WATER DROPWORT is only likely to be a problem after deep ditching.

Ornamental hedges and trees such as **LABURNUM, LAUREL, PRIVET** and **RHODODENDRON,** are more often found in the hedges or boundaries of fields adjoining parks and gardens. It is seldom possible to have them removed, so the area surrounding them should be fenced off. Laburnum seeds are particularly lethal. However, horses are unlikely to eat them unless very hungry.

ACORNS and **CRAB APPLES** Fields containing oak and crab apple trees should be carefully checked in the autumn when the ripe acorns and fruit have fallen. Horses eat both acorns and crab apples. A small quantity does no harm, but a large quantity of acorns causes poisoning, and of crab apples severe colic.

Weeds such as **DAISIES, DOCKS, CHICKWEED** and **BROAD-LEAVED PLANTAIN** are of no food value. Their presence in a pasture reveals poor management. They are rarely eaten, unless cut with the hay. A large quantity of docks can then cause a digestive upset.

NETTLES, THISTLES and **DANDELIONS** are all relished by the horse when cut, but their presence signifies poor grass management.

LITTER

Fields which have roadside hedges or which are crossed by a public footpath should be regularly inspected for tins, broken bottles and other litter. In some areas, such fields require daily checking.

Fields newly acquired should be checked for abandoned farm implements such as harrows, which can be covered with grass and go unnoticed. Binder twine from used hay bales, or on occasions from rotted-down old bales, can also be very dangerous and should be picked up as it can entangle an animal's legs.

RABBIT HOLES

Fields in which rabbits are known to live should be regularly checked for holes. This is particularly important if foals or young stock are grazing them. Unsafe areas may have to be fenced off.

SHELTER

This is essential to protect horses from cold winds, driving snow and rain and in summer from hot sun and flies. A field bounded by thick hedges, stone walls or banks provides shelter in winter. Any trees give shelter from the sun.

Field Shelter

The only effective escape from insects is a field shelter. This is more likely to be used as a refuge from flies in summer than from cold in winter. If well positioned, artificial wind breaks made from stone sleeper walls or heavy plastic mesh, give welcome shelter in winter.

A field shelter should be placed either against the fence or well away from it, with the back of the shed to the prevailing wind. If placed well out in the field it gives some shelter from several directions. Open-fronted L-shaped sheds are the most practical, and they ensure that no horse can get trapped inside by another horse.

In poor draining areas, a concrete floor may be laid, with an apron extending well beyond the shed. A cheaper alternative is to put down loads of stone, rubble or rammed chalk, according to availability. The floor area should always be kept well strawed down. Concrete can be slippery in freezing weather. On good draining soil, with the shed on a slight slope, a base may not be necessary.

Shelter sizes vary from a pony-type structure 3m x 3m (10ft x 10ft) to buildings of 12m x 18m (40ft x 60ft) which hold a number of animals. They must be strongly constructed to withstand horses rubbing against them. Some field-shelter manufacturers insist on a concrete base. In certain districts, planning permission is necessary.

Shelters should be kept regularly free of droppings and once or twice a year they should be thoroughly hosed out and disinfected.

FENCES

It is the responsibility of the stock keeper to ensure that his fields are securely fenced. Animals which get out on a road can cause accidents, damage to life and property, and often fatal injury to themselves. They can also damage crops and.gardens. It is possible to insure against these risks, but insurance companies rightly expect suitable care to be taken. An habitual escaper who causes damage may be responsible for his owner being taken to court. Subsequent insurance may be difficult to obtain and expensive. Animals which make a habit of getting out can rarely be contained against their will. A 1.5–1.8m (5–6ft) gate or fence is no deterrent; however, electric fencing will usually contain them. (See Electric Fencing, page 313).

If cattle or sheep are kept, either to run with horses or to provide alternate grazing, fencing must be strong enough to contain 'gadding cattle' (when pestered by flies in summer), and with the rails sufficiently close to contain sheep.

Owners of valuable stud and competition horses should take greater care with the choice of fencing. It must be secure and safe. If the animals are kept in adjoining fields it is best to separate them with a double line of fencing approximately 1.8m (6ft) apart, which prevents squealing and playing over the fence. An electric fence alongside the permanent fence will safely separate animals.

Hedges

A well-trimmed, thick hedge is the best type of field fencing. It should be kept at a height of 1.2–1.5m (4–5ft) and should be cut back annually to encourage

A good field shelter.

growth at the bottom. This work can be carried out by arrangement with a neighbouring farmer or a contractor. If it is done regularly by mechanical means the hedge trimmings will be chopped up by the machine and can be left to rot. Older, overgrown hedges have to be cut by a large circular-saw type of machine, and the large hedge trimmings should be collected and burned. Weak, long and straggly parts of a thorn hedge can be cut and laid. Elder should be dug out or cut down, as it weakens the other parts of the hedge-line; its only use is as a protection from flies, and for this purpose two or three large bushes can be left. Holly should be encouraged, as it gives good winter shelter. Yew, because it is poisonous, should be completely cut out.

Hedges under trees are often weak. Any gaps or weak parts should be filled with solid, creosoted posts and rails. Plain wire is inadvisable. Barbed wire should not be used.

It is natural for horses to browse. They may well take great pleasure in barking any trees left unprotected. Plastic mesh laced round the trunks, or a protective guard of well-creosoted posts and rails, will keep horses away from trees: note that link fencing or wire netting can be dangerous if horses strike or stamp at flies.

Walls

In certain parts of the country, stone walls are used as field boundaries. If they are of a suitable height and are kept in good repair they are stock-proof and safe, and provide good shelter. A low wall may need a rail above it to prevent animals jumping out. In the case of cobs and ponies, a cheaper alternative is one or two strands of well-tightened, plain wire or an electric fence.

Banks

Banks are found in Ireland, the west of England, and in some other areas. They give good shelter, but to be stock-proof low banks require topping with a hedge, rails or wire. Some banks have a deep drainage ditch on one side, which can be a danger to young stock and should be fenced off. Steep banks can present a hazard,

Post and rail fencing with safe corner.

particularly to foaling mares and young stock, and it is often safer to fence them off. It may also be necessary to fence across a ditch or stream at the field boundary.

Post and Rail

If in good repair, these make safe, stock-proof fences. Though expensive, they have a life expectancy of over twenty years, but can be damaged by chewing, rubbing and leaning. They should be regularly creosoted or painted. Broken rails must be promptly replaced as the split timber and any exposed nails can be a danger; morticed rails, though less easy to replace, are safer, as no nails are used.

Fence posts should be sawn off at an angle, flush with the top rail. If horses try to jump out, projecting posts can cause damage.

Stud Rails

These make a safe, secure fence: they are more durable than wood, will not split, will not be chewed, and require less upkeep. In many cases they are also cheaper. From an environmental aspect, however, some people consider them to be less attractive.

Stud rails consist of plastic strips 10cm (4ins) wide, incorporating high-tensile wire. The strips come in rolls and are erected either on wooden or plastic posts. Three or more lines of rail strips are used. A cheaper but less substantial alternative is a plastic rail at the top with two or three lines of plain wire below. This fence requires very exact straining. It stands up to considerable impact and also has some resilience. In time it stretches and requires tightening.

High-Tensile Wire Mesh Fencing

This is a heavy-duty wire mesh, preferably topped with a wooden or plastic rail to give clear visibility. It is cheaper than stud rail, and is escape-proof against any stock, but it presents the risk that animals – particularly foals or small ponies – may put their feet through the mesh and get caught. Small, diamond-shaped mesh will considerably lessen the risk, but horse owners should be aware of dangers. Mesh fencing must be well maintained and checked for any holes and tears.

Plain Wire

If well erected and firmly strained, this can be the economic answer to many fencing problems. Topped with a rail, it makes a more solid, clearly sighted fence without too much expense. The drawback is that cobs and ponies may loosen the wire by putting their heads through and leaning on it; this can be prevented by using electrified wire or a line of barbed wire between two lines of plain wire. High-tensile steel wire is more expensive than ordinary wire; it is less easy to work with and requires greater skill to erect, but it lasts twice as long. However, if horses cut themselves on it the wounds tend to be very severe.

Barbed Wire

Barbed wire is not recommended but it is cheap and effective, and if it is well maintained and regularly inspected the risk is diminished. If barbed wire must be used, the safest place to put it is on the second line of wire with a single strand of plain wire above and two strands of plain wire below.

When renewing a fence, all old wire must be taken up and removed. To reduce the danger of a horse or pony getting a foot over the wire, the bottom strand should be of plain wire and never be lower than 28cm

Two strands of plain wire safely fitted below a plastic rail.

(15ins) from the ground. Another hazard is that of animals in adjoining fields squealing and stamping at each other through the fence; this can be dealt with only by careful daily observation .

ERECTION OF FENCES

Post and Rail and Wire

Posts should be at a minimum height of 105cm (3ft 6ins), preferably 120cm (4ft) above the ground and 46cm (18ins) below the ground. Posts may be of timber or concrete. All timber should be treated with preservative. Droppers (lightweight wood or steel slats) can be used between posts to keep wire taut. Posts should face outwards with the rails inwards towards the field.

Heavy strainer posts should be placed at all corners, at changes of direction, and/or at intervals of 50m (54yds). They should be dug 75cm (2ft 6ins) into the ground. Stud rails and stud fences can usually be erected by the suppliers. For further information, consult the Ministry of Agriculture – Pamphlet No 711 Wire Fences for Farms.

Electric Fencing

This is cheap and easily erected. Its main uses are:

- To keep horses away from a weakened hedge-line, fence or boundary.

- To provide safe separation from horses in adjoining fields.

- To permit strip-grazing of a field.

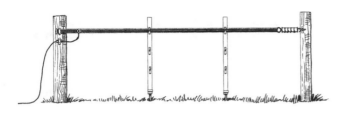

Electric fencing.

- To divide fields temporarily so that areas can be rested.

- To help prevent horses/ponies jumping out of their field.

It should be placed 1.5–1.8m (5–6ft) from the boundary fence.

The fencing usually runs off a portable unit containing a 6-volt battery. However, if it is close to a mains electricity supply, it can be wired to this, with a transformer which will reduce the voltage to approximately 6 volts.

The fencing material consists of:

- Iron or plastic-coated posts with insulators at the top. These can be heeled into the ground. The latter are less likely to cause injury.

- Wooden posts with insulators attached. This type is preferred for permanent fencing. Iron or plastic posts are easier to erect if the fence is temporary, or has to be moved each day as in strip-grazing.

- Plain round wire or wire strips stretched between the posts and attached to insulators at either end. The latter can be more easily seen. It is essential to cut back any long grass or overhanging hedge so that there is no risk of earthing the current, which – particularly in wet weather – will very quickly run down the battery. It is important to check the wire each day and to test the strength of the current with a fence tester. Horses are quick to take advantage of a weak or dead battery.

When erecting a fence, sharp corners and angles should

be avoided, unless the posts can be efficiently stayed. The fence should not be placed underneath or parallel to overhead electric cables, as there can be some risk of the current being diverted down the wire. When moving a fence it is important to ensure that all posts are well heeled in, that insulators are sound, and that attachments to hedges and gate posts have an insulator to break the current.

Plain wire, preferably of the breakable type, can be more easily seen by animals if pieces of coloured cloth are tied between posts.

When first introducing animals to an electric fence, it helps to make them respect it if they are given a shock by feeding them close to the wire. Some horses never gain that respect, and happily jump out. Ponies with thick manes often run underneath. Such animals can sometimes be controlled by putting up two parallel lines of electric fencing. If this fails, they should not be turned out in a field with electric fencing, as they will encourage others in the field to imitate their bad habits

UNSUITABLE FENCING

For safety reasons, the following fencing cannot be recommended, as there can be a risk of horses getting caught up:

- Wire netting.
- Sheep or pig wire.
- Chestnut fencing.
- Old iron rails. These are likely to spring out of line, and the iron spikes can cause nasty injuries to horses.

GATES

Wooden Gates

Oak is the most suitable type of wood. It is expensive, but it does not warp, and it lasts for forty years or more. Soft wood gates, though much cheaper, are inclined to warp, and the timber breaks more easily.

Gate-posts should be of oak, set 90cm (3ft) in the

ground. They should be hard-rammed with stone filling, or set in concrete. Old railway sleepers make a suitable alternative.

All timber should be treated with preservative. Hinges should be of galvanised steel and of a strength suitable for the gate.

Fastenings should be of a secure type, with no sharp projecting edges on which a horse can get caught. When fastened, they should help to take the weight of the gate, preventing it from dropping.

Metal Gates

If of heavy-duty construction, metal gates are satisfactory. Lightweight metal has a limited life; it is easily bent and the spars can fracture or come apart from the framework, exposing dangerous edges. Posts are usually of metal set in concrete, but wooden posts can be used. All metalwork should be treated with rust-resistant solution and kept regularly painted. The hinges and fastenings used for wooden gates are suitable.

Hanging Gates

If gates are to last, they must be well hung, so that when opened they clear the ground and do not have to be lifted or dragged.

Padlocks

Owing to the increase of horse thefts, it is advisable to padlock any gates opening on to roads. The hinge end of the gate should be secured by a chain and padlock, or by a metal bolt driven in above the top hinge so that it cannot be opened by lifting it off its hinges.

Gateways

Gateways can become very muddy in winter. They can be improved with loads of stone, chalk or rubble, or by a roughened concrete apron. Surface water should be directed off the surrounding area by drains or ditches.

FEEDING

Summer

In summer, many horses and ponies at grass become too fat. If they are resting or only in light work (four hours a week), they are unlikely to require extra food. If they are on good grass, their grazing time may need to be restricted. If they are in heavier work and if the grazing is poor or limited, they may need a daily or twice-daily ration of hard feed. For amounts to feed, see Chapter 21.

Winter

In winter, supplementary feeding is essential. Hay should start being fed as soon as the horse shows interest when it is put out in the field. As long as there is grass to eat, they prefer this to hay. At first, they may only require hay at night, but as the grass diminishes, hay should be fed in the morning as well. If good hay is fed and not cleared up, this means that too much is being fed, and the ration should be cut down until all is cleared up.

If the hay is of inferior quality, it will often be left. This should not be taken as a sign that too much is being put out, but rather that the quality of the hay needs improving.

In wet weather, some hay is bound to get spoiled and to be trodden into the mud. In frosty weather, when the ground is hard, the horses should clear up all hay put down.

Methods of Feeding Hay

If one or two animals are kept, hay may be fed in haynets, which must be securely tied to a fence or tree and placed at least 3.5m (12ft) apart. In wet weather all animals prefer to eat with their heads down and their backs to the wind, so haynets may then be unwelcome.

If more than two animals are turned out together, hay should be fed on the ground. It should be placed in the most sheltered part and on the least muddy area, even if this involves taking it to the far end of the field. Hay should be put out in slices, according to weight, and not shaken up as it will then be less likely to be spoiled. Heaps should be at least 3.5m (12ft) apart, and there should be two or three more heaps than horses to allow for them moving about. If there are ten or more animals and if the weather permits, it is better to put the hay cut in a larger circle. To check any fighting or bullying, wait and watch until all horses are settled and eating.

Quantities

Good-quality meadow hay is suitable for feeding in the field.

If there is little or no grass, the following feed scale is suggested:

Small ponies	4.5–5.4kg (10–12lbs) hay
Medium ponies	4.5–6.8kg (12–15lbs) hay
Large ponies and cobs	6.8–9kg (15–20lbs) hay
Large cobs and horses	11.3–13.6kg (25–30lbs) hay

When the horses and ponies are of mixed sizes the hay ration at the beginning of winter should be about 4.5kg (10lbs) per animal. As winter progresses, this ration can be increased to 6.8–9kg (15–20lbs) per animal, depending on the weather and if it is being cleared.

In snow or freezing weather, extra hay must be given; it will not be wasted. Animals require extra food for warmth, and the process of eating and digestion helps to keep them warm and content. Their daily ration cannot be equated with that of a similar-sized horse kept in a warm stable. More food is necessary.

If hay supplies are short, oat or barley straw can be fed on its own, or the straw can be fed mixed with hay. Well-bred animals will then require a high protein supplement.

Concentrates (see also Chapter 20)

Healthy, resting ponies or those in light work (two to four hours per week) do not necessarily need a daily feed.

Ponies under 12.2hh rarely require extra feed.

Ponies and cobs in heavier work require a twice-daily ration of hard food suitable for their size.

Horses wintered out should receive a twice-daily feed.

All animals in light or poor conditions should receive as much food as they will consume.

When feeding concentrates, if only two or three animals are kept they may be fed in the field, providing an attendant stays to prevent bullying and fighting. This is not safe with a large number, and they should be brought in and fed separately.

Feed Blocks

These are manufactured specifically for feeding to horses and ponies wintered out. Together with rough grazing and/or hay, they provide a balanced and nutritious diet. They are weather-resistant and can be placed on the ground in a container or suspended from a fence post or tree. They should be placed well away from the water supply.

For groups of horses, allow one block per four to five horses and follow the maker's instructions. Generally speaking, horses only take up as much as they need and there should be no problem. However, certain horses develop a craving and finish a block in a day or within days. This makes it an expensive feed, and in some cases it has been known to cause severe colic.

COMPANY

Horses are by nature herd animals and they are much happier if part of a group. In summer, by standing head to tail they are able to ward off flies. If it is not possible to find another horse for company, then a donkey or even a goat can be a successful substitute.

To avoid fighting and possible injury, it is important that groups of horses turned out together should agree. It is often arranged that mares and geldings are kept in separate fields. For this reason, many riding schools only keep either mares or geldings. If groups are mixed, be sure that you do not put two aggressive leader-type animals in the same field. Generally speaking, once animals know each other and have sufficient room they sort themselves out and there should be no trouble. However, care must be taken when adding a new animal to an established group.

DAILY CARE

Horses should preferably be visited twice a day, and on one of these visits they should be checked at close quarters. An observant person quickly notices if all is not well, particularly if the same person checks the horses each day. Any unusual behaviour or stance will then be noticed. A horse who is standing away on its own, or who is unusually slow or unwilling to come up when food or hay is put out, should be more closely examined. In areas where horses have been vandalised, a daily check of the whole horse, especially the dock area, is advisable.

Horses should look well and healthy with:

* A bright, alert expression.

* Weight evenly taken on all four feet, or a hind leg only rested. Well-bred horses and young stock sometimes flex a knee when grazing. This is normal.

* No discharge from eyes or nose.

* No stiffness or unevenness of stride when moving off.

* In summer, the coat lying smoothly and showing a gloss. In winter the coat may be rough, but it should not be dull and staring. A horse with a tight staring or standing-up coat is cold and may be ill. He should be caught, stabled and his temperature taken.

Resting horses or young stock should be caught up at least twice a week and their feet, eyes, nose, skin and coat checked.

When observing a group of horses, a tit-bit may be handed out from a pocket, but a feed or catching bowl encourages competitive aggression, with a risk of injury to all concerned.

Water troughs and fencing should be checked daily.

New Zealand Rugs

These are designed to keep the horse warm and dry when turned out or living out.

Two rugs are necessary for a horse living out all the time, so that one is on the horse and one is drying. Even in dry weather the rug should be removed daily, the horse's back massaged to restore circulation, and the shoulders and legs inspected for rubbing.

Rugs with surcingles are only suitable for horses being turned out for short periods, as surcingles may shrink in the wet and also cause pressure on the spine.

HEALTH

All horses should be immunised against tetanus, and a regular worming programme should be carried out. A careful worming strategy is essential if horses are to remain healthy and the level of larvae on the pastures kept to a minimum. The tactics are to use a suitable (veterinary advised) anthelmintic at correct intervals and according to the time of the year (see Chapter 15). Teeth should be checked

In **summer** many horses and ponies become grossly fat, with a consequent risk of laminitis, and if they are ridden the extra weight puts strain on the legs, heart and lungs. In many cases it is necessary to restrict their diet. Unless the paddock is bare of grass, one or two hours' daily grazing is ample. The animals may be stabled at night, bedded on shavings, and turned out in the day; sun and flies will then often deter them from eating. A small amount of hay can be given at night: a pony 1–1.4kg (2–3lbs); a cob 1.8–2.7kg (4–6lbs). If a secure yard is available, it can be used instead of, or in conjunction with, a stable.

In **winter** a pony's thick coat can disguise poor condition. However, if the crest, backbone and rib areas are felt, any lack of flesh will be apparent. The crest in particular is a reserve of fat and it should feel firm and solid. Well cared for healthy ponies should look plump and well throughout the winter, whatever the weather. Ponies with staring coats and bony outlines are a sign of poor management, inadequate worming and insufficient food (see also Chapter 21).

Seasonal Problems

In Summer

LAMINITIS Ponies subject to laminitis must be kept on a strict diet. If there is any sign of 'footiness', heat in the foot, or standing with weight on the heels of the forefeet, take the pony off grass and consult your veterinary surgeon immediately.

SWEET ITCH The signs are continual rubbing of the mane and tail until they become raw. Only certain horses – usually ponies – appear to react to the saliva of the midge which causes the allergy. The condition is improved if the horse is stabled before dusk and until well after dawn. Fly repellent should be used in the stable and in the field. It has been found that if the wind is over 3mph midges will not appear.

BLISTERED MUZZLE Horses with white muzzles are particularly prone to sun-blistered mouths and noses, a condition which can be caused by an allergy to a particular plant. To prevent this, apply barrier cream. If blistering occurs, antiseptic ointment should be used. It may be necessary to stable the animal.

RUNNING EYES Flies settle round a horse's eye to feed on 'the tear'. Sensitive horses can wear a headcollar with a fly guard (browband with a string fringe). More sophisticated protectors, such as browbands impregnated with fly repellent, are now available from tack shops. Some fly repellents are safe to use round the eyes, and can be applied with a small sponge. Horses can be kept in during the day. The eyes should be sponged clean twice daily. If eyes become infected, veterinary advice may be necessary. Light-skinned horses are more subject to this problem.

BOT OR GADFLY From June to September the bot fly lays its eggs on a horse's front legs, particularly just below the knee. These hatch to larvae, which the horse licks. The larvae then penetrate the skin and the mucous membrane around and inside the lips, resulting in the start of the bot fly cycle. The flies cause great irritation, and although they do not land on the skin, horses will gallop about and kick to get away from them. They may then injure themselves and others.

WARBLE FLY These flies are becoming much less common, but any swelling on the horse's back in spring and early summer should be suspect. When the larvae hatches out it must be killed. Some modern wormers kill bots and warble larvae before they reach their final sites.

Various **fly repellents** are now available, which in warm weather help to make horses more comfortable. Many repellents have to be applied twice daily, but some are longer acting and in dry weather are effective for several days. A paraffin rag wiped on the legs can be effective against bots, but should not be used on horses with sensitive skins as it may cause blistering.

In Winter

The following problems are likely to occur, particularly with fine-coated horses in a very wet season. Horses should be closely observed each day and preventive measures taken. At any sign of lameness, obtain veterinary advice.

RAIN SCALD The skin of the back, loins and/or quarters is softened by incessant rain, thus weakening the natural defence mechanism against skin germs. Inflammation and dermatitis follow; the hairs clump together and fall out in patches.

Prevention: Sensitive-skinned horses should wear New Zealand rugs. In autumn, liquid paraffin or barrier cream can be applied to the horses' backs when the coat is free from mud. In severe cases, consult your veterinary surgeon.

MUD FEVER This is similar to rain scald, but occurs on the legs and sometimes the heels. If on the latter, continual exposure to mud increases the infection. Severe cases may spread to the belly and inner thigh.

Prevention: In the autumn and during the winter months apply barrier cream regularly to the legs, after they have been cleaned and dried. Wet, muddy legs should never be brushed, but thatched and left to dry (see Bandaging, Chapter 29). Then they can be brushed clean.

CRACKED HEELS These are inflamed sore horizontal cracks occurring in the heel region. Cracks in the heel and pastern area become intermittently infected, which can cause acute lameness, particularly after the animal has been standing still.

Prevention: As for mud fever.

THRUSH This is a disease of the frog. Most cases are associated with poor foot hygiene or even foot disease. However, some horses will contract a mild form of thrush in winter through standing in mud.

Prevention and treatment: As soon as the ground becomes muddy, apply Stockholm tar in and around the frog area. Repeat it as necessary. The foot and frog must first be thoroughly washed clean and dried. The veterinary surgeon may recommend an antibiotic spray as treatment.

LICE These insects can be found in the coat at any time in the winter, but usually in early spring. Any horse rubbing its mane and tail should be suspect.

Prevention: Apply louse powder regularly, as directed by the manufacturer. If this proves ineffective, obtain veterinary advice, as more sophisticated treatment may be necessary.

RINGWORM AND OTHER FUNGAL INFECTIONS These can appear at any time of the year, but in winter the fungus in its early stages can go undetected, as the long coat covers the scabby area.

Horses at grass, or those brought in from grass to be stabled, should be closely examined for any raised area of hair or circular scabs. The fungus can appear anywhere on the body, but it is more likely to be in places where the horse has rubbed against a gate or fence post, i.e. head, shoulders and quarters.

Prevention: Particularly if cattle have been in the field, all gates and fences should have a thorough coating of creosote. Ironwork should be washed with a strong solution of soda or repainted.

SUDDEN ACUTE LAMENESS This can usually be attributed to a punctured foot or a small flesh wound, which has gone septic. It may be necessary to transport the animal to the stables. The foot and leg should first be thoroughly cleaned and the leg examined for injury. Veterinary help should be obtained.

CHAPTER 46

WORKING THE GRASS-KEPT HORSE OR PONY

Grass-kept horses and ponies can be used for hacking and light to medium work. Highly strung or 'gassy' animals may then be more amenable.

Horses needed for hard work can be kept at grass in the summer, as long as they are given regular exercise and increased concentrates. If the horse starts to get too fat, grass intake on a good pasture may have to be restricted. In winter, if hard, fast work is required, the horse should be stabled at night and regularly exercised each day.

Summer

- Horses and ponies on good grazing should preferably be brought in two hours before being ridden, so that their stomachs are not gorged with grass.

- For grooming and care after work, see Chapter 10.

On a wet morning, the horse's back should be scraped with a sweat scraper and semi-dried with a stable rubber or straw. If a dry stable rubber or numnah is put on and secured by a surcingle or saddle, the back will quickly dry. The horse will not be harmed if ridden with a wet back, although this is not recommended. He will probably react by first hollowing under the weight of the rider; a sharp horse may even buck.

Winter

- If on hard food, the horse should be fed early in the field, or caught about 1½ hours before being ridden and given his feed.

- For grooming, see Chapter 10.

Wet and muddy backs should be scraped with a sweat scraper or plastic curry comb, dried as above, then brushed clean. All lumps of mud must be removed from the saddle and girth area, and a fresh numnah or stable rubber should be placed under the saddle.

If on returning from work the horse/pony is warm, any resulting sweat patches should be wiped clean or sponged off. Feet should be picked out and the animal should be turned out as soon as possible, so that he can roll and be less likely to catch a chill. If he is wearing a New Zealand rug, straw can be put under it to allow more air to circulate, and he can be given hay. His hard feed can be given in the field, or he can be caught later and fed in the stable.

If it is essential to keep the horse or pony in after work, put on either (a) a sweat rug and roller with straw underneath, or (b) a sweat rug and light blanket and roller. The horse/pony may then be fed. It is important that ponies with thick coats are not allowed to sweat profusely.

THE COMBINED SYSTEM

This is a system whereby in winter the horse is stabled at night and put out in the day. In summer, the process is reversed: the horse is in during the day and out at night, thus obtaining protection from hot sun and flies. If time and labour are short the advantages are:

- Daily exercise is not necessary.

- If only ridden at weekends, the horse will be quieter and more manageable.

- Daily grooming is reduced.

- Less bedding is used and at times less hay.

- Mucking-out time is reduced.

- The horse can be clipped and turned out in a New Zealand rug.

- The horse can be made ready for work quickly, as he will be dry and comparatively clean.

- In winter, less food is wasted in the field, and the food ration can be reduced, as the horse will be in a warm, dry stable at night.

- In summer, if necessary, grazing time can be more easily controlled.

YARDING PONIES

This is a system whereby ponies are wintered in a covered yard or large barn, which may or may not have an adjoining railed exercise area. It is a system very suitable for riding schools. Pony mares and foals can also be kept in this way. Young stock are increasingly kept in this way during their first winter after weaning and often as two- and three-year-olds.

The yard is kept strawed down on the deep litter system and is cleared out at the end of the winter. Hind shoes should be taken off and, if feasible, front shoes also. Mares and geldings may have to be separated, or only a single sex kept or accepted for winter keep.

An area of 20m x 20m (65ft x 65ft) is sufficient for ten ponies. A hay and feed manger should run along two sides. Preferably there should be two separate water troughs.

ADVANTAGES

- Ponies will be clean, dry and warm.

- Less time will be needed to groom before being ridden.

- They will be available when required.

- They will need less feed and there will be less waste. (See The Combined System, page 320.)

- If access is suitable, the yard can be cleared by a 'muck lorry' .

- Grassland can be spared unnecessary damage through winter poaching.

DISADVANTAGES

- Certain ponies do not settle and may fight and upset the group. Fights may occur, resulting in injuries. At feed time, it is sensible for someone to be present to prevent squabbling.

- Labour and suitable equipment will be needed in the spring to clear the yard, unless access is suitable for a muck lorry.

CATCHING HORSES/PONIES

- Do not be in a hurry. Careless procedure can easily upset a horse, spoil him, and make him difficult to catch.

- The headcollar should be placed over the left arm. Difficult horses may be encouraged to come to a handful of food

- With the rope in the right hand, approach the horse from an angle towards the shoulder so he can see you, and as he accepts the food, move closer towards his left shoulder, placing your right hand with the rope on and then over his neck.

- Avoid any sudden or jerky movements, and talk quietly to the horse throughout the procedure.

- Quietly put on the headcollar.

Procedure with Difficult Horses/Ponies

• Take a large feed bowl, half fill it with food of a type that will rattle when the bowl is shaken. Tuck the headcollar and rope under your left arm, or alternatively hold a rope under your arm or inside your coat.

• Always allow the horse to see you approach.

• If he moves away, make a big circle and again approach from the front.

• Talk to him as you approach.

• Shake the bowl, and then lower it to the ground, which may gain his attention.

• Walk up to him with the bowl in the left hand.

• As he takes the food, the right hand can quietly move up his jaw and towards his neck.

• If he moves away, do not try to hold him but try to go with him, but if he shoots off, do not follow. Reorganise your approach as from the start. When he stands and eats, move quietly round to his shoulder, taking hold of the rope with your right hand.

• Place the rope under and then up or over and down his neck. When the horse is held by the rope, you should be able to put on the headcollar. On some occasions, a length of string or thin rope may be less obvious.

If difficult horses are caught and fed each day, it becomes a routine, and when the animals are wanted for riding, they will be easier to catch.

Difficult horses are usually more easily caught if only one person makes the attempt and any onlookers are sent away. Other horses in the field should be caught and held at the gate or in the corner of the field. At first do not take them out of the field, as a horse left on his own will be encouraged to gallop about. However, if there is a convenient passageway or track to the stable yard, then the horse will probably follow the others along it, or he may walk in amongst the group and allow himself to be caught. If this occurs, it may be more sensible to quietly move alongside him from the rear. Talk to the horse and place a hand on his quarters (taking care he does not kick you), moving it up his back and neck as you walk alongside him. Secure him by placing a rope round his neck, and then quietly put on the headcollar. Horses must not be allowed to escape on to roads.

If the horse still will not be caught, leave a quiet horse or pony in the field with him and return in half an hour.

Horses are usually easier to catch in hot weather and more difficult in wet and/or windy weather, or when there is spring grass.

The catcher can often put horses off by wearing an unusual hat or raincoat, particularly if it is plastic.

A better contact is made with the horse if gloves are not worn.

Horses who are difficult to catch are best turned out wearing a well-fitting leather or nylon headcollar. If of nylon, it should have a 'panic snap', which will give way if the horse catches the headcollar on a post etc. A 16cm to a maximum of 20cm (6–8ins) length of knotted string can be attached to the centre 'D' and used when catching the animal; the string should never be in a loop. The horse's head should be regularly checked for sores and rubs.

Staff (e.g. an instructor or groom) catching a horse and riding it to the stables should first put a headcollar on the animal and then fit the bridle on top of it. If catching the horse with only a bridle, do not put the reins over first in case the horse pulls away and takes the bridle with him.

On no account should animals be chased about or rounded up. This is self-defeating, and can only make them upset and even less willing to be caught. A sufficient number of helpers (a minimum of six) can form a straight line and walk towards the horse, guiding him into a corner. Even then, if he is determined, he can easily outwit the plan.

Remember that the 'boss' is responsible for the safety of any helpers. Although horses and ponies are usually less upset by children and will accept them if they move near, fear of being caught may alter this attitude.

Unaccompanied small children should not be allowed to catch ponies, particularly if there are several ponies in the field.

TURNING OUT HORSES AND PONIES

It is important that safe procedures are followed. Horses will often kick out when first released, and can injure people or other animals.

When leading a horse through a gateway you must be on the side nearest the gate, to control and stop it swinging back and hitting the horse.

Procedure for One Horse

- Bring the horse through the gate.

- Turn his head towards the swinging end of the gate.

- Shut the gate.

- Take the horse several yards into the field and once again turn his head towards the gate and yourself.

- Pat him, take off his headcollar, and step back.

- Watch the horse as he moves off. Do not turn away.

- Never slap him on the neck or quarters, as this may encourage him to gallop.

- Try to proceed quietly and avoid any fuss. If he is wearing a bridle, follow the methods listed above, then undo the throatlash and noseband (and the curb chain if you are using one).

- Slide the reins up to the headpiece and ease the whole bridle off staying close to the horse so that he releases the bit without pulling away. Give him a pat on the neck and move away from him so that he realises he is free.

Procedure for Several Horses

- All horses should be brought into the field, taken well away from the gate, with at least 3m (12ft) between each horse, and turned to face the gate.

- Shut the gate and proceed as directed above.

- Finally, release all horses together.

Horses who are unfamiliar with one another or who are likely to gallop off, should be taken to the field in pairs. When the first pair have settled, the next pair can be taken out and released.

ROUGHING OFF A FIT HORSE

This is the preparation of a fit horse for a holiday and a rest out at grass. The rest is important both from the physical point of view and to relieve mental stress. Rests are given to:

- Hunters at the end of the season.
- Competition horses.
- Injured horses, to allow the repair processes time to work.

The roughing-off process takes about two weeks and the procedure is:

1. Remove one blanket. After seven to ten days remove the top rug.

2. Thorough grooming should be discontinued and the natural grease should be allowed to accumulate in the coat, forming a protection against cold and wet.

3. Diet should gradually be adjusted, reducing the hard feed and increasing the hay.

4. The horse should be allowed out daily in a small field or led out to graze.

5. Exercise should gradually be reduced and should now consist mainly of hacking out at the walk.

6. Veterinary checks should be made, and any recommended treatment should be completed.

7. Shoes should only be removed if the horse is to have a long rest of more than four weeks. Feet should then be rasped to prevent the horn splitting. Horses with brittle feet who are to be turned out on flinty or very

hard fields may need to have their front shoes left on.(NB: Feet should be rasped or shoes should be removed every four to six weeks.)

8. In cold weather, the horse will be more comfortable if a New Zealand rug is used. For fitting and procedure, see pages 226–8.

Hunters are often turned out at the end of the season. This can be as early as the end of March, when they will need a New Zealand rug, a well-sheltered field, and hard feed and hay to help keep them warm.

Preparation for Turning Out

1. Check the field.

2. If you are turning out a horse on to rich pasture, he must be given short periods of grazing for several days beforehand.

3. Choose a mild day.

4. Make sure the horse has suitable company – preferably not a highly strung galloping type of horse.

5. Do not give a breakfast feed. If the horse is hungry he is more likely to graze than to gallop.

6. Turn the horse out early in the day, so that he will have time to settle in the field before night. He should be watched until he is quiet, and checked again during the day.

It is preferable to allow the horse several short 'holidays' in a year rather than a major let down of several months. Where a complete reduction of muscle tone is allowed the horse may revert to very soft condition and carrying excess weight. This total let down is now recognised as not being in the interests of the long-term welfare of the working horse.

BRINGING A HORSE UP FROM GRASS

It is better for the horse if the change is made gradually over a period of about ten days, as this will allow his microbial digestion time to adjust itself to the change of diet (see also Chapter 21).

Depending on the time of year and quality of grazing, some horses may be receiving hard feed and hay in the field. In this case, a sudden change of diet will present no problems.

Procedure

- Bring the horse into the stable for several hours a day. Give a small feed of 0.5–1.4kg (1–3lbs) of horse and pony cubes or equivalent foodstuff or 1–1.4kg (2–3lbs) of grain, plus a 1.8–2.3kg (4–5lbs) net of dampened hay. Allergic horses should be given soaked hay (see also Chapter 23).

- If the horse is turned out on his own or with one or two others, it is possible to feed them in the field. Always stay to watch the feeding, so that there is no bullying or kicking. (There is no need to feed hay.) This practice is not recommended if there are more than two or three horses in the field, because of the risk of kicking.

- The horse may have to be brought in and kept stabled. In this case, try to arrange for him to be turned out in a paddock for a few hours each day, or, alternatively, graze him in hand for 20 minutes or so.

Feed ration should be two or three 0.5kg (1lb) feeds of 'horse and pony' cubes or their equivalent, and plenty of dampened hay, about 9–11kg (20–25lbs). Do not feed any grain for a week; grain fed too soon can be the cause of filled legs. If you are planning to use grain, change over to it gradually.

Usually if a horse comes in after a summer at grass, weight will need to be lost. Gradually accustom him to a smaller total quantity of feed, but one which includes more concentrates. As more concentrates are fed, hay can be reduced. If the diet is restricted to encourage weight loss, it is essential to put the horse on shavings or a similar bed, and not straw. For amounts of feed, see Chapter 21.

Recommended Treatments

WORMING A suitable dose should be administered. (See also Chapter 15, Internal Parasites.)

FLU AND TETANUS INOCULATIONS If due, these should be arranged. (See Chapter 33.)

TEETH The teeth should be checked by the veterinary surgeon and rasped if necessary.

FEET Arrangements should be made with the farrier to have the horse shod as soon as he is permanently stabled and starts exercise. Four weeks before he is due to come in, depending on the condition of his feet have him shod with light front shoes. This allows some growth of horn and makes it easier to shoe him for work. Valuable horses should never be turned out in company with hind shoes on, as a kick from a shod foot can cause severe injury.

Precautions

With careful feeding and regular walking exercise, the problem of filled legs should not arise. Coughs and colds easily develop during this period, so ensure a plentiful supply of fresh air and take care if the horse sweats when exercised.

Dangers From and Precautions Against Gaining Weight

Turning horses out for long periods in the spring/summer may result in their putting on too much weight. The increase in weight puts strain on a horse's legs and must be reduced before he is fit for active work. Often the time allowed for fittening is insufficient, and horses are asked to work too early, which strains the tendons, joints, heart and wind, with the possibility of permanent harm.

Horses can be turned out for a shorter period of six to eight weeks. They can then be either yarded or stabled at night. In very hot weather, when there are many flies, they can be brought in during the day and turned out at night.

Suggested Exercise for an Unfit, Overweight Horse After Several Months at Grass

1ST WEEK
Walking exercise on the level. Half an hour on the first day, increasing to one hour by the end of the first week.

2ND WEEK
Increase walking up to 1 hour by the end of the week.

3RD WEEK
Slow trotting, starting on the level, together with walking up and down hills.

4TH WEEK
As third week.

5TH WEEK
1½ hours exercise a day, to include some trotting up hills and short, slow cantering on good, level ground. Sweating will help to hasten weight loss. 20 minutes school work may be included.

6TH–8TH WEEK
1½–2 hours exercise a day, to include trotting up hills and longer cantering periods. School work may include jumping and canter circles.

7TH WEEK
If the ground is hard, cantering can be performed in a school. Start with 1 minute on either rein, building up gradually over successive weeks. As this can be very strenuous, care should be taken. Big, heavy horses unused to circle work should not be asked to do this type of work.

For the schooling, work boots should be worn all round. If, when hacking out, roads are slippery, or if the horse is fresh, knee caps and boots are advisable.

All exercise periods should include a good percentage of walking, as this helps to harden legs and to develop muscle without unnecessary strain.

Horses who are turned out for shorter periods – or in the winter, when grass is poor, who have received regular hard feed and hay – come in very much fitter. In these cases, preparation for active work may be shorter. The owner or stable manager must decide how fit a horse is and for what work he is ready.

Unfit, overweight horses should not be lunged, because working on a circle puts extra strain on muscles, tendons and joints. Lungeing may be included after two weeks of walking, but 5 minutes on either rein will be sufficient at first. This should have a beneficial and suppling effect on the horse, as long as a suitably prepared surface is used. Hard or slippery fields are not advisable.

Timetable for Fittening

If the horse is grossly fat the time taken over fittening must be increased.

- **RIDING CLUB** – after 6–8 weeks.

- **HUNTER** – can be brought up in August ready for

gentle cub hunting by mid-September; more serious cub hunting in October; and hunting by November.

- NOVICE EVENT HORSE/SHOW JUMPER – 10–14 weeks

- ENDURANCE HORSE 40–MILE RIDE – 8–12 weeks

- THREE-DAY EVENTER – 14–16 weeks

- GOLDEN HORSE SHOE HORSE – 16–20 weeks

Prevention of Sore Backs, Girth Galls and Sore Mouths

- Check the fit of the saddle.

- Ensure that the bit fits the horse's mouth.

- All leather work must be soft and supple.

- Tack should be cleaned each day.

- Use a numnah under the saddle. Natural cotton or sheepskin are preferable to man-made fibres. In hot weather, a padded cotton numnah helps to absorb sweat without overheating the saddle area.

- Numnahs and girths should be regularly washed – if necessary, each day.

- Use a 'Cottage Craft' type girth.

- Harden saddle and girth areas with salt and water or methylated spirit.

- The horse must be thoroughly cleaned each day; sweat marks quickly become sore if left. In warm weather, the horse should be sponged down; surplus water should be removed with a sweat scraper; then the horse should be walked in the sun until dry. In cold weather, sweat marks can be sponged off with warm water. In both cases the horse should be dried, and thoroughly groomed.

- During grooming the horse should be checked for any signs of saddle sores or girth galls. The first signs are rubbed hair, followed by tenderness and heat.

Treatment

- Harden the area with salt and water. Stop riding, and either lead the horse out in a cavesson or bridle alongside another horse, or, if he is fit enough, lunge him. If the saddle sore is damp (scaled back), spread cold kaolin over the area and top it with a plastic covering. This will help to draw out the inflammation before hardening the skin with salt and water.

SECTION 7

THE STABLE YARD

CHAPTER

48 CONSTRUCTION OF STABLES

Before constructing a new stable yard:

1. Consider your financial situation.
2. Obtain outline planning permission from the local authorities.

SITE

Any local byelaws relating to drainage and construction should be checked. If there is a choice of site, consider:

ACCESSIBILITY If an access road has to be built, this will increase the overall cost of the project. Installing services such as water, electricity and telephone, even over short distances, can be very expensive.

DRAINAGE Stabling requires a solid, dry, well-drained base. If the soil is sand, gravel or chalk, this should present no problem. With clay or a high water level, the sub-soil has to be removed, replaced, and built up with hard core and gravel.

CONVENIENCE Ideally the stables should be sited so that they are visible from the house or staff living quarters. Regular checks can then easily be made.

YARD AND ACCESS ROAD

Foundation

A good foundation is essential. If building a new stable yard, the access road may be put down first to give a firm base for lorries. It is usual to excavate down to the sub-soil, but should it be an area of poor drainage and clay, it may be necessary to go deeper and remove the underneath soil. The whole area is back-filled with hard core or similar material according to availability locally. It is compacted with a vibratory roller. This base should extend at least about 1m (3ft) beyond the area of the yard and 60cm (2ft) on either side of the access road.

The underground water main and electric cable which come from the public road or neighbouring buildings should be put in before the top surface is laid. They should emerge in the tack room or office area. If the water main is to cross an area used by lorries, it is likely to be fractured by pressure. To avoid this and to prevent the water freezing in winter, the main should be laid in a 60cm (2ft) deep trench to the side of the roadway.

The roadway should be edged with a concrete kerb, to prevent the sides from being undermined by surface water and giving way under pressure. The access road should have a width of 3.6m (12ft), and preferably should open out to a distance of 12m (40ft) as it reaches the road. This entrance should be surfaced to a width of 6m (20ft); to save expense the remainder may be left. The 12m (40ft) clear expanse should give good visibility for both riders and traffic.

Drainage

Surface water off the yard will collect at the lowest point, and should be directed off and away from the buildings. If the original area is flat, it must be arranged that the top surface should have a slight slope approximately 1 in 60 towards an open shallow drain. This surface water can then be directed:

(a) away across the field by means of a 'French drain';
(b) to a soak-away; or
(c) if (a) and (b) are not possible, then to a septic-tank system installed to take the stable drains.

A '**French drain**' is a ditch of varying lengths dug to a suitable depth according to requirements, shallower at the yard and deeper as it gets to the field. It is filled with reject stone.

A **soak-away** is a large circular hole 3–6m (10–20ft) deep, filled with reject stone. This works successfully in a sand, gravel or chalk area, but should not be considered in a clay district.

If surface water is to join a drainage system, then one or more gullies will have to be constructed to collect surface mud and straw. They must have strong, removable steel covers, and be large enough to cope with heavy rainwater.

YARD AND ROAD SURFACES

Types

(a) Gravel, stones or chippings are attractive to look at and non-slippery, but are difficult to keep clean and tidy, and are not suitable for a yard with large numbers of horses.

(b) Concrete and tarmac are practical, although tarmac is inclined to be slippery. They are relatively easy to keep clean and tidy, and are of good appearance. If put down to a suitable thickness they will last for many years, providing they have good drainage. Frost can be damaging to concrete and tarmac. Also, they are slippery in icy weather, and suitable precautions have to be taken, e.g. salt or sand put down, or straw tracks made.

The access road can be left untopped, but will then eventually break up. A top surface of concrete or tarmac makes a sound, long-lasting surface, and enhances the approach to the yard.

All yard and road surfaces need to be well maintained, and repaired as necessary. If cracked or broken surfaces are left, the deterioration becomes more rapid and more expensive to put right.

All yards must have a good gate and fences to prevent livestock getting on to the public highway. The gate must be wide enough for easy access, well hinged and have a sound fastening device. The yard gate should be kept shut at all times.

SHELTER

If a barn-type stable is selected (see Types of Stabling, page 334), shelter from cold winds need not be considered. If conventional lines of loose boxes are planned, try to select a sheltered area protected by other buildings, or by hills or trees. If building in a high, exposed position, an 'L'- shaped design, three sides of a square, or a rectangle, will give more protection.

New Ideas

Research is constantly being carried out with regard to flooring, ventilation, and labour saving. Anyone considering the building of new stables should study carefully all the various possibilities before going ahead with any plans.

Design Considerations

TACK ROOM AND FEED SHED These should be near the stable block, and preferably should be connected by an overhang. As the tack room is a security risk, it should be in view of the staff living quarters.

OFFICE, LAVATORIES, CHANGING ROOMS, STAFF AND LECTURE ROOMS should be adjacent to each other and with access to the car park.

MUCK HEAP This should be sited away from the car park and the yard entrance, but should be within easy reach of the stables. The road leading to it should have a surface firm enough to bear heavy lorries. The muck heap should cover an area which is practical for the size of the particular yard. It should have a concrete base and there should be a concrete path leading to the stables. The surrounds should slope so that the surface water will flow to an open drain and will be channelled away – but it should not be directed across fields because of the worm load. One side should be open and the other three sides built up to a height of 1.8m (6ft). Railway sleepers or concrete blocks can be used for this purpose. The path can be raised or the muck heap dug out so that manure can be thrown down rather than up. Good access for removal is an essential consideration.

HAY BARN This should not be adjacent to the stabling because of the risk of fire. Preferably it should be connected to the yard by a hard surface wide enough to accommodate a large hay lorry.

POWER LINES If these are above ground, they must be high enough to give clearance for any muck and hay lorries.

YARD SURFACE This can be of tarmac or concrete – but remember that all surfaces other than gravel are slippery in icy weather. Gravel is not practical if there are large numbers of horses in the stables.

CHAPTER 49

DESIGN OF STABLES

The design of stabling will be determined by:

- Planning authorities.
- Local conditions, e.g. lie of the land, shelter, drainage and other buildings.
- Personal preference.
- Cost.

TYPES OF STABLING

Individual Looseboxes

These are the most popular choice. Each stable is complete in itself and can stand alone or be built in

(a) parallel lines,
(b) an 'L' shape,
(c) a square, or
(d) an open-ended square.

Types (c) and (d) are considered to be more liable to facilitate the spread of disease, as there is less circulation of fresh air. On the other hand, on exposed sites they give more protection. An overhang should be incorporated along the front of the boxes to give shelter to horses and grooms.

American Barn

Barn stabling consists of a large covered area containing two rows of looseboxes facing each other across a 2.5–3.7m (8–12ft) passageway.

ADVANTAGES

- Pleasant working conditions for the staff.
- Labour saving.
- Cost and space effective.
- In cold or exposed areas they provide more warmth and comfort for the horse.
- Less likelihood of water pipes freezing.
- Horses are happier when able to see one another, i.e. herd instinct.
- Horses, feed shed, hay store and tack rooms can be under one roof.
- Horses are more easily observed should anything be wrong.
- The work of student staff is more easily supervised.

DISADVANTAGES AND PRECAUTIONS

- **Fire** This spreads more quickly and there will be greater problems from smoke. It is more difficult to get horses out of a barn-type building than out of individual looseboxes. To reduce the risk of horses being trapped, double doors should be placed at both ends of the building. These can be either sliding or hanging doors, and if the latter they should open outwards. The doorways must never be obstructed. In a very long building, doors should be placed half-way down each side to make it easier to lead horses out. Fire drills should be held regularly – see Prevention and Control of Fire, page 373. A hay store inside the building, although convenient, is an added fire and dust risk.

- **Infection and contagion** These are always a problem where large numbers of horses are stabled together and there is a greater risk of this in barn stabling than in a yard of looseboxes. A design which gives good ventilation, large partitions and high roof, reduces the risk. Barn stabling with low ceilings, a warm damp atmosphere and insufficient ventilation is more vulnerable to the spread of disease.

- **Ventilation** In all weathers and at all times there must be a sufficient supply of fresh air. The requirement for internal stabling is 56 cubic metres (2000 cubic feet) of air space per horse.

  A high roof gives ample air space and ventilation per horse and keeps the stables cool in hot weather. If a ceiling is installed, then the diminished air space necessitates the use of some form of mechanical ventilation. Double doors give a through-draught. Windows can be a source of fresh air, but these should be set high in the outside wall of each box and should open inwards, hinged at the bottom. If the building is roofed with Onduline corrugated sheets, these can contain continuous roof vents and ridge vents where the roof meets the supporting wall, and louvre boards can be installed in the outer walls to assist the extraction of used air.

- **Cough allergies** These can be a problem if some horses are bedded on straw, and other susceptible horses are on shavings or paper in the same building. It is impossible to create a spore- and dust-free atmosphere under these conditions, and the susceptible horses are bound to be affected. Alternative outside accommodation must be available for these horses.

- **Boredom** This is rarely a problem as horses can see each other and watch work going on, but, if finance allows, windows or doors 90cm (3ft) square can be put in the back wall at a height of 1.5m (5ft), which when opened allow the horses to look out. They may

be constructed of wood or unbreakable clear plastic and covered with a grid. When closed they must be draught-proof.

Conversion of an Existing Building

Most farm buildings can be converted into looseboxes or stalls, provided that the building is sound and the roof is of sufficient height. Low roofs are difficult and expensive to alter and roof height is important, both for ventilation and clearance for the horse's head. Few buildings provide sufficient space to incorporate feed and tack rooms. The ventilation may be more difficult to organise, as farm roofs are often tiled or slated. The other advantages and disadvantages are similar to American barn stabling.

Stalls

Stalls are individually partitioned stables in which the horse is permanently tied up, traditionally facing a blank wall. Water, feed and hay are placed in front of the horse.

Some horses do feel restricted and will not settle in stalls, but the majority are reasonably content. Although their movement is very limited, they can still lie down and get up without trouble.

The major advantage of stalls is that they allow more horses to be housed in a smaller space. They are warm, practical, save labour and bedding and are the cheapest type of stable to build. They are specially suitable as day standings for horses and ponies brought in from the field and can be very useful in riding schools.

The conventional design is one line of stalls positioned along the outside wall, with a passageway behind leading to either one or two outside doors. Given sufficient space, two parallel lines of stalls can be built and separated by a 3.7m (12ft) passageway. The more modern design for stalls is two parallel lines facing each other across a feeding passageway. Horses are tied up to the front of the stall, but can see each other through the rails. They are watered and fed from the central passageway. Two wider passageways running behind the horses give room for mucking out and bedding down.

SIZE OF STABLES

Considerations
- Cost – the larger the box, the higher the cost for ground, structure and bedding.

- Comfort of the horse – needs room to lie down.

- Ease of management.

- Larger stables can be used for smaller animals, but the reverse is not possible.

- Within limits, the larger the box, the more comfortable for both horse and groom.

- Looseboxes are preferably square, but if rectangular the longer distance should be from door to back wall.

Dimensions
- 3.7m x 3.1m (12ft x 10ft) is adequate for most horses of 16hh and under.
- 3.7m x 3.7m (12ft x 12ft) or 4.2m x 3.7m (14ft x 12ft) is a comfortable and suitable size for a horse over 16hh.
- 3.1m x 3.1m (10ft x 10ft) is adequate if under 15hh.
- 2.4m x 2.4m (8ft x 8ft) is suitable for small ponies.
- 4.6m x 4.6m (15ft x 15ft) is needed for foaling boxes.
- 1.8m x 3.3m (6ft x 11ft) for stalls for animals under 16hh.
- 1.3m x 2.4m (4½ft x 8ft) for stalls if under 14.2hh.

Roof Height
The roof must be high enough to provide sufficient air space, and to ensure that there is no danger of a horse hitting his head.

Horses in outside looseboxes require 42 cubic metres (1600 cu ft), in barn stabling 56 cubic metres (2000 cu ft) and in stalls 28 cubic metres (1000 cu ft) each.

MATERIALS

Materials used in construction must be strong enough to withstand kicking or leaning against walls or attempting to chew accessible wood. Although ponies can be housed in a more lightly made building, it rarely lasts as long, and can be a false economy.

Types and Their Advantages

Bricks
- Pleasing appearance.
- Warmth.
- Longevity.

Using bricks is very expensive because, for a strong and weather-proof surface, the walls must be cavity built, i.e. double walls with a cavity between.

Concrete Solid Blocks
- Acceptable appearance.
- Hard-wearing and long-lasting.
- Minimal maintenance.
- Comparatively quick and easy to build.
- Cheaper than bricks.

Concrete Cavity Blocks
The advantages of these are the same as those for concrete solid blocks with the additional one of good insulation without extra expense. Although not as strong as solid concrete blocks, cavity blocks are of an acceptable strength for most stables. They can be reinforced with metal strips if they are likely to be subjected to unusual stress.

Concrete Breeze Blocks
These are of an acceptable appearance, are fairly hard-wearing and long-lasting, comparatively quick and easy to build and relatively cheap. These blocks are porous and only suitable for inner linings. For exterior use they need weather-proofing. They are not as strong as solid concrete or cavity blocks, and can sometimes fracture under stress.

Wooden Prefabricated Sectional Stables

- Pleasing appearance if well maintained.
- Readily available.
- Easily erected.
- Can be moved.
- Warm.

There is, however, an added fire risk. Also, wooden, prefabricated sections are more difficult to disinfect, the upkeep costs are higher and they have a comparatively short life compared to bricks or blocks.

Cedar wood is the most expensive material, but lasts longer. A foundation of two rows of bricks, to which the structure is bolted, is required. All wood should be pressure-treated with a wood preservative. Boxes should be lined with suitable hard wood or chipboard.

Whilst good-quality wooden boxes last thirty years and more, ones of inferior quality show wear very quickly. They can be damaged by horses kicking or chewing the wood, and the resulting cracked or splintered boards can be dangerous. Thus when selecting wooden stabling, it is advisable to buy the best that can be afforded, and even then to check on the quality of the timber used and the method of erection.

Concrete Prefabricated Sectional Stables

In recent years there has been some use of this material.

Composite Material Prefabricated Sectional Stables

Although not yet used as much as wood, composite materials may well grow in popularity as their design and structure become better known and proven. They are likely to be cheaper than wood.

Solid Wood

A log-cabin type construction is attractive to look at, strong and durable, but very expensive.

INTERNAL WALLS

Barns

The external structure of this type of stabling is a steel- or concrete-framed barn with a clear span of Onduline

sheets. External walls can be made of Onduline, plastic sheeting or concrete blocks. The internal partitions making up the boxes or pens can be constructed of brick, blocks, wood or prefabricated sectional walls. The last are purpose-built, heavy wooden partitions topped with grilles or heavy galvanised mesh. Mesh is safer as there is no chance of a horse catching his foot when rolling and kicking. These partitions are usually 2.3m (7½ft) in height. Solid partitions between boxes can reduce the risk of infection and contagion. They will reduce draughts. The choice of internal partitions is a question of personal preference and cost. The range is from basic but sufficient, to very elaborate designs.

Separate Boxes

These are usually of wood and may go up to the roof or to the eaves.

LIGHT

Corrugated perspex roof panels can be placed in the roof above each loosebox. They will give ample daylight.

In hot sun the panels on the side of the roof exposed to the sun will increase the temperature in the stable. Those on a north-facing roof cause no problem.

More light can be provided with translucent panels in the outside walls, under the eaves, provided that the walls are high enough for the horses not to reach them.

DRAINAGE SYSTEMS

To install a drainage system, planning permission is required, and must accord with local bye-laws. Professional advice should be taken. The main systems are as follows:

Main Drainage

If this is available, it can be used only for lavatory and tack-room drains. Other arrangements will have to be made for stable drains, roof and surface water.

Septic Tank

This consists of a system of tanks and can be used for the whole of the establishment's drainage. Bacteria present in the first tank work on the material entering it. The liquid then moves on and is eventually piped out under the surface, by which time it is clean.

On occasion it may be necessary to empty this system, by arrangement with a local firm specialising in drain cleaning. Some systems require emptying every six months, while others will function efficiently for many years.

Cess Pit

This is a system of one or more tanks which have to be emptied when they are full. A regular contract should be arranged for this.

Soak-away

If already installed, this system will probably be maintained as long as neighbours do not complain. Planning permission is unlikely to be granted for a new installation. The principle is a tank system with bacteria that work on the solids. The liquid is allowed to drain away across the field by means of a ditch running downhill away from the building.

This system works well in a chalk or gravel area, but not in a clay area. Lavatory and tack-room drains must enter the drainage system via a system of sealed pipes.

Open Drains

Looseboxes and barn-type stables can be drained either to the front or to the back of the box. Drainage to the back results in less air contamination. This is particularly important in barns. Drains to the back of buildings may be more easily forgotten, whereas front drains are more visible and attract more regular maintenance.

Covered Drains

If possible, these should be avoided, as they are not practical, becoming easily obstructed with manure and straw. If already installed, they should have accessible drain traps, which should be fitted with a removable wire tray to catch any solid material. Alternatively, small-gauge wire netting can be used for small traps, but must be cleaned daily. It is essential that any lengths of covered drain can be easily cleaned with rods. If that is not possible, it should be easy to take up the top covering, i.e. in old-fashioned stalls. A covered drain often runs down the length of the building and must be cleaned out each day.

Traps and Gullies

The principle on which gully traps are designed is for water to separate the outside air from the sewer or main drain air, and so prevent the return of the latter. Stable yard and surface drains should have efficient traps between the drainage system and the stable yard, so that solids can be collected and removed. There must be a sufficient body of water between the main drain and the air to prevent sewer gas from emerging. Drains should be well ventilated on the sewer side with a long pipe going well above the building. This should act as a double insurance against unpleasant smells.

DRAINS

In Looseboxes and Barns

These should have a shallow slope of 1 in 60, towards either the front or the back of the box. A small opening at floor level in the centre of the wall can direct any liquid into a shallow open drain running along the outside of the stable wall. Traps for collecting straw and solids should be positioned so that no clogging material can reach the main drainage system. The removable trays in the traps should be regularly cleaned. To avoid any draughts from the drain opening, straw should be bedded up the wall.

The **advantages** of drains at the back are:

- No smell of urine, so more opportunity for the horse to breathe fresh, untainted air.

- Drainage is out of sight but still easily checked.

- There is less chance of draughts, as bedding is always banked up.

The **disadvantages** are:

• Difficult to arrange if looseboxes are on a boundary line or attached to other buildings.

• If not easily accessible they are likely to be neglected.

• It is less easy to wash the box down and sweep out water because of smaller opening and no doorway.

Barn-type Stabling

There should be a shallow open drain, preferably along the outside wall to avoid contamination, but if this is not possible, along the inner walls of both lines of boxes. This should end in a gulley to trap straw and manure.

All drains should be washed clean daily. In a very long building, drains should be of the hog-back type, sloping from the centre towards both ends.

Stalls

Stalls should have a shallow open drain 60cm (2ft) behind the back post of the stall.

Deep Litter

Horses bedded on deep litter or any form of solid bed should have any covered drains sealed off. Drainage is not required if this type of bedding is used, but it is convenient to have an outside shallow drain to take away the liquid after washing the floor and walls.

Yard

This should have a gentle slope (according to the natural lie of the land) towards one or more mud gullies or traps which allow surface water to join the drainage system. All drain covers should be of steel or reinforced concrete, capable where necessary of taking the weight of a heavy lorry.

GUTTERING

All stable buildings must have efficient guttering which is well maintained and regularly cleaned. The

downpipes should be of a sufficient number; and shaped to shoot the water away from the base of the building, to open or covered drains. Broken and leaking guttering gives a bad impression of a stable yard, makes buildings damp and shortens the life of any wooden structure.

FLOORS

For the comfort of the horse, and to save strain on the legs, stable floors should be as level as possible. To assist drainage, a slope of 1 in 60 is suitable, except for stalls, which should be slightly steeper. The necessary floor space is 11–14sq m (118–150sq ft).

Materials

CONCRETE is the most widely used surface and is very satisfactory. A minimum depth of 15cm (6ins) is placed on top of a well-drained prepared base of hard core and shingle. A damp course must be included. It is easily put down, but it must be of an extra-hard mixture to withstand urine and the horse's weight, and 'roughed' to provide a non-slip surface.

CHALK AND CLAY FLOORS are non-slippery and quiet. They are cheap to put down if material is available locally, but they require more maintenance, having to be topped and rammed every year. A clay floor tends to be damp.

COMPOSITION FLOORING is very satisfactory, but very expensive.

RUBBER MATTING on concrete can be used and saves time mucking out. The comfort of the horse is not considered. When the horse is out of the box, manure is swept off and the floor is washed clean. No bedding is used. The smell of ammonia is unacceptable and may be damaging to the horse. Rugs become stained and smelly when the horse attempts to lie down.

POROUS FLOORS need a well-drained base. The surface is dug out and filled with rounded pebbles. There can be a smell problem.

WALLS

Height

RIDGE ROOFS The walls of most prefabricated looseboxes range in height from 2.1–2.3m (7– 7½ft) at the eaves to 3.3m (11ft) at the ridge. This is adequate for a box with a ridge roof and which slopes up from both front and back walls to a ridge.

When building with brick or concrete blocks, it is advisable for the walls to be 2.4–2.7m (8–9ft). This will allow for a 2.4m (8ft) doorway.

PITCH ROOFS A roof sloping from the back down to the front is known as a lean-to or pitch roof, and stables with such roofs require a minimum wall height of 2.7m (9ft) at the front, and 3.3m (11ft) at the rear. In areas likely to have heavy snow, a steeper pitch is recommended. The roof can also be sloped from the front of the box to the rear, when the height of the back wall should be a minimum of 2.7m (9ft).

FLAT ROOFS Stables with a flat roof or a ceiling with store rooms or lofts above require walls at a minimum height of 3m (10ft), but the measurement is dependent on the ground area of the box. See Ventilation, pages 335 and 342.

Damp Course

All walls should have a damp course; tarred building paper or a strip of plastic paper is effective. This is laid on the foundations, at the base of prefabricated stabling, or a little above ground level for bricks or concrete blocks.

Insulation

The insulation of stable walls is not essential, but adds greatly to the warmth and comfort of the horse. It will also increase the ventilation rate. It is recommended that all prefabricated stabling should be fully lined, to increase the stability, strength and warmth. Effective wall insulation is then achieved by placing a sheet of foil-lined building paper between the outer cladding and the inner lining.

Cavity-wall blocks provide their own wall insulation. Brick buildings constructed with cavity walls, i.e. two layers thick, have full insulation.

A single layer of brick needs lining with wood, and reinforcing with supports on the outside every 1.2m (4ft).

Surfacing of Walls

WOOD STABLING. To give sufficient strength and added warmth, all walls should be lined. This is essential up to 1.5m (5ft) high because of the risk of kicking, and it is recommended that it is continued up to the eaves. Material should be exterior plywood, or a wood of similar strength.

Both exterior and interior surfaces of the wood should be treated with wood preservative. Regular painting of the outside with creosote keeps it in good order. If interior woodwork is painted with colourless preservative, the top half of the box can then be painted with white emulsion. For a shavings or sawdust deep litter, the bottom 60cm (2ft) of the wall, if of galvanised metal, should be painted with black bitumastic paint; if of wood, it should be treated with tar.

PRE-CAST CONCRETE OR PREFABRICATED COMPOSITION Such stabling requires lining with wood or chip board. The appearance of the exterior surface is improved by a coat of Snowcem or similar preparation and the wood treated as above.

SOLID CONCRETE BLOCKS AND CAVITY BLOCKS Exterior walls require weather-proofing with a proprietary plastic preparation.

BRICK Cavity walls of brick-built buildings are weatherproof.

The interior walls of both types of blocks and also brick walls are rough. If a horse gets cast, it may damage itself on the rough surface. The walls may be plastered and painted. It should be remembered that lead-based paints are poisonous and should not be used. Plastic paints are now available which can provide a smooth, washable surface.

ANTI-CAST RIDGE This is a ridge standing out approximately 25mm (1in.) from the wall surface and 1.5m (5ft) above the floor. For ponies, the height should be 1.1m (3½ft). The ridge enables a horse to right itself, should it get cast, by stopping the feet slipping up the wall. Wooden stabling can include a reinforced strip of wood fixed firmly to the wall with screws. Block or brick buildings can incorporate a row of blocks or bricks set out from the wall, or a metal strip set into the cement between the blocks and bricks. Whatever is used, it should be shaped so that a crib biter cannot seize it with his teeth. Alternatively, a groove can be made in the wall surface to give a foothold to a cast horse.

ROOFS

Roofs should be:

- Weatherproof.
- Durable.
- Noiseless.
- Non-flammable.
- Warm in winter, cool in summer.

Types

ONDULINE CORRUGATED SHEETING fulfils all the above requirements and is inexpensive. In winter it may need lining to prevent condensation, and for this, roof felting is suitable. Maintenance is nil, unless cracked by a falling bough or heavy blow, e.g. from the horse's head. Perspex roof panels can be placed on the north side to provide more light.

RUBBEROID, GLASS-FIBRE OR PLASTIC TILES AND SLATES have a pleasing appearance, are long-lasting, light in weight, suitable for wooden stabling, and inexpensive compared to conventional tiles and slates.

ROOFING FELT is the cheapest material available, but it does not last, and should only be considered if it is of high quality. Wooden strips should be fitted at frequent intervals to prevent it being torn by strong winds.

TILES AND SLATES are satisfactory, but expensive and heavy. They are not suitable for wooden buildings.

THATCH is not recommended because of the fire risk.

GALVANISED SHEETING is hot in summer and noisy in wet weather. It is satisfactory for day stalls if correctly ventilated. Although cheap to install, maintenance can make it as expensive as other roofing materials, as it requires regular tarring or painting to control rust.

RIGID PLASTIC ROOFING SHEETS are light to handle, reasonably long-lasting, require little maintenance, and compare favourably in price. They are noisy in rain and high winds. They become brittle in time and are then likely to split.

Overhang

This is a short extension of the roof over the door and front wall area of outside looseboxes. It gives protection to both horse and groom, and is a great boon with no drawbacks.

With prefabricated stabling, the overhang is usually an extension of the roof and built of the same material. With brick or block stabling, its material depends on construction. It can be built as a continuation of the roof, or else as a separate structure. Translucent plastic sheets can be used, but these can raise the temperature in the box in hot, sunny weather.

CEILINGS

These can be installed:

- If storage space or living accommodation is required above stables.

- In a high building, to decrease air space and make stables less draughty.

The material used should be fire-resistant and, to avoid condensation, be capable of absorbing moisture. For living accommodation strict local authority building regulations regarding health, safety and fire will need to be adhered to. Professional advice is essential.

VENTILATION

To keep a stabled horse happy requires a constant supply of fresh air, but without draughts. Except in very cold weather, the temperature in a stable should not be warmer than outside. Horses keep fit and healthy if well rugged. It is preferable to keep a horse warm with extra food, blankets and stable bandages rather than by restricting his supply of fresh air. At the same time, he must not be subjected to a draught, as this can easily give him a chill. Ventilation is of prime importance when stables are being built or converted. Any deficiencies will adversely affect the health of the horses. Some prefabricated boxes and purpose-built modern barns are not designed with sufficient ventilation. As long as draughts are avoided, the supply of fresh air should never be restricted.

Ideally, horses should have a permanent opening of 0.3sq m (3sq ft) (inlets) per horse in the walls and 0.1 sq m (1.6sq ft) in the roof (outlets). The acceptable air space per horse is 45.3 cu m (1600 cu ft), and this must be borne in mind when deciding on area and height. Means of providing ventilation are as follows.

Open Top Doors

These are the main source of fresh air for individual outside looseboxes. Except in exceptional circumstances, e.g. driving snow or a very cold direct wind, they should be kept open. The lower door should be 1.4m (4½ft) to 1.5m (5ft) high, to give sufficient protection and discourage a horse from jumping over it.

If stables are in lines, then they are best positioned facing south. If the stables have to be built facing north or east, it may be necessary to shut top doors at times in winter. If so, other methods of ventilation must be used, such as opening the windows or fixing a short metal hook externally to keep the door just ajar. The hook must be short enough for the horse to be unable to insert his head between the door and the door post.

Windows

Windows are necessary for light and ventilation. If two are installed, they should be placed in the front and rear walls. In very windy weather, if the top door has to be shut, the window at the rear can be left open. In very hot weather, a back window allows a through-draught of air. If only one window is to be installed, it is probably better to have it at the rear.

Windows should be placed as high as possible in the wall, hinged at the bottom or centre according to size, and preferably opening inwards. Cold air is then directed up and above the horse's body, before descending and mixing with the warmer air below. Windows should be fitted with unbreakable clear perspex and protected with a grid which opens with the window. Bars are not suitable as these can be dangerous, easily trapping a horse's foot should he kick up or rear. Glass, if already installed, should be replaced with perspex.

Windows fixed at too low a level and opening inwards can be a hazard for the horse. If opening outwards, they must, when fully open, lie flush with the wall. They must be protected by a metal grid so that the horse cannot put his head through.

Louvre Boards

These are broad, overlapping boards, set at an angle to prevent the entry of rain or snow, and with an air passage between. They should be fixed in outside walls at least 2m (6½ft) from the ground. They extract stale air, and if correctly fitted, will assist in preventing condensation on the roof and walls.

Eave Vents and Ridge Vents

These are found in Onduline-sheet roofs, at the eaves where the sheets meet the walls, and at the ridge of the roof where the sheets are capped. They are efficient ventilators. They should also be installed in other types of roofs.

Ventilating Cowls

These are placed on the ridge of the roof. They are manufactured in several designs, and can be used for most roofing types. They allow exit of stale, warm air and ingress of fresh air. It is usual to install one to a box or one to two boxes.

Air Bricks

Air bricks can be put just under the eaves in brick or concrete block stabling. They should not be placed lower down because of draughts.

Power-driven Air Extractors

These can be fitted on the roof and should be turned on when the air is very still and humid.

When buying prefabricated stabling, it is advisable to check on ventilation. Some firms make better provision for it than others.

In **summary,** it is reasonably easy to provide sufficient ventilation in individual boxes but much more difficult in barn stables.

DOORS

There are two types: conventional hinged doors and sliding doors.

Size of Doorway

The ideal size is 2.4m (8ft) high and a minimum of 1.2m (4ft) wide, although for mares, foals and high-class young stock, doors should be 1.4m (4½ft) wide. As an extra precaution against horses knocking their hips, rollers can be fitted on the inside of both door posts. The distance between the posts should then be 1.4–1.5m (4½ to 5ft). Many prefabricated stables have doors 2.1– 2.3m (7–7½ft) high; 2.3m (7½ft) is acceptable, but 2.1m (7ft) is only suitable for smaller horses.

Conventional Hinged Doors

This is the usual type of door for outside loose boxes. They are divided into two, the bottom door being 1.4–1.5m (4½ to 5ft) in height.

FITTING They should be positioned to the side of the box to reduce direct draughts. They should not adjoin the box next door, as this encourages biting and

bullying. They should open outwards so that should a horse get cast against the door it is possible to open it. Outward opening doors also reduce disturbance of the bed. The bottom door should fit firmly down to the floor, so that there are no draughts, and no danger of a horse getting his foot caught under the bottom of the door.

The top edge of the bottom door should be protected from chewing by the horse with a firmly fixed heavy metal strip, which is 8–10cm (3–4ins) deep on both sides. If there is no overhang, then some protection is needed from the weather for the tops of both doors.

Sliding Doors

Sliding doors suspended from an overhead gearing are usually made the same height as the partitions of the boxes or pens of interior stabling, and consist of solid wood to a height of 1.5m (5ft) with grilles above. A sliding door can also be used for outside looseboxes. The horse can look out, but cannot put his head out. A light top door is sometimes fitted to seal the top half in case of need.

Sliding doors cause less obstruction than hinged doors, and are a great asset for interior stabling where horses and people are moving and standing in the central passageway.

FITTING Sliding doors are heavier than a hinged door, and require substantial posts, usually made of metal. They hang from an overhead gearing, and the door is guided at the bottom either in a groove or on a rail. As a groove easily becomes clogged with dirt, a round rail on which the door glides along a groove in the bottom of its lower edge is more efficient.

Doors at the ends of barn stables, because of their size (approximately 3m/10ft wide and 3m/10ft high) must be very heavy, so sliding doors are easier to handle than hinged doors.

Construction

Stable doors receive constant and often rough usage. To withstand this it is essential that they are constructed of strong material. They should be lined with exterior plywood, galvanised sheet metal, or a similar hard, smooth surface. This should also cover the 'stays' on the door, which easily damage a horse's leg or knee if left exposed. Inferior wood, particularly if left unlined, is unsuitable and will not last.

Hinges and **bolts** should be of heavy-duty galvanised iron. Door hinges should be capped to prevent the door being lifted off, and they must be long enough to support the weight of the door.

The bottom door requires a top bolt, and a lower fastening which can be of the kick-over type. Bolts should be of a simple design, and free from anything which could injure the horse or catch on a headcollar.

The ideal door fastenings are those built into the frame work, as the horse cannot get caught on them or touch them with his teeth. There are also special horse-proof bolts on the market, which are excellent but expensive. Some horses are able to open doors, so if the bolt is not of the horse-proof type, it should be able to take a spring hook for extra security. The top door requires one bolt. Both doors should have the necessary catches to fasten them back, flush against the stable wall securely. Because of the danger of fire, stable doors should never be locked.

Window latches must be out of reach of the horse and, if at a lower level, must be protected so that the horse cannot play with them.

STABLE FITTINGS

For safety reasons, stable fittings should be minimal, and in many yards, fixed feed and hay mangers, hay racks and automatic watering systems are not installed.

Tying Up

For tying up, the ring should be placed to the side of the back wall at a height of 1.5m (5ft) (lower for ponies) and adjacent to the haynet ring, which should be nearer the corner at a minimum height of 1.8m (6ft). A short piece of string may be looped through the tying-up ring, so the horse is tied to the string and not directly to the ring. The loop should be small and neat to discourage

the horse from chewing it. A second tie-ring in the front by the manger can be useful.

A tie-ring should also be fixed outside the box so that the horse can be secured during mucking out and grooming. Care should be taken that adjoining horses are not tied up at the same time, as they may well kick at each other.

Rings should be of heavy galvanised steel. In brick or block stabling, they should be bolted through the wall with a holding plate on both surfaces. In wooden stables, the ring should be on a metal plate secured by four heavy screws and fastened through on to the vertical cladding of the box.

Watering

There are a number of ways in which water can be supplied, and these are covered in detail in Chapter 17.

Feed Troughs or Mangers

There are a number of fittings/containers suitable for short feeds:

FIXED MANGERS

These should be solid, with a smooth surface and must be easy to clean. The lip should be broad enough to prevent a horse seizing it with his teeth in case he starts crib biting. An overhanging inner lip prevents food being spilt, but it is difficult to clean. Deep mangers conserve food but encourage a horse to take too large mouthfuls. Shallow mangers are easy to clean, but a horse can easily waste food by pushing it over the edge.

Mangers can be built at ground level. These are safe and create the natural angle for a horse to eat, but they are likely to get soiled with dung and need constant washing.

Mangers are usually positioned at a height of approximately 1m (3–3½ft). To prevent a horse catching his foot underneath or knocking his head as he gets up, they can be boxed in, down to and flush with the floor. Alternatively, they can be fixed across a corner with a sloped but solid boxing-in by boards. It is essential that there are no projecting edges or angles on which the horse could damage himself.

Mangers should be fixed on the same wall as the door to reduce the risk of anyone being kicked as they leave the box after giving the feed.

In stalls, fixed mangers for food and hay are usually installed at breast height.

REMOVABLE MANGERS

These can be made of metal or plastic with hooks which fit either over a door or into metal fixtures. They can be fitted on the door or on to the walls of the front corner and can be fixed on to any type of wall. These removable mangers are easy to wash and cannot be knocked over. The problems are that if fixed on the door they are high, and at an unnatural angle for a horse to eat. Also, they can be lifted out and thrown round the box.

NOTE: Metal fixtures are available which run across a corner to hold the manger. These can be dangerous and are not recommended.

FEED BOWLS

- **Metal feed bowls** are heavy, less likely to be knocked over, but can bruise a horse's leg should he strike or paw at them. They are easily cleaned, but are heavy to carry when taking round feed.

- **Plastic feed bowls** are light and easy to clean and handle. They can get knocked over and thrown around the box, or stamped on by a playful horse. They are easily broken. Lightweight, shallow plastic bowls are comparatively cheap, but do not last, and should be removed when the horse has finished feeding. The heavier, deeper rubber feed bowls without handles are more expensive, but more worthwhile. Feed bowls with handles should **not** be used as a horse can trap a foot in the handle. This is particularly applicable to young or playful horses.

- Feed bowls fixed in a **rubber tyre** are safe, but should be removed from the box after the horse has finished eating. They are unlikely to be knocked over and cannot damage a horse's leg, but they are heavy to move about.

Feeding Hay

Hay can be fed in the following ways:

LOOSE HAY is a safe and natural method of feeding hay and is labour-saving. The disadvantages are that it can be wasteful if the horse is an untidy feeder, and does not clean up. Also, the yard is more likely to be untidy, and this method necessitates tying up the hay for weighing, which is difficult if the hay has to be soaked.

HAYNETS are the most popular way of feeding hay as they are economical, tidy to handle, easy to weigh and easy to damp or soak. The disadvantages are that they are time-consuming to fill, tie up and take down and they can be a danger if not tied up correctly (some horses play with the cord, chew it and get caught up); and they are an expense to buy.

FIXED HAY MANGERS are built up from the floor to a height of about 1m (3½ft). They must be boxed in, and wider at the top than at the bottom. Although practical, they need regular cleaning, which is not easy, and it is more difficult to weigh loose hay.

FIXED HAY RACKS are made of galvanised steel rods or a lighter weight steel mesh. They can be fixed above head level, but dust and seeds tend to fall into the horse's eyes and eating is at an unnatural angle. These problems can be prevented if the rack is fixed breast high, but a horse can then catch his foot or knock his head or eye in the steel rods or mesh.

OPTIONAL STABLE FITTINGS

Grids and Grilles

These are fixed to the top door of looseboxes. They may be a permanent fixture or removable. They are made of steel rods or heavy mesh, and are made in several patterns:

- Covering the entire area of the top door, which will prevent horses biting at people or at other horses passing by, help to prevent banging of the door with the front feet, discourage weaving and prevent crib biting on the door. If there is no overhang, it allows horses to see out and yet keep dry.

- A V-shaped grid allows the horse to put his head and neck over the door, but prevents him weaving, unless he is the type who stands back in his box and weaves. It also helps to prevent crib biting. Care should be taken that the opening is large enough for the horse to draw back without catching his head.

Salt-lick Holders

These should be positioned in a corner. The danger is that a horse can knock his head or eye on one. Salt blocks can be used as a safer alternative, and placed in a corner where the horse is fed. Salt causes a wet or damp patch, which eventually rots through an unprotected metal or wood wall surface.

ADDITIONAL UNITS

Utility Box

In many yards one or more boxes are set aside as utility boxes. The number depends on the size of the yard. The box may be used for clipping, shoeing, washing or veterinary work.

REQUIREMENTS

- High ceiling, minimum of 3.6m (12ft), preferably 4.3m (14ft), to avoid the risk of the horse hitting his head if resisting when being clipped.
- Non-slip, well-drained floor. Rubber flooring is excellent and gives added protection when using electrical equipment.
- Daylight from well-protected windows.
- Efficient electric light.
- Two power plugs.
- Water available near by.
- Tie-rings on each wall.
- Haynet ring on one wall next to tie-ring.
- There should be no fixed mangers or other fittings.

If heavy clipping machinery is to be used, an overhead trackway is necessary at a height of 3.2m (10½ft) and running along the back and side walls. It is then out of the way, should a horse rear. A corrugated perspex roof panel can be installed to give added light. An infra-red light can be fixed and used to dry a horse after washing.

Isolation Box

To be effective, this box must be at least 100m (100 yds) away from the main stabling. It should face away from the stables, with the prevailing wind blowing away from the stables.

REQUIREMENTS

- Size: 4.3m x 4.3m (14ft x 14ft).
- Average height.
- Insulated walls and roof.
- Well-fitting, draught-proof windows.
- Well-fitting, draught-proof doors.
- Electrics: 13-amp plug and fitting for heat lamp.
- Adjacent water supply.
- Adjacent weather-proof shed to store isolated horse's feed, mucking-out tools, grooming equipment, buckets, rugs, medical equipment, etc.

From a practical point of view, it is doubtful if an isolation box is really either worthwhile or effective against infectious diseases, but a box that is quite a way from others is useful for ill horses. For this purpose it needs to be easily accessible from the house. A contagious condition (e.g. ringworm) can be successfully contained by a strict isolation procedure.

Forge

Stables with a resident farrier require a forge. This should be sited away from the stables, hay and straw barn, because of fire risk. It must be on a hard surface and convenient to other buildings for supplies of electricity and water. In a large yard it is usually adjacent to the carpenter's shop (shed). Most yards now use travelling farriers, with a specially equipped van containing a portable forge.

Useful Additions

A retractable hook outside each box is useful for hanging a headcollar or bridle. It should not be possible for the horse to reach it. A folding saddle bracket can also be an asset.

For looseboxes bedded with shavings, sawdust or paper a 90–1.2m (3–4ft) length of light chain fastened to the outside wall of the loosebox makes a convenient place on which to hang rugs and blankets.

Custom-made rug rails or folding brackets are useful.

ELECTRICITY

- A 'trip-switch' system is essential so that if there is any fault in wiring, fittings or machinery the electric current will immediately be cut off, and cannot be restored until the fault is located and repaired.

- All electric wiring and fittings should be installed by a qualified electrician, who should also attend to all servicing and repairs.

- It should be remembered that horses are much more susceptible to electricity than are humans. Quite a low voltage can kill them.

- All cabling should run through metal pipes and be fixed so that it cannot be chewed or pulled by a horse. Wherever possible, it should run outside the boxes and enter the stables from high in the roof.

- Switches should be of a water-tight design and positioned outside the box, where they are out of reach of the horse.

Lights

These can be fluorescent or bulb fixtures. Fluorescent lighting must be fixed high in the roof, out of reach of the horse. In adjoining stables, if walls do not reach the roof, one fixture gives good light to two boxes.

Fluorescent light is more expensive to buy and install than bulb fixtures, but is cheaper to use, lasts longer, and gives a better light. In very cold weather it may not be so efficient.

Bulb fixtures can also be placed high in the roof and then need not be protected, but this height makes replacing and also cleaning difficult; encrusted dust affects their efficiency. If placed in storage areas, they can be a fire risk if they should come into contact with hay or straw.

A much safer but considerably more expensive fitting is the bulkhead type. If these are used, they can be placed at a more accessible height – 2.1–2.4m (7 to 8ft) – which makes cleaning and replacing bulbs easier.

Low-wattage bulbs are long-lasting and economical to use. Light is not instant but progressively strengthens when switched on. This can be helpful in avoiding startling the horses.

Power Points

One 13-amp power point should be installed and available for approximately every six boxes. They must be out of reach or protected from the horse. They are used for clipping, grooming and heat lamps. They must never be used without a plug-in circuit breaker.

The utility and isolation boxes will need good lighting and an extra socket for the clipping machine, heat lamp, etc.

Yard Lights

Efficient yard lights are essential if work is to be carried out after dark. Good lighting will not only make work areas safer, but will save time and be cost-effective. High-intensity tungsten-halogen lamps or high-pressure sodium lamps give excellent light over a wide area. They are cheap to run, although expensive to install. They are very much more efficient than the traditional type of bulb fixture, and require less maintenance. The latter give limited light and are expensive to use.

Yard lights left on at night can be a deterrent to thieves. In this case low-wattage bulbs should be used for economy.

MAINTENANCE

Most manufacturers of prefabricated wooden stabling treat the wood used with a wood preservative, to make it weather-proof, and to protect it from attack by wood beetle and fungus. The woodwork will require regular treatment to ensure that it stays weather-proof and also to improve appearance. Creosote is a cheap and efficient preservative for other woodwork, and also acts as a good disinfectant and fungus killer.

Wooden doors and windows in brick or concrete block stables can either be treated in the same way, or painted. This is a matter of personal choice – painting requires more maintenance and is more expensive. The wood requires a primer, an undercoat and a top coat. Lead-based paint is poisonous and should not be used. Maintenance is important – cracked or peeling paint is a bad advertisement for any yard, commercial or private.

Metal doors and windows are now likely to be of a rust-proof metal and need no attention. Older fixtures require regular painting, with an anti-rust undercoat and a suitable top coat.

All door, window and stable fittings should be of galvanised iron or steel. These will not rust.

Efficient maintenance of buildings, equipment, facilities, fencing, yard and drive are an essential part of successful business management. This is why it is always better to buy the best that can be afforded. This does not mean that money should be wasted on unessential buildings or equipment, or luxurious extras which are not necessary. The successful manager plans ahead, and knows what is essential and what is not.

Repairs to buildings and fencing, if carried out in good time, save further deterioration. Allowing surroundings to remain in bad repair invites accidents to animals and staff.

The impression gained on entering a stable yard is very important. It is possible to tell immediately whether the yard is efficient. Stable managers must be methodical and observant, to ensure that necessary repairs and regular maintenance work are carried out.

FEED SHED

REQUIREMENTS
- It is best constructed of brick or concrete blocks to discourage vermin.
- A dry, light, secure room, convenient to the stables.

A good door is essential, so that there is no danger of a loose horse getting into the feed shed.
- Large yards may have either a small feed shed for daily use, and a larger feed store next door for storing bulk supplies, or separate feed sheds for each area.
- The floor should be of concrete, level except in the door area, where it should have a slight slope and shallow drain to the side of the door.
- See also page 361, on feed storage.

THE TACK ROOM

Horse tack is a constant security risk. Insurance companies have individual rules about siting and structure and tack rooms. It is advisable to consult your own company before either building a new tack room or converting a building for use as a tack room.

Burglary Precautions
- Site to be in view of living quarters and adjacent to or part of the stable block.

- Structure to be of brick or concrete blocks with Onduline or tiled roof.

- Windows to be of glass, protected on the inside with steel bars.

- Doors to be of solid wood or steel, with a normal lock for use during the day and a Yale lock for security at night.

- Communicating doors should have bolts on the inside.

- Yard lights should be left on at night.

- Burglar alarms can be fitted.

- Dogs have been proved to be the best deterrent, either sleeping in the tack room or loose in the yard. If the latter, they should have a well-insulated kennel and the yard must be secure so that they cannot wander.

- Tack can now be marked with a registered number.

This makes identification easier, and discourages theft. In conjunction with the British Horse Society, one scheme is run by Farm Key of Banbury, Oxfordshire, who supply the marking tools and keep a national register.

- Insurance companies often require to see security arrangements, and may give advice on further precautions.

- It is worth noting that many burglaries take place during the day.

If the yard is to be left, tack rooms should be locked.

Layout of Tack Room

- Wash room for cleaning tack. Main tack room for putting tack up when clean.

- Combined wash and tack room.

- Two combined wash and tack rooms, one for school tack and one for livery.

Whichever system is chosen, the main requirements are:

- Floor: concrete is practical and should slope into one corner to drain. If covered with non-slip composition flooring, it is more comfortable for standing and working.

- Lighting: fluorescent strip lighting is the most efficient.

- Heating: there must be no direct heat to leather. All heating should be well above any working areas to avoid any risk of fire and injury. Electric wall heaters fixed high on the wall are suitable, and the fan-heater type is efficient and safe. The temperature suitable for working, and to keep tack in good condition, should be 13–16°C (55–60°F) in winter.

- A sink with an extended draining board for scrubbing. A deep sink makes cleaning easier.

- Hot and cold water: electric water-heaters of the geyser type are suitable. It can sometimes be arranged for room heating and water heating to be on the same system, and connected to other areas requiring heat and hot water.

- Saddle-horses for cleaning saddles should be sufficient in quantity for the number of staff. They can be 3–4.2m (10–14ft) long and must be strongly built.

- Telescopic bridle-cleaning hooks: these should be sited with care, as they can be a hazard if placed near a door or where others are working.

- Saddle racks: preferably of plastic or wood, as iron may mark the leather. Slip-in holders for horses' names make for efficiency.

- Bridle holders: preferably semi-circular in shape so that the top of the bridle is kept rounded. Slip-in holders for the horses' names are advisable.

- Hooks: for lungeing tack, spare girths, stirrup leathers, martingales, breast plates, etc. Name plates help to avoid disorder.

- Bit board for spare bits.

- Cupboard for tack-cleaning equipment.

- Cupboard with shelves or compartments for bandages, boots, shoeing tools, clipping machines, disinfectant, shampoo, hoof oil, Stockholm tar, etc. The top of the cupboard can be flat to provide a sound working surface.

- Medicine cupboard for first aid: thermometer, kaolin poultice, salt, drying-up lotion, Epsom salts, cotton wool, surgical bandages, paraffin gauze, wound powder, wound ointment, methylated spirit, etc.

- Drugs, poisons and worm doses must be kept in a locked cupboard in the office.

- Cupboard for grooming kit.

- Notice board for yard rules, daily ride sheet and work sheet.

- Rug and travelling boxes around walls.

- Several bar stools can make convenient and space-saving seats.

- Lockers for staff. Alternatively, these can be in the staff room.

- One or two roller towels.

- Battery clock.

- Dust bin.

- Fire appliance.

NB: The Health and Safety at Work Act (1974) should be checked and adhered to.

ADDITIONAL PREMISES

Wash Room
Large yards will benefit from an equipped wash room.

REQUIREMENTS
- Drying racks.
- Heavy duty, commercial washing machine and spin dryer.
- Heating to assist drying.
- Large sink with hot and cold water, and a draining board.

Store Room
If space is available, a store room is useful for spare rugs, travelling gear, spare tack, replacement goods for the tack room, feed shed and yard, etc.

Staff and Lecture Room
The need for separate staff and lecture rooms depends on the number of staff, and the size and the functions of the yard.

REQUIREMENTS
- Efficient heating.
- Floor covering, which is easily cleaned but provides some warmth.
- Sufficient upright chairs for students and working pupils.

- One or more tables.
- Small easy chairs.
- Electric kettle, small serving table, power point.
- Blackboard and chalk.
- Wall charts to cover points of the horse, skeleton, muscles, main organs of the body, bits, etc.
- Specimen shoes.
- Specimen bones of the horse's leg below the knee.
- Notice board announcing coming events, competitions, etc.
- Battery clock.
- Long mirror by the door for use of staff.
- Fire appliance.
- Kitchen and small electric stove and sink with hot and cold water (not essential in a small yard).
- Staff and working pupils' individual shelves or lockers.
- Hot drinks machine (not essential in a small yard).
- Video camera and power point (not essential in a small yard).

NB: The facilities must comply with the Health and Safety at Work Act (1974). The minimum requirements are stated in the Horse Riding Establishment's paper – 'Guidance on Promoting Safe Working Conditions' (HM Agricultural Inspectorate Health and Safety Executive 1988).

Changing Rooms and Lavatories
These must be sufficient for the size of the yard; and flush lavatories must be easily available for clients.

REQUIREMENTS
- Changing room.
- Efficient heating.
- Floor covering which is easily cleaned.
- Basin with hot and cold water.
- Shower room, but this is not necessary in a small yard.
- Towels, soap and paper towels.
- Hooks and coat hangers for clothes.
- Boot jacks.
- Mirror.

- Equipment for cleaning basin, lavatories and floors, stored safely in an appropriate cupboard.

NB: If areas requiring hot and cold water are close together, e.g. feed room, kitchen, changing rooms and lavatories, it is more economical as one water heater can be used.

Tack and Clothes Shop

Many yards run a tack and clothing shop. This can be limited to a few items, or cover a large selection of goods. It can be a useful service for clients, and a source of extra income to the school.

REQUIREMENTS
- A trading licence.

- A suitable room, which is dry and essentially burglar-proof.
- A member of staff with sufficient time to cope with customers.
- Cupboards, shelves and counter.
- Cash box.
- Saddle racks.
- Bridle and tack hooks.
- Boot jack.
- Mirror.
- Tape measure.

However, such a shop may alienate local saddlers, who also do the yard's tack repairs. Thus it might be advisable to arrange for them to supply the goods at a discount rate.

50 RIDING AREAS

INDOOR SCHOOLS

With the increase in the number of riders, the demand has grown for all-weather teaching areas and competition facilities, especially during the winter months. An indoor school has become a necessity for any progressive commercial establishment.

The construction will be governed by the amount of space and money available, and also the purposes for which the school is required.

Planning permission is necessary. If a large, competition school is planned, understandably, local residents often object to the inevitable increase in horse traffic. The tactful canvassing of local support before applying to the planning authorities can possibly avert the refusal of a planning application.

There are now many well-established and experienced firms who will advise on construction, and also deal with the planning authorities. Before a decision is made, it is advisable to investigate a variety of types of schools. Owners are often happy to show their schools and discuss design and structure.

Size
20m wide x 30m long
This area is suitable for teaching small children, adult beginners, for lungeing horse and rider and training horses on the lunge. The minimum width should be 20m, both from a functional point of view and so that, if at a later date a school of the size of a dressage arena

is required (20 x 40m, or 20 x 60m), extra bays can be added.

20m wide x 40m long
This size of school is suitable for general teaching, schooling of horses and small dressage and show-jumping competitions. It is the most convenient and economical size for practical work. It is also the minimum size acceptable to the British Horse Society if the school wishes to be recognised as an examination centre.

30m wide x 90m long
This size of school is suitable for all competition work, and also divides comfortably to accommodate two class lessons. The increase in width adds considerably to the expense, as the supports of a wider clear-span roof have to be so much stronger. There is also the inevitable increase in rateable value. A large school, fully enclosed with a seated gallery, has a much higher rateable value. A smaller school with semi-enclosed walls, small unfitted gallery and limited competition facilities should warrant a much lower rating.

Construction
Most modern schools are of clear-span, steel-framed construction with a ridge roof of Onduline sheets. Translucent sheets are placed at regular intervals in the roof to provide light. The ridge of the roof is capped,

and this should be ventilated. The height to the eaves is 4.3–5.5m (14–18ft) – 4.3m (14ft) may be taken as a safe minimum. When floor material is added, the actual roof height will be less. The outer cladding of the walls of the building can be breeze blocks, galvanised metal sheeting, rigid plastic sheeting, asbestos or brick. Brick is very expensive, but sometimes insisted upon by planning authorities.

If the walls are built up to the eaves, translucent sheets should be used on the top section to give light. These can be put on to frames that can be opened in hot weather to give essential ventilation. A fully enclosed school can become airless and an unpleasant place in which to work. A cheaper and more pleasant alternative is to have walls 2.4m (8ft) high with an open space to the eaves. The walls may then be of solid timber, and an outer cladding is not required. This system gives sufficient shelter in most winters, is lighter, and allows a greater feeling of freedom for both horse and rider. In the summer, the school will be cool and pleasant to work in, even in hot weather. Schools used regularly for winter competitions, or situated in heavy-snow areas, may have to be fully enclosed, as do those where it is desirable to shut out external noise.

Kicking boards on the inner side of the wall cladding should be of solid timber, with a minimum of 1.5m (5ft) showing above the floor. They can be sloped from the top at an angle of 12–15° (wider at the top and sloping towards the floor), which will reduce the risk of a rider knocking a knee. To give added light they should be painted white. Dressage markers on the walls should be painted black.

If funds are limited, an alternative to solid walls is straw bales or woven fencing protected by a guard rail. Care must be taken when erecting these to ensure the rider cannot catch his foot, and a watch should be kept for any loose baling string. If a rider or horse falls against these they provide a more cushioned landing than a solid wall.

School doors should be a minimum of 3.7m (12ft) wide and at least 2.4m (8ft) high to avoid any danger of a horse jumping out. If regular topping up of the surface is envisaged, then doorways must be high enough to take the transporters. They should preferably be of the sliding door design, suspended from an overhead gearing. Hinged doors are heavy to handle and a problem in high winds. If used, they should open outwards, as inward-opening doors are a danger to horses working in the school.

Drainage

The roof needs efficient guttering and drainage. If other buildings allow, soak-aways can be dug at intervals round the outside of the building to take this roof water. The soak-away should be a minimum of 4.5m (15ft) deep and filled with stone or pebble. Alternatives are a 'French drain' along both walls or a shallow open drain to direct the water away, possibly to the yard drainage system. A 'French drain' starts at a minimum depth, and is dug out and sloped towards a deep soak-away. The drain and soak-away are filled with stone or pebble. Alternatively a water collection system for roof water may allow the water accumulated to be re-used to irrigate the school surface.

The floor area, unless at a high water-table level, should only require levelling before being covered with the selected surface. It is necessary to take advice, but the extra expense of a prepared base and a plastic membrane is often avoidable. It should be checked that surface water from an upper level cannot seep under the school floor.

Galleries

These can be either at the end or along the side of the arena. For watching and judging jumping a side gallery gives better viewing. For dressage judging an end gallery or box is essential. For competition centres, a tiered gallery is recommended. For smaller centres where the gallery may be used for several purposes, a level floor is of more practical use. If regularly used in winter, some form of heat is needed. Fire regulations must be observed.

Lighting

This can be a very expensive item, and research is advisable as to the cheapest and most efficient system to install, maintain and use. Strip lighting has been the most popular, but improved alternatives have now become available.

Teaching Aids

A full-length mirror or mirrors in the corners and on the centre line and another, wider mirror at the half marker on the long side are excellent instructional aids.

A video camera is an expensive but useful asset. A loud-speaker system is of assistance, for competition and instructional sessions where both the spectators in the gallery and the riders need to be able to hear what is being said.

Jump Store

It is convenient to have jumps readily available, and essential that they are stored in a secure, dry area. If space is available, the area underneath the gallery can be used, with a door in the school wall. For safety this should open inwards to the store and away from the school track.

Surface

Choice depends on the proposed use of the school and the cost. Before deciding, it is advisable to visit other schools to inspect and ride on the surface they have chosen, to check on maintenance, and to consider whether it will stand up to the type of work required.

SAND AND SHAVINGS make a pleasant surface on which to ride. It is suitable for heavy work and jumping. The sand is laid to a depth of about 10cm (4ins). This is then topped with the same depth of shavings, which are rolled level. In time the two surfaces will mix, and can be topped up as required with either shavings or sand, according to the texture of the floor. Sand provides more substance and a heavier surface on which to ride. Shavings make it lighter.

SHAVINGS are at first laid to a depth of 15cm (6ins) and then rolled to make a solid base. As they settle they should be topped up to a final depth of 30–60cm (1–2ft), to ensure that, when jumping, there is no danger of getting through to the ground surface. They make a good general-purpose surface at a reasonable cost, particularly if the shavings can be bought or collected locally. However, there can be a problem of dust in summer – see Control of Dust below.

MANUFACTURED WOOD FIBRE is now supplied by a number of firms specialising in riding surfaces. Many will insist on a plastic membrane, particularly if a prepared base has been put down, as this will prevent stones working up. If the base is topped with very small pebbles, this problem should not arise. In time, plastic membranes usually work up and can wear through, particularly with regular jumping use. They are often an unnecessary expense.

The wood fibre is spread, levelled and saturated with water. It is then compressed by a vibratory roller. The resultant surface is pleasant and resilient to ride on, particularly for working on the flat. It is not so suitable for jump training and lungeing, as the surface can be slippery, and will also get cut up and need considerable maintenance.

SYNTHETIC MATERIALS Many new types of surface are now available. They are more expensive but are harder wearing. They are specifically designed for riding surfaces. Pasada, PVC, Gel-Track and Fibre Sand are just a few.

Maintenance of Surfaces

All surfaces, given sufficient use, will require daily levelling. This can be done by hand, or by a tractor pulling a harrow and leveller. Specially designed machines for this are now available. Particular attention should be paid to the outside track, the lungeing area and any part where jumping regularly takes place.

Some surfaces, particularly after several years' use, may become compacted and slippery. They will require scarifying and then levelling, and in damp weather this may have to be done regularly. They should also be topped up with fresh supplies. A scarifier is a deep type of harrow which digs up the under-surface. It should not be used if a membrane has been put down.

Control of Dust

In time, school surfaces break down into smaller particles, causing a constant dust problem in summer, and in dry, windy spells in winter.

Preventative measures

- Surfaces made from wood will need regular dressings of agricultural salt, approximately 1 tonne per 20 x 40m arena. Salt is usually supplied in sacks of 50kg (110lbs) and can be most easily spread if carted on the back of the tractor and thrown out with a shovel. Care must be taken that not too much salt is applied as it can have a drying effect on horses' feet, with resultant problems. It may also affect the leather boots or shoes of anyone standing in the school for a length of time.

- Daily watering by hose pipe, either by hand or by use of a garden sprayer.

- Installation of an overhead watering system, which is expensive but efficient, quick and labour-saving, and will be a good long-term investment.

- Sump oil spread with a fine spray effectively lays the dust for many months. The surface is oily for a few days, but the oil is then absorbed and appears to cause no further problems. The main difficulty is in spraying the oil so that it is evenly spread over the surface. This treatment is invaluable in drought years when watering is forbidden.

OUTDOOR SCHOOLS

Many firms now specialise in the construction and design of outdoor riding surfaces. If requested, they can provide a complete package deal, and some give a guarantee. It can be a short-sighted policy to cut costs, and finish up with a school only usable in dry weather.

Considerations When Choosing the Site

DRAINAGE Efficient drainage is essential. A site on a slight slope with natural drainage to a lower level can be a great asset. Sub-soils of chalk, gravel, sand or stone all drain well, which makes the site preparation cheaper.

Schools built at the bottom of a slope, or on a sub-soil of clay or other poorly draining material, need considerable work and extensive foundations if the final top surface is to be satisfactory and remain usable in wet weather.

Experienced advice should be taken, and no expense spared on this basic preparation of the site.

ACCESS Preferably this should be convenient to the stable yard, with a hard or firm access to a roadway. A muddy track makes horses, staff and spectators unnecessarily dirty. Mud carried on and in the horses' hooves will not improve the school surface.

SHELTER In exposed areas, neighbouring buildings and/or barns can give welcome protection from high winds and driving rain, but they should not be so close as to interfere with drainage plans.

A tall hedge makes an excellent wind-break.

The surrounding fence can be built of wooden boards or plastic strips, and to a sufficient height to give shelter. As long as there are gaps between the boards, it will be unlikely to be blown down.

In summer, trees can give welcome shade, but if large they may interfere with drainage plans.

Fencing

The school must be safely and securely fenced according to its intended use. A suitable height for most purposes is 1.4m (4½ft).

Posts should be on the outside of the fence; the inside should have no protruding surface liable to cause injury. For economy, plain wire can be used below a top rail of wood or plastic. Barbed wire must never be used.

The entrance should be wide enough for a tractor and any maintenance equipment; slip rails or a gate can be used. The gate should open outwards, and have no protruding fastenings.

The school surface will need to be retained by a ground-level edging, possibly of railway sleepers or other strong timber.

Choice of Surface

This depends on intended use, and the amount of work done on the school. Suppliers of surfaces should be given full information, so that they are in a position to offer the most helpful advice. It is also sensible to see different surfaces under the working conditions similar to those envisaged. Most firms will supply a list of schools they have constructed, and owners are usually happy to discuss matters. It is a help either to ride on the surface, or to see horses working on it. Jump training and lunge work are likely to cause the most problems, making greater demands on the security and durability of the surface.

It is essential that the correct type of foundation is laid to suit the top surface. This is one reason to arrange for the work to be done by a specialist firm, who will also have experience of blending different materials to produce surfaces suitable for particular work.

All surfaces must be put down on a properly prepared base, which should be built up to be well above the level of the surrounding ground surface.

Sand

This has been the most widely used all-purpose surface, and probably in the long run remains the most satisfactory. If laid too deep, it can be heavy to ride on and tiring for the horse. A starting depth of 8–10cm (3–4ins) is recommended, and this can be topped up as required.

In summer, it can be dusty without regular applications of salt and daily watering. In winter, if drainage is faulty, it may freeze, but salt helps to keep the school usable.

The sand used should be clean, and free from soil or debris which could make it compact under pressure.

Types of sand

- Sharp sand, double-washed, provides a workable surface.

- Silicon sand is the best. It is more expensive, and in some areas may be difficult to obtain. It is clean, dust-free, drains well and is therefore less likely to freeze.

- Grade 4 industrial sand is also recommended.

- Black sand provides a good surface, but tends to be dusty in dry conditions and unpleasant when wet. It must be well screened to remove all foreign matter.

Mixtures of sand with shavings and plastic fibre are also workable surfaces, the shavings/fibre making the surface lighter to ride on and less likely to freeze.

Shavings

These are put down on top of a solid base. They are then soaked by hose and compacted with a vibratory roller. The resulting surface is suitable for light work, but lungeing and constant jumping are likely to break it up as is heavy commercial use of the school.

Manufactured Wood Fibre and Wood Chips

Used on their own they are suitable for work on the flat. If mixed with sand, they form an improved all-purpose surface. In freezing conditions, shavings, wood fibre and chips, if well drained, will soften when ridden on and should not freeze.

All forms of wood-based surfaces will eventually break up and start to rot; they will then have to be removed and replaced – a costly operation.

Plastic

Various forms of plastic granules are now obtainable. They form a clean, dust-free, long-lasting surface which drains well and is frost-resistant. They are not so suitable for jump training.

Rubber

Shredded rubber products give a resilient non-skid surface requiring little topping up and are frost-resistant. The shredded rubber is laid on a base of washed stone aggregate which mixes with it to form the surface.

Membranes

A membrane has two distinct uses:

- **Non-porous type.** This is laid down before the

foundation. It prevents water coming up, and on a clay soil stops the foundation and school surface from gradually being absorbed into the clay.

• **Porous type.** This is laid on top of the foundation and below the top surface. It allows surface water to drain through and prevents stones from working up. In time, porous membranes will clog and cease to drain effectively.

All membranes eventually disintegrate and their original function cease.

Tarmac Base

Some surfaces are now laid on a prepared base of tarmac. This obviates the need for a graduated foundation, internal drains and a membrane. A site with good natural drainage is essential. Professional advice should be sought if considering this type of base.

51 BUYING FODDER AND BEDDING

POSSIBLE ARRANGEMENTS

A verbal contract can be arranged with a reputable corn merchant for regular deliveries throughout the year. This system is recommended for yards with limited storage space, or for inexperienced buyers. The latter have to rely on the good faith and reputation of the merchant to ensure that they receive goods of high quality. It should be agreed in advance that if any delivery falls below standard, the goods will be replaced at no extra expense to the buyer. This arrangement is likely to be more expensive than buying from a local farmer, but should ensure guaranteed good-quality supplies throughout the year.

A verbal contract can be made with a local farmer to supply a given quality of fodder and straw throughout the year. Deliveries should be by arrangement, and as convenient. Prices are cheaper than when buying from a merchant. The quality of hay, straw and corn should be checked at harvest time, and then again before delivery, to make sure that there has been no deterioration.

Bulk buying can be arranged direct from the field, thus avoiding extra handling, transport and storage costs. The practice of buying from a neighbouring farmer has much to commend it. There will be many occasions during the year when goodwill on both sides is needed. It is important to be on friendly terms.

Manufactured feedstuffs have to be bought either through a corn merchant or direct from the manufacturer. When bought by the tonne or tonnes, a considerable discount should be obtained. Bought by the sack, goods can be collected and paid for on a cash-and-carry basis.

Bulk Buying

It is always cheaper to buy in bulk, rather than in small quantities. Suitable storage facilities with a sound access surface are essential, as is the cash necessary to pay for the goods. If money is short, it is often possible to borrow from the bank; even with interest added, this may still make a considerable saving.

BUYING OF HAY/HAYLAGE

- If hay is bought off the field at harvest time, extra handling, transport and storage costs are avoided. Even if it is bought at a later date, but in sufficiently large quantities, a favourable price can usually be negotiated.

- If bought in sufficient quantity, a continuing supply of good-quality hay is assured. There is no sudden change in quality content and, with good management, enough old hay will be left to carry on until the new season's hay is ready for use.

- According to the condition of the hay when baled, a minimum weight loss of 25 per cent may have to be allowed for. This occurs as the hay dries out.

BUYING OF STRAW

The same points apply as with the buying of hay except that there is little weight loss as straw is dry when baled.

STORAGE OF HAY

Suitable storage is essential. Poor or inadequate facilities mean hay wasted and money lost.

Purpose-built Hay Barn

This is the most expensive but also the most satisfactory method. The necessary money can often be borrowed from the bank, and the yearly saving in hay costs should make it worthwhile.

The barn should have a ridged roof, with air vents to allow ventilation at the top of the stack and the escape of fumes from the new hay. If sufficient money is available, the north and west – or most exposed – sides of the barn should be filled in with slatted boards as a protection against snow and rain. The two most convenient sides should be left open to facilitate the unloading and stacking of the bales.

Conventional Barn of Brick or Stone Construction

This usually holds only a small quantity and is often more awkward to stack because of enclosing walls; however, the sides cannot fall out even if stacked by inexperienced workers. Although this type of building is useful if already in the yard, it may lack ventilation, and there will thus be more chance of the hay heating up if it has not been sufficiently 'made'.

Covering for an Open Stack

- Plastic sheeting.
- Canvas rick sheet.

Both of these can be successfully used to cover the stack, but there is more chance of hay and straw being wasted.

Building the Stack

BASE To avoid spoiling the bottom bales of a stack, it is usual to put down a base of old hay or straw. This can be loose or in bales, but must be thick enough to prevent any damp getting up into the new hay. Small amounts of hay (10 tonnes or less) can be stored on wooden pallets. These are not practical for the larger stacks because of the weight of hay.

BUILDING THE STACK IN A PURPOSE-BUILT HAY BARN If there are no outside walls, the stack must be built with great care. One bay should be filled at a time. The front should be filled first, and then the sides and back. The middle should be filled last. The bales should be positioned so that each layer locks the layer below and holds it steady. This work requires experienced workers or careful supervision. Should there be any tendency for the side of the stack to bulge out, this can usually be checked by supporting the sides with long poles (jump poles will do). The same principles apply when loading hay or straw on to a lorry or trailer. If transporting it on a road, always rope the load for greater safety.

Heating in the Stack

If carted from the field immediately after it has been baled, hay heats in the stack. To avoid overheating, it is advisable to allow for air passages in each bay. Stand a bale upright at base level in the centre of the bay. As the layers of hay are stacked, pull up the bale with them, thus leaving a square ventilating shaft. At the top of the stack, leave a space between the top layer and the roof so that air can move freely. If the weather looks settled, hay often does better if stood up in the field for a few days before carting.

Stack to be Covered by Plastic Sheet or Canvas Rick Sheet

Build the stack as described above, but with even greater care, as there are no uprights or boards to support the sides. The hay is stacked in a square or rectangle with a flat top. The sheet is put over the top, and should be large enough to hang over the edge. If heating is likely to be a problem, build accordingly (see

Heating in the Stack, above), but also put six bales on top of the last layer. When the sheet is placed over the bales, air can circulate underneath. When heating has finished (two or three weeks), the stack can be secured for the winter. Remove the six bales so that the sheet lies nearly flat; a very slight slope is a help to chute the rain. If using a plastic sheet, place spare bales or heavy-duty tyres on the top to hold it down. This is not necessary with a canvas sheet. The sides are tied down with rope or nylon string, or pegged to the lower bales. This arrangement should be weather-proof. Once it becomes necessary to start using the hay or straw, there is always the problem in re-securing the sheet – and wastage and spoiling of either hay or straw ensues.

Plastic sheets are cheap, and with a stack in a sheltered area and treated with care, they can be a success. If they are not well secured, or if the weather is very rough, they rip and tear very easily: so they are not suitable for windy districts. They usually last only for one year. Large-mesh nets can be bought or made out of baler twine. If these are put over the sheets they give added security and reduce the chance of tearing.

Canvas rick sheets are more expensive, last longer, and are easier to secure to the stack. When the stack is opened, they are easier to replace, and stand much more rough handling and rough weather.

Big Bale Silage/Haylage in Polythene Bales

These are easy to store outside, and will be weather-proof. The area must be vermin proof and with no risk of puncture due to horse interference (chewing over the fence!) or implement damage. Damage to bales will cause spoiling of the contents rendering the feed useless.

Opening the Stack

This should be carried out by experienced staff, but if they are not available, the workers must be well briefed and supervised. When opening a stack, take hay from the top and work down. If individual bales are pulled out from the bottom this can unbalance the rest of the stack, which may fall and injure anyone standing below. It will also spoil many of the bales. When removed from the stack the bales should be stored in a hay shed adjacent to the stables where they will be ready for use.

Storage Areas and Quantities

Hay and straw can be bought either by the tonne or by the bale. When buying by the tonne:

- Each load can be weighed on a weighbridge. This is costly in time and fuel.
  or
- 10–20 bales can be weighed, and from that an average weight per bale can be worked out.

A barn of dimensions 13m x 9m x 5m(45ft x 30ft x 18ft) houses approximately 100 tonnes of hay or 4000 bales.

100 tonnes of hay is sufficient for approximately 50 stabled horses for a year.

65 tonnes of straw is sufficient for approximately 50 stabled horses for a year.

Hay bales can weigh 13–26kg (30–60lbs). Straw bales can weigh 13–18kg (30–40lbs). The weight of the bale depends on:

- The condition of the hay when baled.
- The setting of the baler. This can be altered to make light or heavy bales. An all-female staff will appreciate the lighter bales, even if there are more of them to the tonne.

BUYING LOOSE SAWDUST AND SHAVINGS

If you are in a wooded district, these (according to demand) can be:

1. Free. A verbal contract is agreed whereby all available supplies will be regularly collected.
2. Delivered by lorry. Transport has to be paid for.
3. Bought by the sack. Provide your own transport or pay for delivery.

BUYING BALED SHAVINGS AND PAPER BEDDING

These are both much cheaper when bought by the lorry load. They can be stored outside without a cover, as they are packed in polythene bags. Choose a level site convenient to the stables.

BUYING AND STORAGE OF GRAIN

Bulk buying of grain presents more problems, and is not really practical for the small yard. The problems of good quality and an assured and regular supply can often be solved by contracting to buy an agreed amount from a local farmer. He will be responsible for storage and delivery as requested. Payment can be arranged monthly. Corn is always in short supply by July, and there can be a problem until the new grain is ready to feed. It is important to remember this when buying in bulk earlier in the year, or when making arrangements for regular deliveries. Make sure that the supplier knows the approximate total quantity needed for the year.

If keeping high-performance horses, this arrangement may not be satisfactory, as the necessary quality of grain may not be available locally. Under such circumstances, and if grain is being bought to last for the year, a silo capable of storing sufficient grain for twelve months can be the most practical proposition.

Quantities of up to 4 tonnes can be stored in purpose-built metal containers. Alternatively, a spare brick or concrete block loose box with a solid door can be used as a store. The roof must be made vermin-proof.

Whole corn can be bought in small quantities and stored in metal bins, or paper or jute sacks (not plastic). Living seed dries naturally, and will not deteriorate or diminish in food value. If losses are to be avoided, the storage area must be vermin-proof or well patrolled by cats.

Bruised, rolled, crushed or cut grain should not be bought in bulk. It is dead, the husk is broken, and its food value deteriorates within a week. Grain is often dampened during processing, which can cause it to heat and spoil if not fed within days.

ORGANISING AND RUNNING A YARD

THE OFFICE

A well-organised and well-equipped office is essential for the efficient running of any equestrian establishment, whether it is a riding school, livery yard, training and competition yard or racing stable. It is needed to give a good impression to clients, and to facilitate the day-to-day running of the yard. The suggested equipment is detailed below. Some of this will not necessarily be required for all establishments.

Equipment

- Flat-topped desk with drawers and chair.
- Some form of heating.
- Telephone/fax machine.
- Intercom system between office, tack room, yard and indoor school.
- Outside telephone bell.
- Telephone answering machine.
- Typewriter/word processor/computer.
- Cash box with float for change.
- Correspondence tray.
- Telephone book.
- Tidy for pens, pencils, paper clips etc.
- Stapler and hole puncher.
- One or two spare chairs.
- Wall chart/year planner.
- First-aid box.
- Wastepaper basket.

- Desk diary with a day to a page.
- Record book or books for shoeing, vet, tack, repairs, injections, teeth, worming, horse hours of work, staff days off and holidays.
- Accident/incident book. This should not be a loose-leaf book.
- Accident forms for use should a person be injured.
- Two loose-leaf books containing daily ride lists and work sheets.
- Filing cabinet.
- Photocopier.
- Index card system (may all be recorded on computer).
  Accounts.
  Details of livery horses, equipment and clothing.
  Riding clients' addresses, telephone numbers, dates of birth, standards and other relevant information.
  Staff particulars, with work contracts.
  Student particulars, with contracts and date of exams.
  Working pupil particulars, with work contracts and date of exams.
  RIDDOR (Reporting of Injuries, Diseases and Dangerous Occurrences Regulations) report forms
- Stationery cupboard containing:
  Printed letterheads and cards.
  Brochures.
  Typing paper.
  Desk supplies.
- Cash books (may all be computer recorded):
  Main cash book.

Yard Management

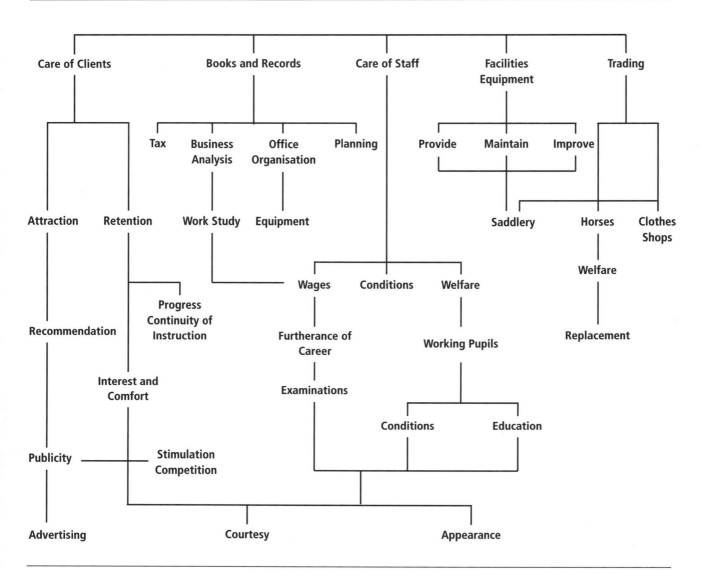

Petty cash book.
Wages book.
VAT book if the business is registered.
PAYE tax tables.
- Index file system for invoices, receipts and correspondence.

- Notice board with:
  Riding school licence.
  Third-party insurance certificate.
  Business licence.
  Employers' liability certificate.
  Fire instructions.

Accident instructions.
Telephone numbers of doctor and veterinary surgeon.
Information on school activities.
Information on local activities.
Staff instructor certificates.

In large yards, the following equipment can increase efficiency:

- Photocopier.
- Tape recorder.
- Word processor.
- Computer.
- Facsimile machine.

ACCOUNTS

Unless the proprietor or secretary has experience of accounting, it is advisable for all businesses to employ an accountant who can:

- Present a true record of the business to the Inland Revenue.
- Claim all available tax allowances.
- Advise as to the present state of the business, future prospects and trends, as well as providing essential information for constructive future planning.

The accountant may be asked to present the books monthly, three-monthly, half-yearly or annually, according to personal preference and the establishment's ability to maintain good records. As he will be paid by the hour, the more that can be done to keep the accounts in good order, the lower his bill will be. He will advise on the system that he wishes to use.

The accountant will require the following books and information:

- Main cash book. Receipts are noted on the left-hand page, payments on the right-hand page. All are totalled monthly.
- Petty cash book for recording small items paid for in cash. The total is transferred to the main cash book each month.
- VAT book. For recording inputs and outputs of Value Added Tax.
- Wages book.
- Bank statements for current and deposit accounts.
- Bank cheque stubs.
- Details of creditors.
- Details of debtors.
- Records of livestock: i.e. list of horses, with details of purchases, sales or deaths.
- Records of stock: i.e. machinery, tractors, cars, trailers and horse boxes, with details of purchase, sale and writing off.
- Stocks of hard feed, hay, straw and bedding.

INSURANCE

Riding is a hazardous sport, and accidents will occur in spite of every precaution taken by the riding-school proprietors. Insurance policies are therefore a must.

Compulsory Insurance

- Third-party insurance is compulsory for riding schools, and is recommended for all types of yard.
- Employers' liability.
- Cars, horseboxes, trailers.
- Tractors used on a public highway.

Recommended Insurance

- Fire.
- Theft of horses, saddlery, machinery, furniture, personal belongings.
- Loss of business due to fire or theft.
- Personal accident and illness to proprietor and/or chief instructor.
- Personal accident insurance for clients.
- Office insurance.

NB: Livery horses and their equipment should be insured by their owners. They should not be accepted by the establishment until fully insured.

Non-essential Insurance

SCHOOL HORSES It is usually considered uneconomic to insure school horses against accident and illness, as premiums are expensive.

EPIDEMIC – causing loss of business. It is a matter of personal choice and type of business as to whether this is considered worthwhile. Several insurance companies arrange umbrella policies, which are worth investigating for riding schools and livery yards. BHS-approved schools receive very favourable terms from the BHS brokers.

CLIENTS' PERSONAL ACCIDENT INSURANCE In recent years there has been a considerable increase in injury claims by clients riding at riding schools. Many proprietors now recommend that clients should take out some form of personal accident policy before either learning to ride or continuing to ride as a hobby. This does not exonerate the school from legal action should an accident occur and negligence be proven.

PRIVATE HORSE OWNER'S LIABILITY The private horse owner is advised to insure against third-party risks (at present this is included for BHS members, in the higher membership categories – Silver and Gold) and they should preferably insure themselves against personal accident.

FREELANCE INSTRUCTORS Independent freelance instructors are also advised to insure themselves against third-party risks and personal accident. As members of the BHS Instructors Register, freelance instructors enjoy comprehensive insurance and the benefit of automatic membership of the National Association of Sports Coaches.

INCIDENT OR ACCIDENT BOOK

It is essential for all types of yard to keep an Incident or Accident Book. This is a compulsory measure for BHS- and ABRS-approved schools. The book must record the time, place and date of any incidents or accidents, an account of what took place, and the name and address of any witnesses. It is often months – even years – before information on an accident is asked for. During this time, important details can be forgotten, staff may leave, and reliable information may thus be difficult to obtain. Suitable accident forms can be obtained from the BHS.

No riding establishment should be left in the charge of a person under sixteen years of age. It is the responsibility of the owner, proprietor or stable manager to decide when a person is sufficiently responsible and capable of being left in charge. The same responsibility applies when allowing riders to be escorted or pupils to be taught by young people without supervision.

At present, litigation is all too easy, and court cases and claims for damages are all too common, so it is essential for responsible proprietors or stable managers to take extra care. For example, a notice stating that: 'Although every care is taken, there can be no responsibility for accidents' gives no protection in law should a claim for accident and injury lead to litigation. The account of an accident should be signed by the person in charge, and, whenever possible, by the injured party and any witnesses. For legal requirements, see Chapter 53.

RECOMMENDED DRESS FOR RIDING CLIENTS

There should be strict rules as to suitable and safe dress for clients riding at any establishment. Correctly fitting BSI-standard hats to PAS 015 or BSEN1384 should be worn, with attached chin-strap securely fastened. Under the Health and Safety at Work Act the wearing of a PAS 015 or BSEN 1384 standard hat is mandatory for all employees and students at Approved riding schools. **The casual loan or hire of hard hats is not recommended.** If, as the result of an accident, the hat should come off or be found faulty, the proprietor of the school could be claimed against for having provided it.

Shoes or boots should be worn with smooth soles and small heels, so that if the rider falls:

- The foot will not be caught up in the stirrup by ridges in the sole.
- The low heel will prevent the foot from slipping through the iron.

The following footwear should not be allowed:

- Any ridged Wellington boots or muckers.
- Buckled shoes.
- Tennis, walking or game shoes with a solid sole and no heel.
- High-heeled boots or shoes.

Floppy plastic raincoats which in a high wind can blow about, making a sudden noise and frightening a horse, should also be forbidden.

ORGANISATION OF FACILITIES

Indoor School

- School walls must be well maintained and of a suitable height and strength. No part of the surface should protrude.
- School doors should be in good working order.
- The gallery must be well fenced off and safe.
- Jumping equipment and jump cups should be kept out of the way of riders, or be placed so as not to be a hazard.

Outdoor School

- Outdoor schools must be kept well fenced. Post and rails or plain wire with a rail above are suitable. Barbed wire must never be used.
- The entrance should have a gate. It should be well maintained and easy to open and shut.
- Broken boards or rails should be immediately repaired or made safe, so that no jagged edges, boards or nails are left exposed.
- Jumping equipment should be as for indoor schools.

Lavatory Facilities

Suitable facilities must be provided. Wash basins should preferably have hot and cold water. Soap and towels should be available.

Catering

Establishments providing food for staff and/or clients must make sure that their arrangements comply with the Food and Hygiene Act (1990). This includes suitable arrangements and facilities for the storage, cooking and serving of food, and for washing up afterwards.

The regulations of the Fire Precautions Act (1974) should also be consulted.

Organisation of Tack Room

- Efficient organisation is essential.
- All saddles and bridles should be clearly marked with the horse's name. Metal discs can be clipped to the near-side front 'D's of the saddle, and to the buckle on the throatlash of bridles. It is then possible to change them if necessary.

  An alternative is adhesive tape placed under the skirt of the saddle or under the cantle to the rear of the saddle.
- There must be an adequate number of saddle racks; saddles placed one on top of the other can be easily knocked to the ground and may break a tree. Racks for school tack can be arranged along a wall in three or more layers.
- Bridles can be hung separately on another wall.
- Tack-cleaning equipment must be provided for each member of staff involved in cleaning. Time is wasted if supplies are short.
- There must be facilities for the drying of numnahs, rugs, staff coats, etc. Old-fashioned drying racks, worked on a pulley system and suspended from the ceiling, are space-saving and efficient, but they must not be sited too close to any upper wall-heaters because of the fire risk.

Livery Tack

Each owner's saddles, bridles, martingales, etc. should be kept together, with the individual grooming kit below them, so that there is no muddle on a groom's day off. Livery owners should keep their spare tack and rugs at home, unless space is available for their own rug box. All livery tack and belongings should be clearly marked by the owner. It is preferable to have a separate school and livery tack room.

Organisation of Feed Room

- Feed sacks, if stored on pallets, should be positioned so that there is space for any yard cats to move round all sides and underneath.

- Floors should be swept daily and kept clean.

- Sinks and feed bowls should be cleaned daily.

- Fresh supplies coming in should be placed behind the sacks, so that the old food is used first.

- Rats and mice can be a serious problem in feed rooms. Apart from the problem of loss of feed, the droppings of rats and mice in the feed eaten by a horse can cause severe colic. Stable managers must be aware of the problem and take suitable action. Some of the poisons used in the past are no longer effective, and farm cats can be far more efficient.

FEES

Fees for Rides and Lessons

The amount charged should depend upon:

(a) Keep of horse feed, bedding, water, electricity.
(b) Labour – mucking out, grooming, tack cleaning.
(c) Instructor or escort time.
(d) Shoeing and veterinary expenses.
(e) Saddlery – upkeep and replacement.
(f) Horse replacement.
(g) Share of overheads – rent, rates, bank interest.
(h) Share of upkeep of box, yard, fields, fencing, water, cultivation.

(i) Share of insurance and office expenses.
(j) Local conditions.

Livery Fees

Items to be taken into account:

- Items (a), (b), (g), (i) and (j) above.

- Exercising time if required.

- Profit should be a minimum of 10 per cent. Livery-only yards must take a bigger percentage if it is to be their main source of income.

Schooling Fees

These depend on the riding time of skilled staff.

Working Liveries

The advantage of these is the use of an extra horse without additional capital expense, but before agreeing to this system it is essential to ensure that the horse will be really useful. Remember, too, that the box taken up by a horse could have been occupied by a full livery.

Clear arrangements must be made with the owner as to:

- Days and time when the horse will be available for use in the school.

- Responsibility for clipping, shoeing, veterinary expenses, tack repairs.

- Responsibility for livery if the horse is ill or off work.

Fees should depend on individual arrangements, but are usually half to threequarters of the full livery fee. The livery owner's insurance company should be notified of the arrangement, as this may affect the terms of the policy.

DIY Livery

Do-it-yourself livery can be an acceptable arrangement for the owner of the yard and the owner of a horse.

Arrangements from yard to yard can be hugely variable. Conditions must be clearly defined, agreed in writing and signed by both parties.

Part livery and DIY are to some degree fraught with potential problems. The ongoing success of such arrangements depends to some degree on the mutual goodwill and endeavour of both parties to make it work.

STAFF

Staff Contracts

Within thirteen weeks of starting work, all staff should have a work contract. This should include:

- Job description.
- Date of commencement.
- Hours of work – but as the care of animals is involved these have to be flexible.
- Days off per week.
- Arrangements for holidays, including statutory holidays.
- Pay per week or month.
- Overtime.
- Sickness arrangements.
- Arrangements for termination of employment on either side, and, disciplinary and grievance procedures.
- Special arrangements: e.g. board and lodging, living-out allowance, keep of own horse, use of car or Land-Rover and trailer, time off to compete, preparation for exams, and keeping of pets.

Employers are reminded that arrangements for terminating employment due to unsatisfactory work or behaviour should fulfil the guidelines laid down in the relevant pamphlets published by the Department of Employment.

Staff Duties

The following are the normal responsibilities of staff in a stable yard.

CHIEF INSTRUCTOR
Typical responsibilities:
- Interviewing prospective clients and students.
- Organising lessons and rides – checking daily ride-list

in desk diary so that it is ready for the secretary to type. Allocation of suitable horses.
- Providing private lessons.
- Preparation for exams.
- Schooling of horses.
- Co-ordinating with stable manager regarding lecture and practical stable-management sessions, and allocation of yard duties etc.
- Co-ordinating with secretary to keep records and book of rides and lessons. Organising lectures and practical stable-management sessions for students and working pupils. These may be taken by a BHS Intermediate Instructor.
- Arranging and conducting regular staff meetings.

STABLE MANAGER
Typical responsibilities:
- Feeding.
- Daily work lists.
- Supervision of stable management of staff, students and working pupils.
- Daily inspections of horses.
- Arranging clipping, trimming, shoeing and veterinary list and saddlery repairs.
- Attendance at veterinary surgeon's visits.
- Feeding and checking of field horses.
- Buying and storage of fodder.
- Executing or arranging for necessary work in fields: e.g. harrowing, rolling, fertilising, topping, fencing, drainage, gates and water.
- Executing and arranging general upkeep of stable yard, fences, gates, drainage, school surfaces (indoor and outdoor), show jumps and cross-country fences.

SECRETARY
Typical responsibilities:
- Answering telephone.
- Welcoming clients.
- Booking rides and taking ride money.
- Typing out daily ride sheets.
- Supervising petty cash.
- Dealing with correspondence.
- Keeping accounts: VAT, wages and record books.

- Liaising with chief instructor and stable manager with regard to record books and client problems or requests.
- Keeping livery and ride accounts.
- Sending out monthly bills.
- Paying accounts.
- Working out PAYE and employees' National Insurance contributions.
- Making up weekly wage packets or preparing wage cheques for signature.
- Making regular visits to bank to pay in cash and cheques, and collecting wages and petty cash money.

INTERMEDIATE INSTRUCTOR
Typical responsibilities:
- Looking after two horses.
- Supervising stable management of students and working pupils.
- Giving stable-management lectures.
- Teaching.
- Schooling.
- General yard work.

NB: If preparing for BHSI, suitable help will be needed from the chief instructor.

ASSISTANT INSTRUCTOR
Typical responsibilities:
- Looking after two to four horses.
- Some teaching.
- Joining in general yard work.
- Helping to catch and groom field ponies.

NB: If preparing for an exam, lessons from the chief instructor and help from the stable manager will be needed.

WORKING PUPILS
Typical duties:
- Looking after three horses.
- Helping with beginner lessons.
- General yard work.
- Looking after and grooming field ponies.
- Education should include:

- Set periods of stable management each week, theory and practical.
- Extra help as the examination date approaches.
- Four riding lessons a week plus teaching practice.

PAYING STUDENTS
Work and instruction according to individual arrangements, but typical duties and education might include:

- Looking after one horse (or possibly own horse and one other).
- Preparation for intended examination.
- Two daily lessons plus one lunge lesson.
- Teaching practice. When sufficiently experienced, taking class lessons or beginner lessons under supervision, or lessons suitable in standard as preparation for intended examinations.

ODD-JOB MAN
Typical responsibilities:
- Liaising with stable manager regarding daily work.
- Stable repairs and yard maintenance. Field work and fence repairs.
- Grass cutting.
- Building and maintaining show jumps and cross-country fences.
- Care of car, horsebox and trailers.

In small yards, the duties of chief instructor and stable manager can be amalgamated, and some of their duties can be reallocated to an Intermediate Instructor or senior BHSAI. Smaller yards are less likely to have paying students, or to do exam preparation, and tend to concentrate on providing instruction and hacking for the general rider.

DOMESTIC STAFF
A cook, a housekeeper and cleaners may be required. If staff and/or students and working pupils are to live in or live in a staff house, arrangements have to be made for:
- Cleaning the house.
- Preparation of food.
- Mature staff living in separate accommodation may

be expected to look after themselves, but young students and working pupils should not be required to do so. It is the responsibility of the school owner or manager to see that young members of staff are well looked after. They should have suitable accommodation, and should be encouraged to care for it and keep it clean and tidy.

Turnout of Staff, Students and Working Pupils

Some yards will have strict rules regarding dress and turnout. All yards should make sure that personnel are suitably dressed. In particular, hair should be tidy; if long, it should be tied back or held in a net. The wearing of jewellery should be firmly discouraged. Earrings are dangerous.

When working in the yard, jeans, sweaters and muckers/paddock boots or wellingtons are suitable. When riding, BSI-standard hats and safe footwear are essential.

GENERAL DAILY ROUTINE

The following tasks are applicable to most stable yards:

1. Check the horse's general well-being:
   - That he is moving comfortably round the box.
   - That he is not pointing or resting a front foot – a sign of discomfort. If the opposite hind foot is also rested, it is a sign of relaxation after work.
   - That there are no signs of injury such as those caused by getting cast.
   - That the bed is in a satisfactory state and that the droppings are normal in appearance and number.
   - That the feed and hay have been eaten.
2. If necessary, adjust rugs.
3. Water, feed and leave horse to eat in peace.
4. Muck out, pick out feet into a skip, bed down, sweep up. If there is time, tidy up the muck heap.
5. Brush off and saddle up. Horses should be tied up:
   - If the groom is working in the stable.
   - If they are saddled up ready for work, in case they try to roll.
6. Exercise.

7. On return from exercise, wash and pick out feet, unsaddle, give small feed of hay.
8. Groom.
9. Skip out box. Tidy muck heap (if not done earlier). Water, feed, lunch.
10. Finish grooming.
11. Clean tack.
12. Stable jobs (see Stable Duties, page 372).
13. Skip out.
14. Water and feed.
15. Sweep yard.
16. Give haynet.
17. Check:
    - Horse rugs.
    - Water.
    - Stable doors.
    - Feed shed.
18. Lock up tack rooms and close yard gate.
19. Later at night look round to ensure that all is well, but try not to disturb the horses.
20. Give water and late feeds if required.

Typical Routine for Competition Horses

This daily routine would be appropriate for a yard of three horses with one groom and help to ride.

0700	Check water. Feed.
0730	Muck out. Bed down. Sweep up. Tidy up muck heap.
0830	Brush off.
0900	Schooling and/or exercise. Lunge one horse if one person responsible for all exercise. **NB:** Hay after work.
1200	Strap one horse. Set fair others.
1245	Feed.
1300	LUNCH
1400	Strap two horses.
1515	Stable jobs. Tack.
1600	Skip out. Water.
1630	Feed. Finish tack.
1730	Skip out. Water. Hay-up.
2030	Water. Fourth feed.

Time	Work	Person Responsible	Lessons, Teaching, Schooling	Person Responsible
0700	Check horses. Water and feed. Horses working at 0800 have small feeds and are fed after work.	Stable manager		
0800	Water. Muck out, bed down (or at 1215).	Staff	Private lesson	Chief Instructor
0900	Brush-off horses. Saddle-up for 1000 and 1100 lessons and for livery owners.Tidy muck heap and sweep yard.	Staff		
1000			Staff lesson	Chief I or II
1100			Client lesson	I or AI
1115	Small haynet all round. Groom horses.	Staff		
1215	Skip out. Water. Bed down.	Staff	Lungeing or schooling	Chief I or II
1230	Feed.	Stable manager		
1230	LUNCH BREAK			
1300			Private lesson	II
1345	Prepare horses for 1400 lesson.	Staff		
1400	Lectures for working pupils.	Stable manager	Lesson	AI or II
1430	Catch and prepare ponies for 1600 lesson. Groom horses.	AIs and working pupils		
1500			Lesson	AI or II
1600-1730	Feed. Skip out. Hay up.	Stable manager & staff	Childrens' lesson	AI and WP help
1730	Turn out ponies.	WPs		
1800	HOME			

Evening Rides

Time	Work	Person Responsible	Lessons, Teaching, Schooling	Person Responsible
1700	Saddle up for 1800 lesson. Prepare hay and leave outside boxes of horses being worked.	Staff on evening rota	Lessons	Staff on evening rota
1800-2100	Prepare horses. Skip out boxes. Water. Check unsaddling and rugging-up. Put away tack. Give haynets.			
2100	Final check or after last lesson.	Stable manager		

General Considerations in a Competition Yard

• Feed one and a half hours before exercise.

• No hay before exercise if working in the early morning, or fast working.

• If working early, give first feed after work.

• Check legs for heat or strain after work and in evening.

• If work reduced, cut down on the concentrate feed.

Typical Routine for Hunter Liveries

This daily routine would be appropriate for a yard of four horses. When getting the horses fit, extra help will be required if other duties are to be efficiently performed.

0700	Check horses. Water. Feed.
0730	Muck out.
0830	Brush off. Saddle-up.
0900	Ride and lead each pair for approximately one and a quarter hours.
1015	Return from exercise. Prepare second pair and take out.
1130	Unsaddle. Haynet.
1145	Strap two horses. Skip out beds. Set fair. Sweep yard. Tidy muck heap.
1245	Feed.
1300	LUNCH
1400	Strap two horses.
1500	Clean tack.
1545	Stable duties. Sweep yard.
1630	Feed.
1700	Skip out. Check rugs. Water.
1730	Finish.
2030	Check horses. Water.

General Considerations in a Hunter-livery Yard

• When the horse is fit and hunting twice a week, daily exercise can be reduced. In some cases threequarters of an hour is sufficient.

• No hay should be given before hunting.

• A warm mash may be given after hunting, but this is no longer thought to be so beneficial as in the past. See Chapter 23, page 178.

• The day after hunting, check legs, trot up, lead out for ten minutes. NB: One experienced (trained) groom should be able to do five horses if he or she is not doing the exercising. A horse walker can be a great boon, and can save many hours spent exercising.

GENERAL STABLE DUTIES

Daily

1. Water.
2. Cleaning out water bowls.
3. Feeding.
4. Mucking out and sweeping yard.
5. Grooming.
6. Exercising.
7. Cleaning tack.
8. Washing feed bowls.
9. Tidying tack room and feed shed.
10. Washing sinks and floors.
11. Cleaning lavatories and basins.
12. Raking and harrowing school floor and track. (Raking done by hand; harrowing done by tractor.)

Weekly

1. Dusting down stables.
2. Checking salt licks.
3. Cleaning drains and traps.
4. Shaking and brushing rugs.
5. Disinfecting stable floors and drains – not deep-litter bedding.
6. Washing lavatory floors and checking supplies.
7. Washing grooming kit.
8. Cleaning head collars and rollers.
9. Oiling tack if it is wet.
10. Levelling school floor; salting and/or watering if necessary.

11. Preparing farrier list.
12. Dealing with saddlery repairs.
13. Washing down barn and stable passageways.
14. Washing stable tools and wheel-barrows.

Monthly

1. Checking door bolts.
2. Checking string on tie-rings.
3. Checking drains.
4. Ordering feed supplies.
5. Checking other supplies: e.g. disinfectant, saddle soap, clipping blades, brooms, stable tools.
6. Checking bandages.
7. Checking saddlery for repairs.
8. Washing stable walls.
9. Checking first-aid box and veterinary cupboard.
10. Ensuring that horses are shod every 4–6 weeks.
11. Worming every six weeks.

Annually

1. Checking night and New Zealand rugs, washing and sending for repair.
2. Sending day rugs and blankets to be cleaned and repaired. Store away.
3. Creosoting wooden stabling and wood surfaces
4. Painting block walls of stables.
5. Painting school walls.
6. Painting show jumps
7. Repairing stabling. Repaint as necessary.
8. Checking roofs and yard surface. Repair as necessary.
9. Checking windows.
10. Buying hay, straw and other bedding.
11. Checking saddlery for replacement.
12. Checking fields, fencing and gates before turning out horses for spring.
13. Checking fire extinguishers.
14. Servicing clippers, groomers, corn mill, water heaters and other equipment.
15. Organising inoculations.
16. Checking teeth every six to twelve months.
17. Mucking out deep litter (twice a year minimum).

PREVENTION AND CONTROL OF FIRE

It is advisable to consult the local fire brigade headquarters about fire regulations relating to stables, and to seek their advice regarding a particular stable yard and its problem areas. For legal requirements, see Chapter 53.

Recommended Layout of Stable Yard and Surrounding Buildings

- Hay and straw barns should be well away from the stable area. Small amounts of hay and straw may be kept easily available for daily use in a convenient covered area.

- Tractor-sheds and horsebox garage should also be away from the stables and the hay barn.

- Petrol and diesel tanks are not in general use, but if you do have them on your premises they should be near the access road and away from all other buildings.

- If supplies of tractor fuel have to be stored, they should be in heavy-duty jerry cans. Not more than 50 litres (approx. 10 gallons) should be stored at a time. Note: Plastic containers are not safe for storing diesel fuel, petrol or paraffin.

Recommended Procedures and Practices

- No smoking in stable and working areas. 'No Smoking' notices should be prominently displayed and no-smoking rules strictly adhered to.

- Fire notices should be displayed in all stable blocks and in all other ancillary buildings. Instructions for procedure in the event of fire must be clearly set out and legible.

- Fire drill should be held at regular intervals, particularly when there has been a new intake of staff or working pupils.

- Suitable fire appliances appropriate to the contents should be placed in all buildings, including covered school and gallery.

- All personnel should be clearly instructed as to which appliances should be used on the following sources of fire:

- Hay and straw.

- Electrical faults.

- Oil and petrol.

- A sufficient number of fire alarms should be available in accordance with the size of the yard. They should be regularly tested to ensure that they are in working order. They should be protected from frost.

- There should be an adequate number of hoses and water-points, so that the hoses reach all areas of stables, barns, and other buildings.

- Unless there is a risk of frost, the main fire hose should always be kept fitted to the water supply, and ready for use. In freezing weather, it should be drained and insulated.

Action to be Taken in the Event of Fire

These instructions should be given to all personnel:

1. Call the fire brigade, or make sure that someone else has called them.
2. Sound the fire alarm.
3. Make a head count of all staff and other persons on the yard.
4. If the stables are on fire, or could catch fire, take or release the horses into a field or designated area. Begin with those nearest to the fire.
5. Tackle the fire with appropriate fire appliances, without putting yourself at risk.

Horses may be frightened and refuse to leave their stables. A coat or cloth put over the head and covering the eyes may make them more willing to move. In the case of a difficult horse, several people may be needed to move him. Inhaled smoke is very dangerous to horses, so make sure that they have access to fresh air. If they have inhaled smoke, it is advisable to seek veterinary advice.

The Most Common Causes of Fire

- Electrical fault in wiring or appliance. Installing a trip-switch, which cuts off the electricity supply in the event of a fault, should avert this.

- Electric or paraffin heaters.

- Clothing or rugs put too close to heaters when drying.

- Smoking. Careless use of matches and throwing away of cigarette ends.

- Children playing with matches.

- Self-combustion of hay in the stack.

- Bonfires placed too close to hay barns or other buildings.

53 THE LAW

This chapter outlines the relevant contents of the Acts which cover horses and stables. The Acts discussed are:

- Protection of Animals Act of 1911 and 1912.
- Animals Act of 1971.
- Riding Establishments Acts of 1964 and 1970.
- Health and Safety at Work Act of 1974.
- Reporting of Injuries, Diseases and Dangerous Occurrences Regulations (RIDDOR) 1985.
- Control of Substances Hazardous to Health (COSHH) regulations.

Only outlines can be given, and they are not a substitute for professional advice.

RIDING ESTABLISHMENTS ACTS 1964 AND 1970

The setting up, and to a large extent the running, of a riding establishment is governed by the combined effect of the Riding Establishments Acts 1964 and 1970. These are applicable in England, Scotland and Wales.

Northern Ireland has separate legislation entitled Riding Establishments Regulations 1980, and riding stables in the Isle of Man are governed by the Riding Establishment (Inspection) Act of 1968.

The Riding Establishment Acts 1964 and 1970 forbid the keeping of a riding establishment in England, Scotland and Wales except under the authority of a licence issued by the local authority in whose area the premises (including land) are situated.

THE RIDING ESTABLISHMENT

The term Riding Establishment covers the carrying on of a business 'of keeping horses to let them out on hire for riding and/or for use in providing instruction in riding, for payment'.

Licences

A licence may be granted annually after application by an individual over the age of eighteen years, or by a company. The normal duration of a licence is twelve months. Licences are not renewable, and a new application and inspection of the premises is required on the expiry of the existing licence. Licences run from the date of issue, but some local authorities issue licences dated from 1 January.

As an alternative, if a local authority is not satisfied that a case has been made for a full licence it has the power to grant a provisional licence for a period of three months. This provisional licence can specify the conditions required, so that the licence holder has time to meet such conditions and will therefore eventually be granted a full licence. The local authority may extend the period for a further three months, but only on re-application by the licence holder. One of the purposes of such an extension is to give the licence holder the opportunity to complete work already begun in order to meet the specified conditions. The local authority may not issue provisional licences to any person or company for more than six months in any one case.

The cost of the licence, either full or provisional, is at the discretion of the local authority, and the total fee includes administrative and inspection charges. No provision is made for a charge to be levied for the three-months' extension to a provisional licence.

The Acts give the local authority complete discretion over the granting or refunding of a licence. However, an aggrieved applicant can appeal to a magistrates' court, in relation to both the refusal and to any conditions that the authority has imposed. The magistrates may then give such directions as they think proper in respect of the issuing of a licence or to the conditions.

There are no provisions within the Acts for third parties to appeal against the granting of a particular licence.

Planning Permission

Before consideration is given to the granting of a licence to a new applicant, the local authority should ascertain that under the Town and Country Planning Act of 1971 permission has been obtained either to erect new buildings, or for change of use of all or part of existing premises.

Qualification for a Licence

No one under eighteen is qualified to apply for a licence, nor is any person who for the time being is disqualified under any of the following:

- The Riding Establishment Acts.
- The Protection of Animals (Cruelty to Dogs) Act (1933) from keeping a dog.
- The Protection of Animals (Cruelty to Dogs) (Scotland) Act (1934) from keeping a dog.
- The Pet Animals Act (1951) from keeping a pet shop.
- The Protection of Animals (Amendment) Act of 1954 from having custody of animals.
- The Animal Boarding Establishments Act (1963) from keeping a boarding establishment for animals.

An applicant for a licence must be a qualified person within the above provisions, and he/she must be able to satisfy the local authority that he/she is suitable, and qualified, either by practical experience in the management of horses or by being the holder of one of the following certificates:

- BHS Assistant Instructor.
- BHS Intermediate Instructor.
- BHS Instructor.
- BHS or ABRS Fellowship.
- Any other certificate prescribed by order of the Secretary of State.

An inspecting veterinary surgeon is required to consider the suitability of the applicant in accordance with the above conditions.

Inspections

Before granting a licence the local authority grants powers in writing to an officer from their own or from any other local authority; this person will be a veterinary surgeon or veterinary practitioner selected from the list specifically drawn up jointly by the Royal College of Veterinary Surgeons and the British Veterinary Association for this purpose. He should carry out a detailed inspection of the premises and submit a report to the local authority, who will consider whether the premises and the persons employed in the management of the riding establishment are suitable to be holders of a licence.

Inspectors are recommended to visit the riding establishment at any time within reason when all horses are likely to be at the stables: giving not more than 24 hours' notice of the visit. The inspection should be made at the time of year appropriate to the type of use of the animal, and to seasonal activities such as trekking or hunting. To determine the standard of care and management, the animals should be in full use.

Conditions of a Licence

Without prejudice to their discretion to withhold a licence on any grounds, the authority shall take particular regard to the following:

Accommodation

- Accommodation should be suitable for the animals in respect of size, construction, number of occupants, light, ventilation, drainage and cleanliness. Stalls should be wide enough and long enough to allow the animals to lie down and to get up easily and without risk of injury. Boxes should be large enough to allow the animal to turn round. Stalls and boxes should be free from fittings, projections or structural features which might cause injury. Doors should always open outwards. These and the following requirements apply to new constructions, and to buildings which have been converted for use as stabling.

  The local authority considers the number of horses kept at an establishment (including animals at livery) in relation to the buildings and land available, and may impose a condition specifying the maximum number of horses of all categories (both for use in the riding establishment or otherwise) which should be kept at any time. Yards should provide sufficient space for every animal kept on the premises.

- Lighting should be good enough to preclude the use of artificial light during daytime. Switches, wires and other electrical equipment should be protected in such a way that horses cannot injure themselves.

- There should be adequate ventilation without draught.

- Drainage must carry away any liquids voided, and be sufficient to keep the boxes and stalls dry. Provisions must be made for the storage and regular disposal of manure and waste bedding.

- Adequate and suitable food, water, and bedding, together with both rest and exercise where required, must be provided, as well as suitable facilities for the storage of reasonable reserves of food and bedding.

Grazing

- Where horses are kept at grass there must be adequate pasture, suitable shelter, and water at all times. When the animals are either in work, or during the winter period when the grass is not growing, the Acts require that adequate supplementary feed should be given. The veterinary surgeon takes into consideration the type of animal, together with the type and location of their pasture. Arabs, Thoroughbreds or hunter types require more protection from the weather than do native ponies.

- It is necessary within the Acts to maintain fences in a safe condition and to keep the grazing free from hazards and rubbish.

- A competent person who can recognise injuries or illness must visit the horses at grass daily.

Horses

- The term 'horse' within the Acts includes any mare, gelding, pony, foal, colt, filly, stallion, ass, mule or jennet. They are required to be kept in good health, physically fit, and suitable for the purpose for which they are being maintained.

- It is not permissible to use animals that are three years old or under, or mares heavy in foal or within three months of having foaled.

- Good, not thin condition is required.

- It is not necessary for horses to be shod, but their feet must be kept well trimmed and in good condition. If they are shod, the fitting of the shoes must be correct.

- Animals must be free from illness, sores, galls, or injuries from the bit, saddle or other sources. Where injury or illness has occurred it is an offence not to provide curative care and treatments.

- On any inspection of the premises by an authorised officer of the local authority, a horse found to be in need of veterinary attention must be removed immediately from work on the verbal instruction of the officer. These instructions are generally confirmed in writing as soon as possible after the inspection. The horse cannot be returned to work until the licence holder has obtained at his/her own expense a certificate from a veterinary surgeon that the horse is fit for work, and has lodged the certificate with the local authority. This is a mandatory condition in the granting of a licence.

- A register of all horses three years old and under and in the possession of the licence holder must be maintained and must be available for inspection.

- Horses at part livery and partially used in the riding establishment are within the provisions of the Act and subject to inspection.

- Horses kept at full livery for private owners should be noted and can be inspected by the authorised officer who has powers under Section 1 (3) of the principal Act to inspect any horse found on the premises.

- It is an offence under the Acts to conceal, or cause to be concealed, any horse maintained on the premises, with the intention of avoiding an inspection of this horse.

- It is also an offence to allow a horse to be in such a condition that riding him would be likely to cause him suffering, as would letting him out for hire, or using him to provide, in return for payment, instruction in riding, or to demonstrate riding, whether for payment or not.

- When instruction in riding is given on a horse that is the property of the pupil receiving instruction, it is not usually considered that this requires a licence, regardless of where that instruction is given.

Saddlery

- All riding equipment must be maintained in good condition and, at the time when it is supplied to the rider, should not be subject to any defect which on inspection is considered to be likely to cause suffering to the horse or an accident to the rider. It also constitutes an offence to supply a saddle which is ill-fitting, causing a sore back; or bits, curbs, etc. which cause injury to the mouth; or bridles, girths, stirrup leathers and irons which might break due to faulty

materials, manufacture, or stitching and thus place the rider in peril or render the horse liable to injury.

- The inspector examines the saddlery when fitted to the horse for which it is intended, and pays particular attention to the correct fitting of Western-type saddles and also to ex-army troop saddles, as many of these are not fitted with safety stirrup bars and might cause a falling rider to be dragged. This type of saddle should be used only with patent safety stirrup irons.

Infectious Diseases and First-aid Equipment

- All reasonable precautions must be taken against the spread of infectious diseases and, so far as it is possible, there should be provisions to isolate an infectious animal.

- Veterinary first-aid equipment and medicines must be kept on the premises in a suitable and clean place set aside for this purpose. It is recommended that before assembling this equipment, consultation as to its contents should be made by the licence holder with the establishment's veterinary surgeon. He will take into consideration the veterinary experience of the licence holder when advising on particular medicines etc., as well as on the standard contents, such as antiseptic solutions, powders, bandages, dressings, scissors and a clinical thermometer. It is also strongly recommended that the name, address and telephone numbers of the establishment's veterinarian and doctor should be prominently displayed.

Fire Precautions

- Precautions must be taken for the protection of the horses in case of fire. Notices prohibiting smoking should be displayed.

- There must be clear access to all stalls and boxes, and where multiple numbers of stalls and boxes are housed within the same building, more than one exit is strongly recommended.

- Fire extinguishers must be regularly serviced, and an adequate supply of water should be available easily.

The local fire-prevention officer should be consulted on the types of fire-fighting equipment most suitable for the premises.

- The 1970 Riding Establishments Act requires that there should be a notice on the outside of the premises, giving the name, address, and telephone number of the licence holder or other appointed responsible persons, together with directions as to the actions to be taken in case of fire. The directions should include particular advice on the removal of animals from the stables.

Management and Supervision

- A person under sixteen years of age must not be left in charge of the business. All horses let out on hire must be supervised by a person of sixteen years or over, unless the licence holder is satisfied that the hirer is competent to ride without supervision. Before he can be satisfied that the rider is competent to ride unsupervised, a licence holder must therefore have either previous knowledge, or have made an assessment of the capability of the rider and must then provide a horse 'suitable for the purpose' of that rider's requirements and ability. A licence holder who has failed to make the correct assessment might be unable to use this as a defence should litigation follow.

- The knowledge of the licence holder is an integral part of the granting of a licence, and should this knowledge be regarded as inadequate, the applicant can appoint a manager or other person to supervise the riding establishment on his/her behalf. The local authority can then issue a licence which will also carry the name of the person with the necessary knowledge or qualifications upon it. If this person leaves the establishment during the currency of the licence, it would appear that the licence becomes null and void.

- Similarly, it would appear that a riding establishment cannot be sold as a licensed business, because without the qualifications or experience of the licence holder the licence is ineffective. Any purchaser therefore has to make an application to the local authority for a licence before proceeding to operate the business, and to satisfy the authority that he/she has the necessary qualifications or experience.

- Provisions are made whereby, in the event of the death of a person holding a current licence, the licence passes to his personal representatives, and can be extended for one year (three months if a provisional licence) from the date of death. Such an extension can be made more than once if the local authority is satisfied that the extensions are necessary to wind up the estate.

Insurance

- The 1970 Riding Establishments Act clearly states that the licence holder shall hold a current Public Liability Insurance Policy which provides indemnity against liability at law to pay damages for accidental bodily injury sustained by the hirer of a horse, or those using the horse to receive instruction in riding. The licence holder must also insure the riders in respect of any liability which they might incur through injuries to any other person, caused or having arisen through the hire or use of the horse.

- The amount of indemnity is not specified, but licence holders are advised to make sure that the amount exceeds the highest awards the courts have made in respect of riding accidents.

- In addition, where the business employs staff, compulsory insurance is required under the following:
  - Employer's Liability (Compulsory Insurance) Act 1969.
  - Employer's Liability (Defective Equipment) Act 1969.

Rights of Local (Licensing) Authorities

- Powers of entry.
- The right to impose conditions upon a licence so as to secure all the objects specified within the Acts.

- The right to dispense with some conditions if they think that circumstances so warrant.

Five conditions in Section 2 of the 1970 Act are mandatory. They are:

1. The removal from work of horses which are in need of veterinary attention.
2. The supervision by responsible people aged sixteen years or over.
3. Leaving the premises in charge of a person over sixteen years.
4. The insurance policy requirements.
5. The register of horses three years old and under.

Any person who operates or has intentions of operating a riding establishment is advised to obtain copies of both the 1964 and 1970 Riding Establishment Acts in order to be fully conversant with their requirements.

THE HEALTH AND SAFETY AT WORK ACT 1974

The provisions of this Act apply to employers and employees engaged in the keeping and management of livestock. Riding schools, livery yards, etc., are not specified, but do come under the Act. Responsible employers should be aware of their duty, both to themselves and to their employees, to see that work conditions are as safe and as healthy as possible.

Any business employing more than five people, including casual workers, must issue a Safety Policy Statement. Instructions to employees should be clearly set out. Assistance in drawing up this statement can be obtained from the BHS Riding Schools and Recreational Riding Office, The Deer Park, Stoneleigh, Warwickshire; or The Association of British Riding Schools, Old Brewery Yard, Penzance, Cornwall TR18 2SL; or the local Health and Safety Executive (address and telephone number in the local Yellow Pages Directory). The local office of ADAS also supplies leaflets and a poster on safety measures and safer working conditions.

In December 1986, HM Agricultural Inspectorate of Health and Safety published requirements extra to the 1974 Act. Many of the following directives are covered, but in addition there are strict regulations for health and safety in relation to food preparation, first aid, sanitation, etc. Responsibility for staff training, instruction, suitability of horses, dress, etc, are also covered. It is therefore essential for persons setting up – or running – a yard to obtain a copy of Horse Riding Establishments Guidance on Promoting Safe Working Conditions by the Health and Safety Executive. They can thus ensure that they are able to comply with the necessary requirements of HM Agricultural Inspectorate.

General Obligations

- Employers must ensure the safety of their employees by maintaining safe systems of work, safe premises, and safe equipment.

- Employees and the self-employed must take reasonable care to avoid injury.

- Employers, the self-employed and employees must not endanger the health and safety of third parties.

- There should be a named person to whom any faults in equipment or other hazards can be reported.

- Employers must ensure that all employees and others on the premises are correctly instructed in any work that they have to do and in any equipment that they have to use.

- Employees should be encouraged to produce new ideas for improved safety measures and methods of working.

- Employers should ensure that their insurance policies cover the use of such equipment and machinery by any member of staff, paid or unpaid.

- Employer's liability insurance is compulsory for all employers, and an up-to-date certificate of this must be displayed.

- Well-equipped first-aid facilities should be available on the premises and a responsible person always on call when required. Leaflets on first aid can be obtained from HMSO Publications Centre, PO Box 276, London SW8 5DT or the local Health and Safety Executive.

Risk Areas

- Employers should give these special attention.
- Employees should be instructed as to any special precautions which should be taken.
- Employees should be instructed in the correct and safe use of equipment and machinery.

Employees likely to be left in charge of the yard, in the absence of senior staff, should be fully briefed as to what to do should an emergency arise. A list of relevant telephone numbers – for doctor, veterinary surgeon, owner or manager, electrician – should be provided.

Employees should be instructed about:

- The working of the trip-switch and how it is re-set
- The whereabouts of all water-main stop-cocks.

The main risk factors are:

COMBUSTIBLE MATERIALS – stables; hay barns; oil, diesel or petrol storage areas; electric, gas or oil heaters; bonfires.

ELECTRICAL EQUIPMENT AND MACHINERY such as clipping and grooming machines, corn mill, etc. All electrical wiring, fittings and equipment should be regularly inspected. Equipment should be regularly serviced. A circuit-breaker plug should be used with all electrical equipment. Electric cable should be inspected before using any equipment, to see that it has not been damaged by friction or by a horse treading on it. Plugs should be checked for faulty wiring and cracked casings. Personnel should be properly instructed in the equipment's use and care. Junior staff should be supervised when using electric clippers and groomers.

The corn mill and/or chaff cutter should be worked only by experienced staff. If it is not fitted with a dust extractor, face masks must be used. The dangers of inhaling dust should be explained.

CHAFF CUTTER If electrically powered, see previous paragraph. If either hand or electric powered, guards must be fitted to protect operator's hands. Warning notices should be fixed on the machine, and the risk of getting hands or clothing caught up should be explained.

BARLEY BOILER If electrically powered, see Electrical Equipment, above. If gas powered, it should be well away from any combustible material, on a concrete floor and be worked only by experienced personnel.

TRACTORS, FIELD MACHINERY, HAY TRAILERS, ETC. Regulations and advice are published in a Ministry of Agriculture, Fisheries and Food pamphlet, which should be essential reading for all personnel.

- No extra persons should be allowed to travel on the tractor, hay trailer, or other farm vehicles.
- Personnel driving the tractor should have a current driving licence.
- Tractors must be suitably insured, and if driven on or across a road, the necessary road tax must be paid. The dangers of working on steep hills, close to deep ditches and boggy areas should be explained; also the risk of allowing children to play in the area where the tractor is kept or where it is working.

LAND-ROVER, LAND-ROVER AND TRAILER, HORSEBOX Drivers must check that children and animals are not near when drawing away or reversing.

HORSE WALKERS OR HORSE EXERCISERS These must be well maintained in a suitably fenced area. They should only be used by experienced personnel. Horses should not be attached to the machine by junior staff.

OUTSIDE STRUCTURES Buildings should be kept in sound order. In strong winds, loose tiles, slates or guttering can be dislodged and cause injury to anyone passing underneath.

STEPS AND STAIRS up to or down into working areas should be sound and with a level tread. It may be advisable to fit a hand-rail or rope.

LADDERS AND STEP-LADDERS must be strong enough for the work involved. Unsupervised junior personnel must not be allowed to use ladders. Two people should be responsible for putting up the ladder and for taking it down and putting it away. Ladders should never be left up against a stack or near a stack, as they can be knocked down, breaking the ladder and possibly causing injury. They may also be an invitation to children to climb up and play on top of the stack. This must never be allowed.

LIFTING LOADS

- When collecting hay from a stack, employees must be instructed to take bales from the top of the stack and work down. Lower bales should not be pulled out, as this may cause the stack to fall, resulting in serious injury.

- When loading hay bales on to a trailer and taking them to the yard they must be stacked with care, so that the load is steady. If it is to be transported on a public road, the load should be roped for extra safety. Extra care should be taken on a steep or uneven surface.

- When lifting bales and sacks of feed, personnel must be instructed to lift weights in the proper manner. If in doubt, they should ask for help. Senior staff should check that the weights are not too heavy for the persons involved. They should not be allowed to risk straining their back.

Instructions for lifting sacks and other heavy weights:

- Estimate the weight and if necessary ask for help.
- Stand close to the sack.
- Square the sack up in front of you.
- Bend your knees; do not lean over and lift up.
- Take hold of the sack. Lift it by straightening the legs.
- Look where you are going. When putting the sack down, bend your knees. Try not to bend your back.

YARD AREA Stable tools must be carefully used and tidily replaced after use. If left about they are a hazard to staff and horses: they can fall down, be tripped over or stepped on, resulting in a cut foot and/or bruises to face and eyes. Rakes should always be left standing upright, with the head at the top.

BALING STRING When the bale is to be used, the string should be cut, tied up and placed in a suitable sack or bin. Left on the ground, uncut, it can cause a fall and injury should either a person or a horse get their feet caught in it.

WHEELBARROWS should never be left about. When in use, they should be placed parallel to the stable wall. Handles should never be left pointing outwards to the yard. When not in use, barrows should be parked tidily in the stable tool area.

YARD SURFACES should be swept clean. It may be necessary to hose them clean should a film of mud and manure form, as this can be very slippery. It is likely to happen in springtime when horses and ponies are brought in from the fields.

NARROW DOORS AND GATEWAYS should have warning notices. Internal swing doors for feed shed, tack rooms or office are dangerous and are not recommended.

COLD WEATHER CONDITIONS In freezing weather salt should be used if there is ice on the yard. (This should be obtained in the autumn.) In deep snow, working paths should be cleared and then kept clear with regular applications of salt. As an emergency measure, sand or coal ashes help to keep a slippery surface usable. In continuing freezing weather, straw or shavings manure tracks, 2m (6ft) wide, give a secure footing for horse and staff. This is more effective than using plain straw. Water buckets must be emptied carefully into a drain, and not spilt on the yard to create an icy surface.

FIELDS, FENCES AND GATES should be kept in good order. Heavy gates on broken hinges can cause a fractured leg for human or horse. Bog and deep ditch areas should be fenced off. If a field entrance on to a road is dangerously placed it should preferably be moved but, if this is not possible warning notices should be fixed, and staff should be advised of the danger.

CAR PARK, AND YARD AND DRIVE ENTRANCES The following warning notices should be fixed:

- Speed restriction 5 mph.
- Children and horses.

Drives and access roads can be made safer by placing 'humps' at intervals to slow down traffic. Cars, horseboxes and trailers should be parked away from the stable yard, preferably in a car park. There should also be a special area set aside for bicycles.

TREES Tall old trees near the stable yard and car park area can be a hazard. They should be regularly inspected, and if necessary lopped or felled.

THE ANIMALS ACT 1971

This Act is not applicable in Scotland or Northern Ireland. It is based on strict liability and deals with the injury that horses may cause, and with damage arising from their straying. Strict liability means that there is liability without proof that the person claimed against was negligent, if he was responsible for the animal, and the animal has caused injury or damage.

Animals

These are divided into animals belonging to a dangerous species, and those which do not. Bulls and wild and unbroken stallions may come within the former category. Liability for injury caused by animals of a dangerous species will be placed on the 'keeper' unless he can bring himself within one or more of the exceptions of the Act. Liability for damages caused by an animal which is not of a dangerous species (such as a horse or pony) also rests with the 'keepers' in the following circumstances:

The damage is of a kind likely to be caused, and when caused is likely to be severe, unless the animal is restrained; and the likelihood of the damage, or of its being severe, must be due to the characteristics of that particular animal which are not normal in animals of this species except at particular times or circumstances: e.g. mares in season, etc.; and those characteristics were known to the keeper or known to any other person who had charge of the animal as the keeper's servant, or where that keeper is the head of a household, or were known to another keeper of the animals, who is a member of that household and under sixteen years of age.

Who is a Keeper?

A person will be treated as a keeper if:

- he owns the animal, or has it in his possession, or
- he is the head of the household of which a member under sixteen years of age owns the animal or has it in his possession.

The keeper must be an individual. The person who has control over company-owned horses would be regarded as the keeper. Liability for damage rests on the person who at the time has possession; he need not be the owner. Anyone who takes possession of and keeps horses to prevent them from causing damage or to restore them to their owner is NOT a keeper.

Damage by Trespassing Stock

In broad terms an animal is said to trespass in the same way as humans: that is, where it is on land on which it has no right to be, or where its owner has no right to put it. This includes animals straying on to a highway or from a highway on to private land. Exceptions exist with regard to animals straying from unfenced common land on to highways.

The only lawful use of the highway (unless they have grazing rights on the verges) is to pass and repass on it. The highway need not adjoin the land; the stock may have wandered through other land from the highway. Bridlepaths are highways.

Section 8 of this Act imposes a duty on the person placing horses on the land to take care to prevent them from straying on to the highway: the only exception being if they stray from unfenced land, and providing the person who placed them on the land had a 'right' to do so. Unfenced land is regarded as land in an area where fencing is not customary; this is common land, town or village greens or some moorlands.

Claims under Section 2 for Injuries and Damage Done by Animals

1. Where a horse strays on to land owned or occupied by another person, the owner of the horse is strictly liable for the following damage and expenses:

    (a) Damage done to the land or to any property on the land which is owned or in possession of another person.

    (b) Expenses which are reasonably incurred by the other person in keeping the horse until it can be returned to the owner.

    (c) The expense of finding out to whom the horse belongs.

    (d) Expenses incurred by the occupier of the land in exercising his right to detain the horse.

    (e) Injuries caused by the straying horse; death of or injury to, a person is included in the word 'damage'.

2. There would be NO responsibility for straying livestock if:

    (a) The damage is due wholly to the fault of the person suffering it, but that person cannot be regarded as at fault just because he could have prevented it by fencing, or

    (b) It is proved that the straying would not have occurred but for a breach of duty imposed on another person who has interests in the land to fence it, or

    (c) The person suffering the damage has voluntarily accepted the risk. But where a person employed by a keeper incurs a risk as a result of his employment he shall not be treated as accepting it voluntarily.

    (d) The livestock strayed from the highway whilst lawfully using it.

3. The occupier of the land on to which the horse strays without being under the control of any person has the right to detain it, unless ordered by a court to return it. The right ceases, however:

    (a) At the end of 48 hours unless the occupier notifies a police officer and also the person to whom the horse belongs, if that person is known.

    (b) If the person claiming the horse offers the occupier sufficient money to cover any proper claim for damage and expenses.

    (c) If the occupier has no proper claim, when the horse is reclaimed by someone entitled to its possession.

4. If a horse has been detained for at least 14 days the occupier may sell it at market or by auction, unless any court proceedings are pending. The occupier is entitled to the proceeds of the sale, but any excess over his claim has to be paid to the person who but for the sale would have been entitled to possession.

5. Occupiers are liable for any damage caused to the horse that they have detained, by failing to treat it with care, or by failing to supply it with food and water.

6. Occupiers who sell the animal have no right to sue for damages. They must choose one remedy or the other.

Fencing

The Act does not define a 'duty to fence', so common-law rules have to be considered. These show that the law has never imposed any obligation on a person to fence his land, but obligations may have been created by formal agreements or may have become established through long usage or custom.

For example, if a person has maintained a fence for, say, forty years and upwards this would indicate that he thought he had a duty to do so. He cannot then deny the obligation, and his actions have created a liability on him and his successors in title.

Where horses stray from common land where they had a 'right' to graze, the position might differ if it can be established that it was the responsibility of the land owner to fence against animals straying from it. Any fences erected must be kept in a proper state of repair so as not to constitute a danger to adjoining occupiers or to people on a highway.

Dogs

This Act also gives protection against dogs either killing or causing injury to livestock. For proceedings and defences, reference should be made to Section 9 of the Act.

PROTECTION OF ANIMALS ACT 1911 – ENGLAND AND WALES

PROTECTION OF ANIMALS ACT 1912 – SCOTLAND

These two Acts are perhaps the most widely used of the many Welfare of Animals Acts currently on the statute book. Both Acts are similar in content. Any reference to magistrates in the Protection of Animals Act 1911 should be read as a reference to the sheriff in the Protection of Animals Act (Scotland) 1912. As the powers of courts in regard to penalties are subject to changes under the Criminal Justices Act, these have not been included.

The Acts consolidate the laws relating to cruelty to domestic and captive animals. The expression 'domestic animal' covers any animal of whatever species which is tame, or which has been, or is being sufficiently tamed to serve some purpose for the use of man. 'Captive animal' means any animal which is in captivity or confinement.

The 1911 Act has 15 sections but this chapter only deals with those sections which are relevant to horses. The expression 'horse' includes mare, gelding, pony, foal, colt, filly, stallion, ass, mule or jennet.

The Offences of Cruelty in Section 1 read that if any person:

(a) 'Shall cruelly beat, kick, ill-treat, over-ride, over drive, over-load, torture, infuriate, or terrify any animal, or shall cause or procure, or, being the owner, permit any animal to be so used, or shall, by wantonly or unreasonably doing or omitting to do any Act, or by causing or procuring the commission or omission of any Act, cause any unnecessary suffering, or being the owner permit any unnecessary suffering to be so caused to any animal, shall be guilty of an offence.'

This long paragraph indicates, among many other things, that if a person carries out any of the above offences, or hires or permits any other person to commit them, it will be an offence. It also means that by unreasonably preventing any of the following from taking place, the owner and/or offender will be guilty within the meaning of the Act. Failure to feed, failure to provide veterinary treatment, or failure to slaughter an animal which is incurably diseased and thereby causing unnecessary suffering may well be offences of omission. The abandonment of animals was not an offence under this Act as it was originally passed, so the Abandonment of Animals Act 1960 was passed to rectify the omission. It is now an offence to abandon an animal (whether permanently or not) in circumstances likely to cause that animal any unnecessary suffering. This would appear to include both liberating animals in order to get rid of them, and leaving animals shut up unattended.

The clause of causing or procuring the commission of certain acts by others, could be interpreted that the parents of children who, acting on the instructions of their parents, perform acts of cruelty, could be guilty. In such a case it would be irrelevant if the child is below the age of criminal responsibility. There is no reason why one person should not be charged with 'causing' or presumably 'permitting' and another with ill-treating in respect of the same act. This would be the case where a horse is ridden in an unfit state (e.g. lame), the stable manager (or the owner) knowing/causing (or permitting) it to be ridden, and the rider for knowingly riding it.

Cruelty and Unnecessary Suffering

The mere infliction of pain is not sufficient to constitute any offences given in the Act. It has been ruled that cruelty is defined as: 'causing unnecessary suffering', and courts have to examine whether the defendant is doing something which it was not reasonably necessary to do. If the reason was of sufficient importance to justify the act done then no offence has been committed. Suffering must not only be disproportionate to the alleged reason for which it was inflicted, but must also be substantial. What is clear from judgements given is that 'cruelly' and 'so as to cause unnecessary suffering' mean exactly the same thing. The inflicting of mental suffering by the defendant's own positive act could be within the category of cruelly infuriating and terrifying animals; it may also be included in the category of torturing animals.

(b) 'Shall convey or carry or cause or procure, or, being the owner permit to be conveyed or carried, any animal in such a manner or position as to cause that animal any unnecessary suffering' . This sub-paragraph refers to all kinds of animals carried, and in regard to horses it is generally assumed to be by mechanical methods. Other legislation exists where offences of carrying can occur both by air or sea, but perhaps the most relevant legislation is the Transit of Animals (Road and Rail) Order of 1975 and amendments 1979.

(c) 'Shall wilfully, without any reasonable cause or excuse, administer, or cause or procure or being the owner permit such administration of, any poisonous or injurious drug or substances to any animal, or shall wilfully without any reason cause or excuse, cause such substances to be taken by any animals'.

This means that the administration of any injurious drug which causes unnecessary suffering, or the owner attempting or permitting an attempt by another person, to poison an animal is regarded as an offence.

(d) 'Shall subject or cause or procure, or being the owner permit, to be subjected, any animal to any operation which is performed without due care and humanity'.

Treatments and operations to agricultural animals (as defined in the Agriculture Act 1947) which are permitted to be carried out by unqualified persons, are specified in Schedule III of the Veterinary Surgeons Act 1966.

The Powers of a Court to Order Destruction of Animals in Section 2

Where the owner of an animal is convicted of an offence of cruelty within the meaning of this Act, the court has powers, providing they are satisfied that it would be cruel to keep the animal alive, to direct that:

(a) the animal should be destroyed, and

(b) the animal should be assigned to any suitable person for destruction, and the person to whom it is assigned shall as soon as it is possible destroy that animal or arrange for its destruction in his/her presence without unnecessary suffering.

The court may order that any reasonable expenses incurred in the destruction shall be paid by the owner of the animal or be recoverable by a summons as a civil debt. It should be noted that unless the owner agrees to the destruction, a veterinary surgeon must give evidence to the court to the effect that it would be cruel to keep the animal alive. If no such consent, or evidence, is given, courts cannot make an order for destruction.

There can be no appeal against an order for the destruction of an animal. However, the magistrates may themselves suspend or rescind an order on complaint.

The Powers of Confiscation in Section 3

If the owner of any animal shall be guilty of cruelty within the meaning of this Act, the courts, upon conviction of the owner, may if they think fit, in addition to any other punishment, deprive such a person of the ownership of the animal, and they may also make an order as to the disposal of the animal as they think fit under the circumstances. They are not permitted to make such an order, unless it is shown by evidence of a previous conviction, or as to the character of the owner, or otherwise, that if the animal is left with the owner, it is likely to be exposed to further cruelty.

The Regulations for Knackers in Section 5

Any police constable has a right of entry to a knacker's yard during hours when business is usually being carried out, for the purpose of examining whether there has been any contravention or non-compliance with the provisions of this Act. For the purposes of Section 1 of this Act relating to offences of cruelty the knacker shall be regarded as the owner of any animal delivered to him. The expression 'knacker' means a person whose trade or business is to kill any cattle including horses for the purpose of the flesh not being used as butchers' meat.

Animal Pounds in Section 7

If anyone impounds, or causes to be impounded or confined in any pound, any animal, they shall while the animal is so impounded or confined, be responsible for supplying a sufficient quantity of wholesome and suitable food and water to that animal, and if they fail to do so will be liable upon conviction to a fine.

The person taking the animal to the pound is therefore responsible for it under this Section, not the pound keeper. The pound keeper may be guilty of an offence under sections of this Act if it can be shown that he was legally responsible for performing that act.

If any animal impounded or confined is without sufficient suitable food or water for six successive hours, or longer, any person may enter the pound for the purpose of supplying the animal. The reasonable cost of the food and water supply shall be recoverable from the owner of the animal as a civil debt.

Poisoned Animal Food in Section 8

It is an offence to offer for sale, to give away or to cause or procure any other person to do so any grain or seed which has been rendered poisonous, except for bona-fide use in agriculture. It is also an offence to place, cause some other party to place, or knowingly be a party to the placing in or upon any land or building any poison or fluid or edible matter (which is not sown seed or grain) that has been rendered poisonous.

The defence can be used that the poison was placed by the accused for the purpose of destroying insects or small ground-vermin where such is necessary in the interest of public health, agriculture or the preservation of other animals, or for the purpose of manuring the land, providing that all reasonable precautions to prevent injury to all domestic animals and wild birds were taken.

Diseased and Injured Animals in Section 11

(i) If a police constable finds any animal so distressed or so severely injured, or in such a physical condition that in his opinion, there is no possibility of removing it without cruelty, he shall in the absence of the owner, or if the owner refuses to consent to the animal being destroyed, summon a veterinary surgeon if any reside within a reasonable distance and providing it appears by the certificate of that veterinary surgeon that the animal is mortally injured, or so severely injured, or so diseased, or in such a physical condition that it is cruel to keep it alive, it is lawful for the police constable, without the owner's consent to slaughter or arrange for the slaughter, with such instruments or appliances, and with such precautions, and in such a manner as to inflict as little suffering as is possible. The constable also has powers to arrange for the removal of the carcasses if the slaughter of the animal takes place on a public highway.

(ii) If a veterinary surgeon certifies that the animal can be removed without cruelty, it is the responsibility of the person in charge to arrange for the removal with as little suffering as possible, and if the person in charge fails to do this, the police constable may without the consent of the person in charge cause the animal to be removed.

(iii) Any reasonable expenses incurred including those

of the veterinary surgeon and regardless of whether the animal is slaughtered or not may be recovered from the owner as a civil debt.

(iv) 'Animal' in this section applies only to horses (as already defined), and to bulls, sheeps, goats or pigs.

The Summons in Section 13

Frequently, the most valuable piece of evidence is the animal itself. If the owner refuses to produce the animal, or to have it examined by a vet, a summons may be applied for ordering the production of the animal in the same way as a summons to an accused person.

Rights of Appeal in Section 14

(i) Any appeal from conviction or order (other than appeal against destruction in Section 2) made by a summary court, shall be made to a Crown Court.

(ii) Where there is an appeal against a summary court conviction or order, the court may order the owner not to sell or part with the animal until the appeal is determined or abandoned, the court may order that the animal be produced on the hearing of the appeal providing such production is possible without cruelty being incurred.

PROTECTION OF ANIMALS (AMENDMENT) ACT 1954

This Act, which applies to Scotland but not to N. Ireland, extends the powers of the court to disqualify from having custody of animals, a person previously convicted of an offence of cruelty to an animal.

Powers of Disqualification in Section 1

(i) Where a person has been convicted under the Protection of Animals Act 1911 or the Protection of Animals (Scotland) (1912), and is subsequently convicted under either of these Acts of an offence, the court by which the subsequent conviction is made, may if it thinks fit either, in addition to, or in substitution for, any other punishment, order the person to be disqualified, for a such period as it thinks fit, from having custody of any animals or any animal of a kind specified.

(ii) A court which has made such a disqualification order may if it thinks fit suspend the order:

(a) For a period of time as the court thinks necessary to enable arrangements to be made for the custody of any animals to which the order relates; or
(b) Pending an appeal.

(iii) Any disqualified person may after twelve months from the date of the order and subsequently from time to time, apply to the court by which the order was made for removal of the disqualification, and the court will consider the character of the applicant, the conduct following the order, the nature of the offence of which he was convicted, and any other circumstances of the case; and either:
(a) Direct the removal of the disqualification from such a date as they specify, or shall vary the order to apply only to animals of a kind specified; or
(b) Refuse the application.

Further applications for variations cannot be made within 12 months after the date of the direction or refusal.

ADDITIONAL EFFECTS OF DISQUALIFICATION . Any person disqualified under this Act from having custody of an animal may not be granted a licence to keep either an animal-boarding establishment or a riding establishment, a pet shop, or dog-breeding kennels.

Further penalties can be imposed for breaches of any disqualification order.

PROTECTION OF ANIMALS (AMENDMENT) ACT 1988

Gives power to the court to disqualify a person convicted of cruelty to animals from having custody of animals.

PROTECTION AGAINST CRUEL TETHERING ACT 1988

It shall be an offence to tether any horses, asses or mules under such conditions or in such manner as to cause that animal unnecessary suffering.

RIDDOR (REPORTING OF INJURIES, DISEASES AND DANGEROUS OCCURRENCES REGULATIONS) 1985

Certain accidents must be reported to the enforcement authority (either the Health and Safety Executive – HSE or the local authority) either immediately or within seven days. Accidents which result in absence from work for more than three days (excluding the day the accident occurred but including non-working days) must be reported (using Form F2508) within seven days of the accident. Fatal accidents or major injury must be reported immediately by telephone with a written report within seven days. Major injury accidents include any fracture (other than a bone in the hand or foot), amputation, loss of eyesight or any penetrating injury to the eye, loss of consciousness or any other injury requiring admission to hospital for more than 24 hours.

Records of reportable injuries must be kept for at least three years and must include:

• Date and time of accident.

• Name, address and occupation of person affected.
• Where the accident occurred, the nature of it and a brief description of the circumstances.
• Additional information (e.g. any witness statements) is always valuable.

COSHH (CONTROL OF SUBSTANCES HAZARDOUS TO HEALTH) 1989

Every employer has a responsibility to assess the risks associated with hazardous substances in the workplace and to take steps to eliminate or control those risks. In the stable yard this would relate to substances like disinfectant (used for floors and drains), creosote (for maintaining fences), or liniment (for treating minor muscular injuries in horses).

The employer must identify all such substances ensuring that the following procedure relates to their storage and use.

• Safe storage in identified area.

• Specified use, when, where, and by whom.

• Necessary relevant training for use with appropriate protective clothing/equipment provided.

• Data sheet maintained on each and every identified substance to provide full information on the nature and relevant severity of its potential toxicity and immediate action to take if poisoning should accidentally occur.

APPENDICES

APPENDIX

1 RATIONING

This appendix is for those who like to know some of the theory and science behind the practice.

Nature is both economical and well balanced so it provides plants that are complementary to animals. Plants store the sun's energy and animals use it to drive their body processes and provide movement. Plants build up natural elements into proteins; animals rebuild these to form their own structure. When devising rations we must be particularly concerned with energy and protein.

In the wild, the horse selects the foods he needs; in domesticity we must provide the right quantity of a suitable mix of foods, particularly for the stabled horse. This is the art of rationing. As already shown, this is achieved partly by experience, partly by observation and partly by careful thought based on sound theory. However, it is normally based on simple calculations which once done, create a starting point from which later changes can be made. Also, calculations allow different foods to be compared so that the ration is not too costly.

The following rationing system for horses involves six simple steps.

EXAMPLE 1

Step 1
How big is the animal? Measure or estimate (see page 159).

Step 2
How much can it eat each day? (See page 159 and the table on page 160).

Allowing, say 2.5% of his body weight, divide the weight of the horse by 100 and multiply the answer by 2.5. A 500kg horse will thus eat 5 x 2.5 = 12.5kg (about 28lbs).

Step 3
Provide for maintenance using mostly grass, hay or other forage (see page 157). This amount varies according to the quality of the food; so calculate what is needed first, then divide that amount according to the quality value of the food proposed and thus find the weight of forage needed. The unit used is Digestible Energy, measured in megajoules (expressed as MJ of DE). As shown on page 162, a horse or pony needs 18MJ of DE plus one more MJ for every 10kg of his body weight. Thus a 500kg horse needs:

18 + 50 = 68MJ of DE.

To provide 68MJ from our average hay which has 9MJ per kg (see table on page 162), divide 68kg by 9. Answer = 7.6kg (about 17lbs) of hay is needed for maintenance. (For fast work provide some of this in concentrates to keep the horse slim.)

Step 4
Provide concentrates for work. As shown on page 162, provide 1–8MJ of DE for each 50kg of body weight. In

selecting between 1 and 8 use the scale points of work as shown on page 163. For example, a 500kg horse in medium work on a scale point 4 needs 10 x 4 = 40MJ of DE to provide enough energy for its work. If this is provided by horse and pony cubes which have a DE of 10MJ per kg (see table on page 162) then 4kg (about 9lbs) of cubes are needed.

Step 5

Check the protein, check the hay/concentrate ratio and check for capacity.

The text on page 162 suggests that medium work needs 7.5–8.5% of protein in the total ration.

The table on page 162 shows that the hay has about 8% protein and the cubes will have 10%.

This gives a ration of 8.6% protein – a little over but near enough.

A ration consisting of 7.6kg of hay and 4kg of cubes has a ratio of 65% hay and 34% concentrates. (The table on page 163 suggests a slightly higher percentage of concentrates for medium work so consider this in Step 2.)

Step 2 showed that the horse could comfortably eat 12.5kg. Thus if we are providing 7.6kg of hay and 4kg of cubes, making a total of 11.6kg, we are well within its capacity and can give more if it seems appropriate.

Step 6

Feed this ration, observe, consider and adjust as necessary.

EXAMPLE 2

Here the system explained in Example 1 is applied to a 16.2 hh advanced event horse.

Step 1

Weigh the horse on a weighbridge (using a trailer, then subtracting the weight of the empty trailer). He weighs 600kg.

Step 2

Capacity at 2.5% of body weight =

$(600 \div 100) \times 2.5 = 15kg$ (34lbs)

Step 3

Provide for maintenance =

$(18 \div 10) + 600 = 78MJ$ of DE needed

Supply only 40% of capacity from hay:

$15 \times 0.4 = 6kg$ (13.3lbs)

Our good hay has a DE of 11MJ, so 6kg x 11 = 66, leaving (78 – 66) 12 MJ still needed.

Step 4

Provide for production. Our horse is on scale point 6 (hard work).

$(600 \div 50) \times 6 = 72MJ$ of DE

Add the 12MJ still needed for maintenance:

$72 + 12 = 84$ MJ of DE

Feed, say, eventing cubes at DE of 13 MJ = 6.5kg (14lbs)

Step 5

From Steps 3 and 4 the suggested ration is:

6kg of good hay at, say, 9% CP
6.5kg of eventing cubes at, say, 11% CP

Check the protein level: it is 10%
Check the hay/concentrate ratio: it is 48% to 52%.
Check the capacity: we are feeding 12.5kg (the capacity is 15kg).

Step 6

Feed the ration, observe, consider and adjust if necessary.

2 MEASUREMENTS

Land-based industries in Britain went metric in 1976 and ever since the feed room has been a muddle of metric and imperial measurements. Now, with our closer ties to mainland Europe, it is important that everyone in Britain should try to use metric measures.

Weight

If in doubt remember: 'two and a quarter pounds of jam weigh about a kilogram.'

pounds (lbs)	kilograms/pounds (kg/lbs)	kilograms (kg)
2.2	1	0.45
4.4	2	0.9
6.6	3	1.36
8.8	4	1.81
11.0	5	2.27
13.2	6	2.72
15.4	7	3.18
17.6	8	3.63
19.8	9	4.00

- Note the weight on sacks and containers.
- Resolve to make every feed board metric.

Typically:
- Horse cubes come in 25kg bags but some straight feeds come in 40kg sacks.
- Fertiliser comes in 50kg bags (used to be supplied in 1cwt bags).
- Hay bales weigh about 25kg (40 to the tonne).
- Straw bales weigh about 16.5kg (60 to the tonne).
- Stacked hay takes up about 7 cubic metres per tonne.
- Stacked staw takes up 10.5 cubic metres per tonne.
- A hay crop of 10 tonnes per hectare used to be reckoned as 4 tons per acre. (A hectare is about 2.5 acres, which is about the size of a professional football pitch).
- A ton and a tonne (metric) are nearly the same.

Volume
- A litre is about 1¾ pints, or remember the phrase: 'A litre of water's a pint and three quarters'.
- We know from filling up with petrol at a garage that whereas we used to take 10 gallons we now take 45 litres.
- A 5-litre container holds just over 1 gallon.

THE BRITISH HORSE SOCIETY

WHAT IS THE BRITISH HORSE SOCIETY?

The British Horse Society is a charity to promote the welfare, care and use of the horse and pony; to encourage horsemastership and the improvement of horse management and breeding; to represent all equine interests.

The Society is internationally recognised as the premier equestrian riding, training and examination organisation in the United Kingdom, and operates an Approvals scheme for all types of equestrian establishment. It incorporates some 390 Riding Clubs and works closely with the Pony Club. The Society is also the national governing body for recreational riding and fully supports the independent sporting disciplines within the British Equestrian Federation.

The BHS plays a major role in equine welfare, safety, provision of access to the countryside, and protection of riding and driving Rights of Way. It represents riders to Government and to the EU in Brussels in all matters, especially those concerning taxation, rates, planning and the law.

Membership benefits include £5 million public liability insurance, personal accident insurance, a yearbook, magazines, special facilities and discounts at BHS functions, and access to BHS advice and support.

On the 30th September 1997 the BHS had 60,378 members. In addition, Affiliated Riding Clubs represented approximately 40,000 members.

By joining the Society you are helping all who ride. For further information and membership details write to: The British Horse Society, The Deer Park, Stoneleigh, Kenilworth, Warwickshire, CV8 2XZ. Tel: 01926 707 700.

WHY CHOOSE A BHS-APPROVED ESTABLISHMENT?

Whatever your reason for seeking out a riding centre – be it for a pleasant hour's hacking, or training up to advanced level – you will want to choose an establishment that is both reputable and well run.

By selecting one of the establishments listed in the BHS publications **Where to Ride** and **Where to Train** , you will not only be able to satisfy yourself on that point, but you will also be able to discover a good deal about the centre before you pick up the telephone to discuss your ride.

Each of the establishments listed in **Where to Ride** and **Where to Train** has been inspected by a fully qualified representative of the British Horse Society, who will have looked carefully at the level of instruction available to clients, the on-site facilities and equipment, and the standards of safety, horse care and management.

A riding centre will not be approved by the BHS unless it has a Local Authority Licence. As a general rule, every riding establishment is supposed to have a

Local Authority Licence; without this licence it will not have proper third-party insurance cover. To hold such a licence, granted annually, the establishment will have been inspected by a council representative who will have been concerned primarily with riders' safety and the welfare of the animals being used. Unlike the BHS inspector, the council inspector will not have assessed the standard of the instructors or whether the horses are well-schooled and forward-going.

It is also reassuring to know that at an Approved establishment you will not be allowed to ride without the correct protective head gear. An Approved centre will be able to supply you with an appropriate hard hat if you do not already possess one of the correct standard. And newcomers to riding can expect to be offered helpful advice on what to wear so that their first lessons are safe and enjoyable.

Another advantage is that many BHS-Approved establishments run the BHS Progressive Riding Tests, a scheme of instruction and assessment designed for riders of all ages and abilities. Further details of these tests – which can be administered only by an Approved centre – will be available from your chosen BHS Approved centre.

The staff at the **Riding Schools Office** are available to answer any queries during normal working hours, whether you want to know more about the Society's system of approving riding establishments, about an individual Approved Establishment or just have a general enquiry about finding your nearest Approved Centre.

The phone number is 01926 707 700, and we are always pleased to deal with any enquiries. Please do not hesitate to contact us.

BHS EXAMINATIONS AND PROFESSIONAL QUALIFICATIONS

The British Horse Society administers an examination system designed to meet the requirements of the riding instructor, yard management, riding holiday sector, and those who wish to further their knowledge in riding,

stable management and horse care. It holds these examinations – aimed primarily at career students – at BHS-Approved centres throughout the British Isles and overseas.

Before embarking on a career with horses it is important to select carefully the place of training. Whilst it is not obligatory for candidates to receive their training at a BHS-Approved centre, it is certainly advisable to do so. Details of approved BHS Training Centres can be found in the publication **Where to Train** available from the Training and Education Office, British Horse Society, The Deer Park, Stoneleigh, Kenilworth, Warwickshire, CV8 2XZ.

Examinations are taken at officially appointed BHS-Approved centres throughout the UK and overseas.

Candidates for all examinations must be members of the BHS at time of application and on the day of their examination.

First Stage

The minimum age for entry is 16 years. The candidate must understand the basic principles of horse care and, working under supervision, must show some knowledge and practice of looking after a well-mannered horse in the stable or at grass. He/she must be capable of riding a quiet, experienced horse or pony in an enclosed space.

Second Stage

Before entering the riding section of this examination the candidate must have passed the BHS Riding and Road Safety test. The candidate must understand the general management and requirements of horses for their health and well-being. Working under regular but not constant supervision, the candidates should be able to carry out the care of a stabled and grass-kept horse during all seasons of the year. The candidate must be capable of riding a quiet, experienced horse or pony in the countryside and on the public highway, as well as in a manege.

Third Stage

The minimum age for entry is 17 years. The candidate must show an ability to look after up to four horses under a variety of circumstances. He/she should be tactful, yet effective, understanding the reasons for his/her actions, both in horse care and while riding. A written test is also included in the examination. A candidate taking just the Horse Knowledge & Care section (Groom's Certificate) at this stage is expected to be a competent enough rider to be able to take horses on ride-and-lead exercises.

BHS PRELIMINARY TEACHING TEST : The minimum age for entry is 17½ years. Candidates under the age of 18 years are required to hold four GCSEs, Grades A, B or C, one subject of which must be English language or literature. Candidates are examined in their ability to give instruction in basic subjects. In order to enter the Preliminary Teaching Test a candidate must have passed the Horse Knowledge and Riding Stage II examination.

BHS PRELIMINARY TEACHER (BHSPTC): This is the first certificate on the instructor's ladder and is awarded to those passing the Horse Knowledge and Riding Stage III and the Preliminary Teaching Test. This certificate qualifies the holder at a 'Trainee Teacher' level. The candidate then needs to log 500 hours of teaching experience (25% of which may be stable management teaching). Once the 500 hours have been completed the log book (which is despatched with the BHSPT certificate from the office) is returned to the BHS, together with a copy of a current Health & Safety Executive First Aid at Work certificate, where random checks are made. Mature candidates (over 25 years) with extensive experience of teaching in the industry may apply for exemption from the requirement of 500 hours (information sheet available from the T & E Office). The candidate is then awarded an International Instructor Level 1 (BHSAI) certificate.

BHS INTERNATIONAL INSTRUCTOR LEVEL 1 (BHSAI): Candidates holding the Certificate of Horsemastership (pre-1986) and passing the Preliminary Teaching Test will become BHSPTC (unless they can prove extensive experience – see BHSPTC).

Fourth Stage

The minimum age for entry is 20 years. The candidate must be capable of taking sole charge of a group of horses of various types in stables and at grass. He/she must be an educated rider capable of training and improving horses in their work on the flat and over fences. Candidates passing the Horse Knowledge & Care section will be awarded the Intermediate Stable Manager's Certificate.

BHS INTERMEDIATE TEACHING EXAMINATION : The minimum age for entry is 20 years. This is open to candidates who hold the International Instructor Level 1 (BHSAI) Certificate. Candidates are examined on their general teaching abilities relating to class rides; individual dressage and jumping lessons; lunge lessons and general discussion, talks and short lectures.

INTERNATIONAL INSTRUCTOR LEVEL 2 (BHSII): This is achieved by passing the Intermediate Teaching Examination and the complete Horse Knowledge & Riding Stage IV examination. Candidates are also required to hold a full Health & Safety Executive First Aid at Work certificate.

INTERNATIONAL INSTRUCTOR LEVEL 3 (BHSI): The minimum age for entry is 22 years. Candidates must hold the International Instructor Level 2 (BHSII). This Instructor's Certificate is obtained by passing the two following examinations:

(a) **BHS Stable Manager's Examination** : This is also an examination in its own right and candidates need not necessarily go on to complete the full Instructor's Examination by taking b) below. The entry requirement for this examination is the Intermediate Stable Manager's Certificate (H K & C Stage IV). The minimum age for entry is 22 years.

(b) **BHS Equitation and Teaching Examination** : On passing this examination, the candidate completes the full International Instructor Level 3 (BHSI) Certificate. The entry requirements are the

British Horse Society Examination System

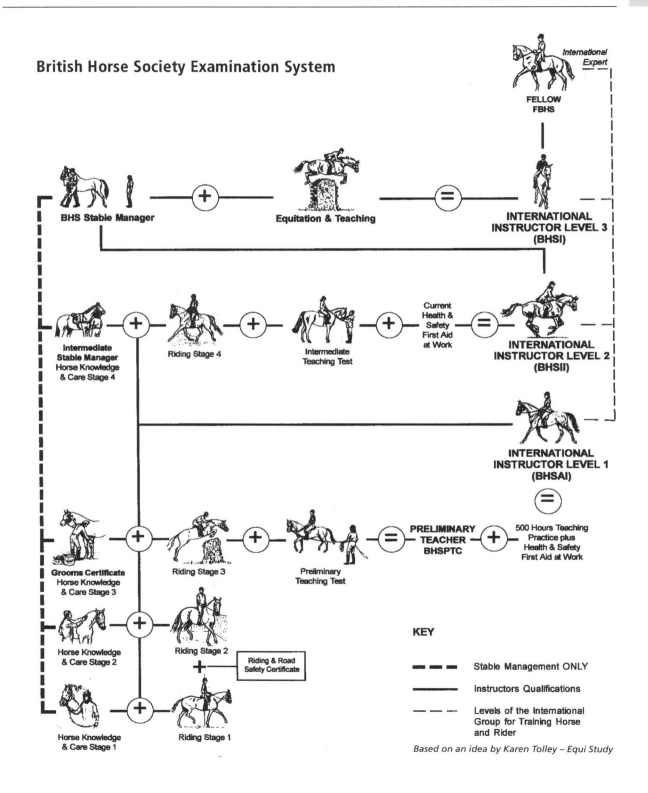

KEY

▬ ▬ ▬   Stable Management ONLY

────────   Instructors Qualifications

─ ─ ─   Levels of the International Group for Training Horse and Rider

Based on an idea by Karen Tolley – Equi Study

International Instructor Level 2 (BHSII) Certificate and the BHS Stable Manager's Certificate.

BHS FELLOWSHIP EXAMINATION : The BHS Fellowship Certificate is the senior qualification. The minimum age for entry is 25 years. The candidate must show a depth of knowledge and effectiveness in all aspects of equitation and horsemanship; proving themselves to be persons to whom others can turn for advice in the various spheres of equestrian activities. The Fellowship is available in five formats to assist those who have followed specific discipline routes or those who have reached a high overall level of ability.

In addition, the British Horse Society is an awarding body for Scottish and National Vocational Qualifications (S/NVQs) in the Horse Industry. Full details on request.

The present structure of the BHS examination system is shown on page 399. If you wish to find out more about the syllabi and requirements of the different exams, contact the Training & Education Office, The British Horse Society, The Deer Park, Stoneleigh, Kenilworth, Warwickshire, CV8 2XZ. Tel: 01926 707 700.

INDEX